# FOOD WASTE

## ACROSS THE SUPPLY CHAIN

## A U.S. PERSPECTIVE ON A GLOBAL PROBLEM

Edited by

**Zhengxia Dou, James D. Ferguson, David T. Galligan, Alan M. Kelly, Steven M. Finn, and Robert Giegengack**

University of Pennsylvania

Food waste across the supply chain: a U.S. perspective on a global problem

Edited by Zhengxia Dou, James D. Ferguson, David T. Galligan, Alan M. Kelly, Steven M. Finn, and Robert Giegengack

373 pages, 15.2 × 22.9 cm

ISBN 978-1532835070

© 2016 by the Council for Agricultural Science and Technology
CAST, Ames, Iowa
All rights reserved

Printed by CreateSpace, an Amazon.com Company

**Council for Agricultural Science and Technology (CAST):**

www.cast-science.org

CAST assembles, interprets, and communicates credible science-based information regionally, nationally, and internationally to legislators, regulators, policymakers, the media, the private sector, and the public. CAST is a nonprofit organization composed of scientific societies and many individual, student, company, nonprofit, and associate society members. CAST's Board of Directors is composed of representatives of the scientific societies, commercial companies, and nonprofit or trade organizations. CAST was established in 1972 as a result of a meeting sponsored in 1970 by the National Academy of Sciences, National Research Council.

The authors named in each publication are responsible for the contents. The CAST Board of Directors is responsible for the policies and procedures followed in developing, processing, and disseminating the documents produced.

# Contents

## Figures

## Tables

## Contributors

**Jonathan Bloom**

Journalist and food waste expert, author of "American Wasteland" and moderator of the blog Wasted Food

**Jean C. Buzby**

Chief of the Diet, Safety, and Health Economics Branch, Food Economics Division, Economic Research Service, U.S. Department of Agriculture

**Alexandra Cirone**

Student, Health and Societies Major, College of Arts and Sciences, University of Pennsylvania

**Nicole M. Civita**

Visiting Assistant Professor of Law, University of Arkansas, and Director of the Food Recovery Project

**Elena Crouch**

Student, Biology Major, College of Arts and Sciences, University of Pennsylvania

**Zhengxia Dou**

Professor of Animal Systems, School of Veterinary Medicine, University of Pennsylvania, organizer of *The Last Food Mile Conference*, and editor-in-chief of this volume

**Barbara Ekwall**

Senior Liaison Officer, United Nations Food and Agriculture Organization Office for North America

**James D. Ferguson**

Professor of Nutrition, School of Veterinary Medicine, University of Pennsylvania, and an editor of this volume

**Steven M. Finn**

Managing Director of ResponsEcology, author of the blog Food for Thoughtful Action, and an editor of this volume

## *Contributors*

**David T. Galligan**

Professor of Animal Health Economics, School of Veterinary Medicine, University of Pennsylvania, and an editor of this volume

**Robert Giegengack**

Professor Emeritus of Earth and Environmental Science, College of Arts and Sciences, University of Pennsylvania, and an editor of this volume

**Elise Golan**

Director of Sustainable Development, Office of the Chief Economist, U.S. Department of Agriculture

**Nora Goldstein**

Editor of the journal Biocycle: Composting, Renewable Energy, Sustainability

**Ethen Grant**

Student in the Philadelphia School District

**Karen Hanner**

Managing Director of Manufacturing Product Sourcing, Feeding America

**Brittany Hartmire**

Student, Health and Societies Major, College of Arts and Sciences, University of Pennsylvania

**Jane Kauer**

Lecturer in Anthropology, College of Arts and Sciences, University of Pennsylvania

**Alan Kelly**

Professor Emeritus of Pathobiology and Dean Emeritus, School of Veterinary Medicine, University of Pennsylvania, and an editor of this volume

**Christine Kim**

Graduate, University of Pennsylvania Class of 2015, Health and Societies Major

**Gomian Konneh**

Student, College of Arts and Sciences, University of Pennsylvania

## *Contributors*

**David Masser**

President of Sterman Masser Inc., a Central Pennsylvania potato farm in the family for over 200 years

**Emily M. Moscato**

Assistant Professor of Food Marketing, Haub School of Business, St. Joseph's University

**Roni A. Neff**

Assistant Professor of Environmental Health Sciences and Health Policy and Management, Bloomberg School of Public Health, Director of the Food System Sustainability and Public Health program, Johns Hopkins University

**Thomas O'Donnell**

U.S. Environmental Protection Agency, Region 3 NAHE, Philadelphia

**Leah Oppenheimer**

AmpleHarvest.org

**Jibreel Powers**

Student in the Philadelphia School District

**Andrew Shakman**

Co-Founder and President/CEO, LeanPath

**Yasmin Siddiqi**

Global Marketing Manager, DuPont Packaging

**Jarrell Smith**

Master of Urban Planning Candidate, Department of Urban and Regional Planning, University of Florida

**Cathy Snyder**

Founder and Executive Director, Rolling Harvest Food Rescue

**Marie L. Spiker**

Graduate Student, Center for a Livable Future, School of Public Health, Johns Hopkins University

*Contributors*

**Brian Spooner**

Professor of Anthropology, College of Arts and Sciences, University of Pennsylvania

**Dave Stangis**

Vice President, Public Affairs and Corporate Responsibility, Campbell Soup Company

**John L. Stanton**

Professor of Food Marketing, Haub School of Business, St. Joseph's University

**Jarrett Stein**

Director of ABCS Partnerships and Student Engagement, Netter Center for Community Partnerships, University of Pennsylvania

**Kathleen Sullivan-Sealey**

Associate Professor of Biology, College of Arts and Sciences, University of Florida

**John D. Toth**

Research Specialist, School of Veterinary Medicine, University of Pennsylvania

**Patricia L. Truant**

Ph.D. Candidate, Department of Public Health, Johns Hopkins University

**Steven M. Waldmann**

Executive Director, Society of St. Andrew

**Hodan F. Wells**

Economist, Economic Research Service, U.S. Department of Agriculture

**Jessica Zha**

Graduate Student, Perelman School of Medicine, University of Pennsylvania

# *Preface*

**Alan Kelly**

The new Post-2015 UN Development Agenda, "Transforming our World," seeks to end poverty, achieve gender equality, and ensure food security in every corner of the globe by 2030. These are ambitious goals with food loss and food waste among the urgent challenges that must be resolved. Each year around 1.3 billion tonnes, a third of all food produced in the world, is wasted with financial costs estimated at $1 trillion per year. According to the Rockefeller Foundation, cutting post harvest food loss in half would yield enough food to feed 1 billion people and go a long way towards accomplishing the UN development agenda.

The problems of food loss and waste manifest themselves in different ways in different parts of the world. Although the UN Development Agenda affects all of us, most immediately the Agenda targets the world's 800 million poor and hungry peoples most of whom subsist in rural areas of developing countries. One out of three adults in rural Sub-Saharan Africa is chronically undernourished. Also targeted are some 2 billion subsistence farmers, mainly women, who maintain livestock, cattle, sheep, goats, and depend upon their animals for survival. Many exist in harsh environs where they face a precarious future as their regions are among the most vulnerable to the effects of climate change.

To escape from hunger and poverty small livestock farmers need help from veterinary and other professions. They are poor because their livestock carry a high incidence of infectious disease and because of primitive farming methods that lead to low productivity. But in their struggle for survival, farmers find their meager profits further diminished by pre-market losses of 30% to 40% due to weak transportation and power infrastructures, inadequate management of the supply chain, food contamination, and decay. The losses are estimated to cost farmers at least 15% of their profits and cement the poverty trap they are unable to escape. Pre-market food losses also create significant environmental consequences since waste food rots and releases quantities of greenhouse gases that advance climate change. Moreover, in order to compensate for their losses, small farmers habitually increase the size of their herds, triggering overgrazing and unsustainable land degradation.

Until inefficiencies of production and the food supply system are corrected the UN Development Goals will not be realized. The problem requires novel collaborations across a range of disciplines since it involves animal, human and environmental health, and increasingly, distant urban markets. Veterinary medicine is well suited to play a pivotal role in this discussion for, in addition to its critical role in animal disease control and food safety, the profession appreciates that human health is intricately connected to the health and well being of all types of animals and the environment. These mature values are captured by the slogan "One Heath."

It was the One Health viewpoint that led us, a group of food animal veterinarians and colleagues, to prepare a text on the growing incidence of food waste in the United States. Food in this country is abundant, inexpensive, and safe; farming is efficient and road, rail, and power infrastructures are all well developed. In contrast to much of the developing world that lacks cold chain and pasteurization facilities and where a gallon of milk sours within a day, milk in this country has a shelf life of up to 3 weeks. But the very success of farming that has made food inexpensive for most Americans has also led to overconsumption, an epidemic of obesity, and prodigious waste, waste that is virtually unheard of among poor farmers of the developing world. American consumers throw away an estimated 35 to 40% of all food that is purchased. According to the Economist, the U.S. loses or wastes 1,520 calories per person per day with enormous consequences for greenhouse gas emissions from landfills, let alone the resources used to produce the food. Nevertheless, in the midst of all our abundance we have growing poverty and food insecurity in this country with nearly 50 million Americans living in food insecure households.

These unacceptable food insecurity figures will rise as the price of food increases due to global population growth, climate change, drought and disease. To combat this, the U.S. must address the problem of food waste more forcefully and begin to convert food waste into a food supply for the hungry. Europe has recognized this need and is already well ahead of us. The present volume is the first to outline the problem comprehensively, to describe current efforts to stem the loss and the crucial activities of charitable organizations to supply our Nation's hungry. We are a nation of over consumption. Our landfills, with wasted food being the largest component, are puffing out pollutants that contribute to extreme weather events globally and add to the struggles of the most vulnerable, the poor farmers of the developing world. The new UN sustainability agenda and the imperative of curbing food waste presents us with the formidable challenge of rethinking our values and for most of us, our way of life. We hope the present text will, in a small way, help to stimulate a dialogue and suitable response.

# *Introduction*

## *Understanding Food Loss and Waste Across the U.S. Supply Chain*

### *Reduction, Recovery, Recycling*

**Zhengxia Dou**

民以食为天 *(min yi shi wei tian)* is a Chinese proverb, meaning "food is a basic necessity of man." Taken in its historical context (over 2,000 years ago, in the Spring and Autumn period in China), the original phrase conveys a more profound meaning, which may be translated as "the King's power to rule is founded upon his ability to feed the people" .[1] Regardless of its origin, human history attests time and again that lack of food (hunger) breeds discontentment which in turn brews unrest and social instability.

These perils are equally relevant today as the world faces the vital challenge of providing enough food to meet the basic necessities of 7.3 billion people. Despite tremendous growth in agricultural productivity, hunger and malnutrition are widespread, and 1 in 9 people go to bed hungry every night. The situation may further deteriorate in the coming decades as the world population continues to grow to a projected 9+ billion by 2050 and food demand increases to an estimated 50-100% above the current level of production. Compounding the challenge of food security are the unrelenting worldwide problems of dwindling resources and environmental degradation, coupled with the overarching trends of urbanization and globalization. These complex and profound issues facing mankind are the subject of wide-ranging debates and discussions.

Various strategies have emerged to address the pressing need to sustainably feed the world. Enhancing agricultural output to increase the food *supply* stands as the most appealing option for many, spurred by our confidence in the seemingly unlimited potential in human ingenuity with science and technology innovation. Strategies have also emerged focusing on curbing *demand*. Striving for a global population with a sustainable replacement rate is considered critical, particularly in regions where a high degree of food insecurity co-exists with birth rates more than double the world average, such as in Sub-

[1] http://www.ofnumbers.com/2013/03/20/chapter-3-food-and-beverage/#identifier_0_22

Saharan Africa. There are no simple solutions. We need the combination of all possible ways and means to tackle the problem.

Recently, there has been a *waste-less-to-feed-more* movement worldwide, largely kindled by the United Nations Food and Agriculture Organization report[2] that about 1/3 of food produced for human consumption is lost or wasted before reaching a human stomach, amounting to 1.3 billion tonnes globally on an annual basis. Halving the amount of food wastage would mean enough food to feed the world's 800 million hungry people. Furthermore, reducing food wastage would help conserve natural resources and lower the environmental footprint that is associated with our food system. Conceivably, reduction in food loss and waste represents a realistic opportunity of significant magnitude with multiple benefits and very few negatives or conflicts.

It was under this global backdrop that *The Last Food Mile Conference* was organized by a group of faculty at the University of Pennsylvania. Supported by a USDA grant and held on the Penn campus on December 8-9, 2014, the *Conference* convened more than two dozen distinguished speakers from industry, academia, NGOs, and governmental agencies to examine issues of food loss and waste across the U.S. supply chain. Panel presentations and discussions covered wide-ranging topics, e.g. the scale of the problem, food waste streams and measurements, consumer food behavior, ongoing food waste reduction, recovery, and recycling efforts, and lessons learned and barriers encountered. More than 200 participants demonstrated a great deal of enthusiasm and engaged in active discussions and networking.[3]

The vitality of the conference persuaded the organizing committee to leverage the growing interest in food loss and waste by putting together this book, the first to explore the scale of the problem involving wastage of food and resources across the U.S. supply chain, to examine ways to reduce what is squandered and, where possible, turn wholesome (otherwise wasted) food into nourishment for this country's growing poor and food insecure families. This book is based on presentations given at the *Conference*, augmented with additional information taking advantage of flexibility in print format. Further, the book's scope is expanded and the content enriched with a few new chapters that bring fresh perspectives to the field or present new data that have not been published previously. The end product is a collage of 19 chapters authored by

---

[2] http://www.fao.org/docrep/014/mb060e/mb060e00.htm

[3] For Conference summary and survey results, visit: http://repository.upenn.edu/thelastfoodmile/followup/

people of outstanding rank in the field and covering key issues with breadth and depth.

Brief descriptions of the chapters are provided below to serve as a guide.

Food loss and waste is a global phenomenon and its implications are far-reaching beyond wasted food per se. The UN FAO recently led a series of studies on the scale of the food waste problem and its environmental ramifications at the global level. **Chapter 1** summarily presents the findings and conclusions from these studies, providing a one-stop reference for a large amount of material and information to interested readers. **Chapter 2**, undertaking an approach to examining global food production at the most fundamental level, discusses the elemental necessities required for the production and provision of food, namely soil, water, nutrients, solar radiation, fossil fuels, plus marine production systems. Here, the reader will find a comprehensive analysis of the nature, state, capacity, and potential limitations of these basic necessities, focusing on the non-renewable resources.

The significance of the United States in the landscape of global food production, consumption, and wastage cannot be overstated. In terms of production, the U.S. produces more food than France, Germany, UK, Canada, Japan, and Australia combined. In terms of consumption, the U.S. is the third most populous country in the world (behind China and India) and ranks number one in per capita food availability. How much of that food is never eaten but lost and wasted at the national level, and at what segments of the supply chain? In what scale and magnitude are natural resources embedded in the wasted food? **Chapter 3** provides authoritative information detailing the weight, calorie, and dollar value of food lost and wasted at the retail and consumer levels of the U.S. food system, while **Chapter 4** presents freshly obtained data, derived from a set of official statistical databases, that quantify the acreage of land, quantity of irrigation water, and amounts of fertilizer nutrients embedded in the wasted food at the national level.

Is food-wasting behavior a new vice that happens to infect modern day consumers? **Chapter 5** examines the issue from a historical perspective, arguing that food waste is nothing new and that continued social changes, cultural mingling, and the mega-trends of urbanization and globalization form the undercurrent that shapes the way we handle food as a people or as individuals.

Certainly, consumers are at the crux of the food-waste matter. In the U.S. and other developed countries, food loss and waste at the consumer level is the single largest component of wastage along the supply chain. Several chapters in this book are devoted to consumer food behavior – **Chapter 6** provides a glimpse of Americans' awareness and attitudes regarding food waste, as derived from a nationally-representative survey; **Chapter** 7 summarizes current understanding of consumer knowledge and factors affecting consumer food behavior, based on literature review and synthesis; **Chapter 8** provides an insightful and in-depth analysis of the cyclical nature involving consumers and retailers that leads to the wasteful culture and practice of discrediting food of "imperfect" appearances.

Many of us are appalled by scenes of food wastage occurring at buffet-style eateries. What constitutes the waste streams? How much can be attributed to individual behavior and how much occurs because of the system (i.e. the business-service model)? What proportion may be recovered to feed the hungry and how much is unfit for the most desirable use? **Chapter 9** provides quantitative information from a carefully-designed study to shed light on these questions. The authors describe a research project that identifies four waste streams at an all-you-can-eat dining hall and the relevant quantities and proportions based on 10-day sampling and measurements. Their data speaks volumes.

Across the kindergarten-to-high school system in the U.S., food programs are in place to ensure that all pupils have access to adequate and nutritious food. Students from low-income families are provided with meals, usually lunch and in some cases breakfast and dinner as well, at reduced or no cost to them. Not surprisingly, a large amount of the food is tossed into the trash bin. In **Chapter 10**, a group of $8^{th}$ graders, guided by adult counselors through a participatory project, presents their findings on the types and quantities of food thrown away and the associated reasons. The problems seen in the eyes of these youngsters and expressed in their own voices are quite thought provoking.

**Chapter 11** discusses key change aspects in reducing food waste by moving from a culture of abundance to a culture of responsibility.

Food losses occur every step of the way along the supply chain, at farms, processing sites and manufacturing plants, packaging and storage facilities, retail stores, foodservice sites, and homes. Opportunities exist at each step to reduce, recover, and recycle (the 3Rs) food waste. Indeed, such 3R efforts have been taking place all over the country in a variety of formats thanks to numerous organizations and individuals. **Chapter 12** showcases some of the measures taken by the *food industry* with four exemplary stories. **Chapter 13** highlights some *charity organizations'* food rescue efforts that are geared toward the noble purpose of hunger relief in America. At the end of the supply chain are millions of *households*; how to reduce food wastage at home is perhaps the greatest challenge. **Chapter 14** describes a pilot program led by the U.S. Environmental Protection Agency with case studies demonstrating 20-35% reduction in kitchen food discards among participating households. Food remnants, generated in millions of tons each year from the food handling and processing sector, are generally unfit for human consumption, but such "waste" can be recovered for beneficial use in *animal feeding*. **Chapter 15** details the types, sources, and magnitudes of such food material and their uses in the animal industry, and discusses relevant limitations and challenges. Finally, *composting* is the last resort in the 3R hierarchy. **Chapter 16** provides an authoritative overview of food waste composting in the U.S. with the most updated information available, while **Chapter 17** discusses lessons learned and barriers encountered on a small Caribbean island where composting food discards and other organic wastes is vital to the health of the people as well as its ecosystem.

What role has the U.S. government played in combating food loss and waste? What laws exist to provide guidance and protection regarding food donations to feed hungry people? **Chapter 18** describes various federal programs and policies related to food waste reduction, recovery, and recycling, and **Chapter 19** covers the legal aspects of food recovery and donation in detail with clear and concise interpretations plus practical guidance.

This book, with its coverage of diverse topics contributed by leading, multi-disciplinary experts, is meant as **a source book for many** – for educators teaching in or outside of classrooms, for sustainability officers looking for practical solutions, for consultants developing strategic and actionable measures, for advocates educating and engaging the public, for hunger relief agents searching for inspiration and creative ideas, for researchers endeavoring to gain insights from solid scientific information, for concerned citizens wanting to make a difference through everyday actions, and for policy makers devising programs and shaping policies that may affect our future.

The USDA and EPA recently (9/16/2015) announced the nation's first-ever food waste reduction goal, calling for a 50% reduction by 2030.[4] It is our hope that this book will help support the policy and advance solutions to the complex problems of food loss and waste and sustainably feeding the growing population in the U.S. and beyond.

[4] http://www.usda.gov/wps/portal/usda/usdahome?contentid=2015/09/0257.xml

# *Part One*

# *Food loss and waste in the context of sustainably feeding the world*

# Chapter 1

# *Global Food Loss and Food Waste and the Environmental Footprint*

**Barbara Ekwall**

## ABSTRACT

Looking at the global context, this chapter will examine major trends in food security and nutrition as the world meets the challenge to feed 9.2 billion people by 2050, 2.3 billion people more than today. Urbanization, changing diets, rising middle class, natural resources and climate change will have a profound impact on food systems.

Reducing food loss and waste is an integral part of efforts in view of achieving a zero hunger world. Every year, the world wastes or loses 1.3 billion metric tons of food, roughly one third of the globe's food production. Food loss and waste, which occurs at all stages of the food value chain, is an indicator that the food systems are not functioning as they should. When food is thrown away, the land, water and energy used to produce, process, distribute and prepare the food is also thrown away. This chapter will look at the environmental footprint of water, land and greenhouse gas emissions. It shows that if food loss and waste were a country, it would be the largest consumer of irrigation water, would be occupying the second largest land area, and would be the third largest emitter of greenhouse gases.

This chapter also highlights the need to make the reduction of food loss and waste an integral part of efforts to eradicate world hunger. FAO estimates that the food produced but never eaten would be enough to feed 2 billion people, and the economic loss represents a staggering 1 trillion USD.

This chapter concludes that reducing food loss and waste is a low-hanging fruit that can make a considerable difference to reducing hunger and malnutrition, preserve the environment and contribute to economic development. In a world of limited resources, we not only need to produce more, we need to produce better and consume more intelligently.

## INTRODUCTION

Rarely has a publication by the Food and Agriculture Organization of the United Nations (FAO) generated so much interest, surprise and maybe even shock as *Global food losses and food waste – extent, causes and prevention*[1] when it was published in 2011. Is it possible that about 1.3 billion tons of food are lost or wasted globally every year? This figure is astronomical (Figure 1.1). How many zeros is that, anyway? And how can this happen while millions of people go hungry?

The FAO was established in 1945 as a specialized agency of the UN system. FAO has 194 member countries plus one member organization, the European Union. Achieving freedom from hunger is at the heart of FAO's mandate – to make sure people have regular access to enough high-quality food to lead active, healthy lives. Our objective is to raise the levels of nutrition, to improve agricultural productivity, to better the lives of rural populations and to contribute to the growth of the world economy in a sustainable manner. We do this by generating and disseminating knowledge and providing technical expertise related to food and agriculture. The publication *Food losses and food waste – extent, causes and prevention* was followed by a number of other publications and information briefs,[2] and the issue of food loss and waste is now an integral part of efforts to eradicate hunger, as shown by the Zero Hunger Challenge launched by the UN Secretary General in 2012, and the Post-2015 process leading to the adoption of the Sustainable Development Goals in September 2015.

**1.1** Global food loss and waste

The present chapter reflects the keynote presentation given by the author at "The Last Food Mile Conference" organized by the University of Pennsylvania in December 2014. The views expressed in this article are those of the author and do not necessarily reflect the views or policies of FAO. The article's objective is to give a broad overview of the global context of food security and nutrition as well as some of the challenges of creating a world free from hunger. Subsequently, it will look at the issue of food loss and waste, and discuss in more detail its environmental impact, focusing specifically on water, land and greenhouse gas emissions. The article highlights the importance of reducing food loss and waste as an integral part of the effort to eradicate hunger, to preserve and protect natural

---

[1] FAO, 2011. *Global food losses and food waste – extent, causes and prevention*, by J. Gustavsson, C. Cederberg, U. Sonesson, R. van Otterdijk and A. Meyerbeck. Rome (http://www.fao.org/docrep/014/mb060e/mb060e00.pdf)

[2] Available at: www.fao.org/save-food/en/

resources, and to promote economic growth. It also calls for political will and strong partnerships across sectors and groups.

## FOOD SECURITY

The publication *The state of food insecurity in the world* (SOFI) by FAO, IFAD and WFP offers, annually, the latest estimates related to global hunger, progress made, and trends at the global and regional levels. The 2014 SOFI publication[3] shows mixed progress in reducing hunger and malnutrition. This progress is measured against two major internationally agreed-upon goals: Millennium Development Goal 1c of halving the proportion of undernourished people in developing countries by 2015, and the World Food Summit 1996 (WFS) target of halving the number of undernourished people by 2015. While the world is on-track to achieve the MDG goal (the red line in Figure 1.2), the absolute number of people who suffer from hunger (the yellow line) remains persistently high. The global community is far from reaching the more stringent WFS target to reduce the number of people suffering from chronic hunger to 500 million.

Our planet produces enough food for everyone. Yet, although trends in the past few years show that we are going in the right direction, there are still 805 million children, women and men who suffer from chronic hunger and malnutrition. About 70% of them live in rural areas and are small-scale farmers or rural landless people. Ironically, those that produce food are also those who are most affected by hunger and malnutrition.

The opposite problem is facing many developed and middle income countries, with obesity rates nearly doubling since 1980. Obesity[4] often goes hand in hand with poverty, unemployment, and low levels of education. It is putting enormous stress on public health systems and will increasingly do so in the future. Obesity is preventable.

## FUTURE CHALLENGES

Figure 1.2 shows the global food security situation as discussed in SOFI 2014. Let us also look at some future challenges.

### Population

By 2050, it is estimated that the world population will increase to 9.2 billion, which is 2.3 billion more than the population today. The highest population increases are expected to take place in poorer countries, especially in Sub-Saharan Africa. For example, Niger, which today has a population of 14

[3] Available at: http://www.fao.org/3/a-i4030e.pdf
[4] Body Mass Index (BMI) greater or equal to 30.

million people, is expected to triple this number by 2050. There is a vast body of evidence on the strong links between poverty and population growth. When households have limited ability to save for the future, they often depend on children for additional income or as a safety net for old age. In addition, low levels of education and lack of women's empowerment, especially when combined with lack of access to health and reproductive rights, further increase population dynamics in poor countries and poor communities.

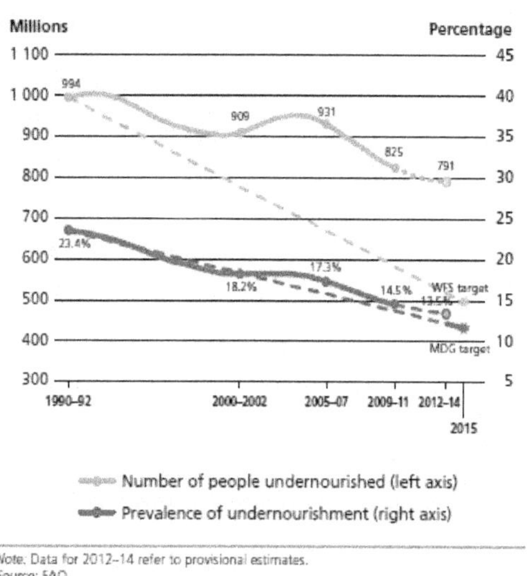


**1.2** The trajectory of undernourishment in developing regions: actual and projected progress towards the MGD and WFS targets. Source: FAO, IFAD and WFP, 2014

## Urbanization

The world of tomorrow will be an urban world (Figure 1.3). In 1950, 30% of the population lived in cities. In 2008, more people lived in cities than in rural areas for the first time in human history, and this trend will only increase. Some estimates say over 67% of the world's population will live in urban areas by 2050. Urbanization has major implications for agriculture and food security. More people living in cities means fewer people working in agriculture, more convenience products (such as fast food), more imports, and changing diets and lifestyles which lead to more obesity. Urbanization also means longer and more complex food value chains and an increase of food loss and waste.

## Middle Class

The world of tomorrow will also see the rise of an important middle class. The UN estimates that at least 3 billion more people are likely to enter the global middle class by 2030. They will demand more resource-intensive meat and vegetable oils.

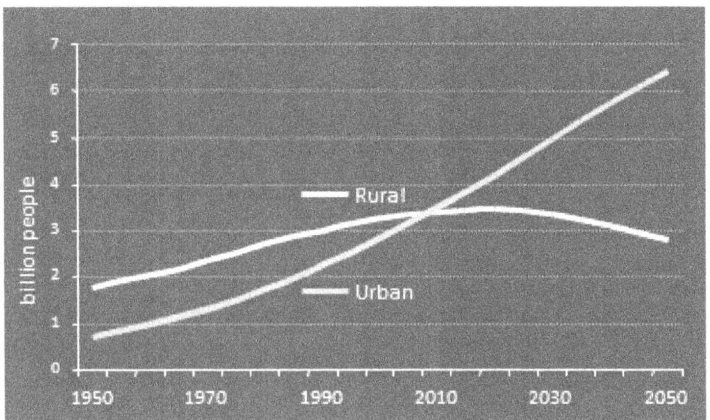


**1.3** The acceleration of global urbanization, 1950-2050

**Increased Food Production**

How much food will the world of tomorrow need? FAO estimates that to feed the world population of 9.2 billion by 2050, the production of basic staple food would have to increase by 60%. This increase needs to take place where the food is most needed: in developing countries. Increasing food production will be a daunting task, particularly against the background of current stress on natural resources and climate change.

## FOOD LOSS AND WASTE

As mentioned in the introduction, it is hard to accept what the figures tell us: that about one third of all food produced is lost or wasted in producing or consuming food.

What is food loss and waste? It is a decrease, measured in mass, at all stages of the food chain from harvest to consumption, of food that was originally intended for human consumption, regardless of the cause. FAO measures food loss and waste by quantifying the physical mass, or weight. Another alternative is to measure food loss and waste in terms of energy, or kcal.[5]

The distinction between food loss and waste is not always clear. Generally, food loss occurs mostly at the production stages, from the moment the

[5] Regarding the latest discussions about the definition of food loss and waste, see FAO. 2014. *Definitional framework of food loss*, Working Paper (www.fao.org/3/a-at144e.pdf)

product is ready to be harvested to the moment it reaches the retailer. Food waste typically takes place at the retailer and consumer side of the food-supply chain.

Harvest can be destroyed by animals, diseases or bad weather. Food can be lost throughout the food supply chain, including harvesting, threshing, and milling, but also during transportation, handling, storage, and retailing.[6] The consumer in developed economies meets a delightful sight at the market, because only the best products are offered. Not so perfect looking food items are discarded at an early stage, although they may be nutritious and safe for human consumption. Finally, at the household level, food is wasted because of lack of planning, misleading food safety information, particular eating habits, a culture of abundance, or simply convenience.

As a result, globally, roughly 30% of cereals, 40-50% of root crops, fruits and vegetables, 20% of oilseeds, meat and dairy products, and 35% of fish are thrown away.[7]

Food loss and waste indicate that the food systems are not functioning as they should. In low-income countries, food losses tend to be a result of managerial and technical limitations, infrastructure problems, lack of storage facilities, transportation and cooling systems. As shown in Figure 1.4,[8] there is little waste at the consumption level in low-income countries. In contrast, medium- and high-income countries present a big share of food waste (about 20% in developed countries) due both to consumer behavior, but also to policies and regulations. Per capita food waste by consumers in Europe and North America is 95-115 kg/year (210-250 lbs), while this figure in Sub-Saharan Africa and South-Southeast Asia is only 6-11 kg (13-24 lbs) year.[9] However, food is also lost in the fields in industrialized countries as a result of surplus production, low price, or lack of labor.

---

[6] Several case studies discuss the causes of food loss and waste and critical loss points throughout the value chain for banana, maize, milk and fish in Kenya. See FAO. 2014. *Food loss assessments: causes and solutions*, (www.fao.org/fileadmin/user_upload/save-food/PDF/Kenya_Food_Loss_Studies.pdf). A more detailed analysis of causes and drivers of food losses and waste can be found in the report of the High-Level Panel of Experts on Food Security and Nutrition of the Committee of World Food Security, June 2014, entitled *Food losses and waste in the context of sustainable food systems* available at www.fao.org/3/a-i3901e.pdf

[7] See FAO, 2014. *Global initiative on food loss and waste reduction* (www.fao.org/3/a-i4068e.pdf)

[8] FAO, 2011.

[9] FAO, 2011.

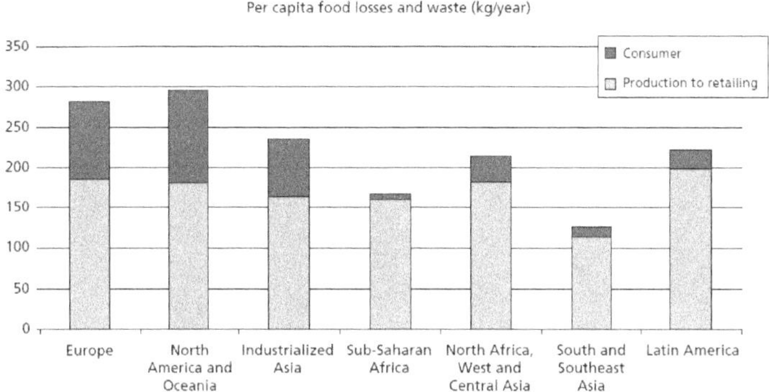


**1.4** Per capita food loss and waste, at consumption and pre-consumption stages, in different regions. Source: FAO, 2011

When food is thrown away, the land, water and energy used to produce and bring this food to the consumer have also been "thrown away." We will now examine food loss and waste from the perspective of three major environmental challenges: water, land and greenhouse gas emissions.[10]

**Is There Enough Water?**

Agriculture is by far the largest source of water usage world-wide, accounting for about 70% of all water withdrawals. Irrigation is crucial for increasing agricultural yields, especially for cereals.

In Figure 1.5, a distinction is made between physical and economic water scarcity.[11] The orange/brown areas are those where there is simply too little water to expand irrigation. In much of Sub-Saharan Africa, there is economic water scarcity, which means that there is enough water to be pumped for irrigation, but there is a lack of infrastructure and economic incentives. Finally, there are areas where water is sufficient, typically in the temperate zones.[12]

Water scarcity will increase in the future. Irrigation in agriculture has the potential to cause severe environmental problems, such as water depletion, salinization, and soil degradation.

---

[10] For a more detailed discussion, readers are referred to *Food wastage footprint – impacts on natural resources. Summary report*, published by FAO in 2013. www.fao.org/docrep/018/i3347e/i3347e.pdf

[11] See also: www.fao.org/NR/Water/art/2007/scarcity.html

[12] FAO, 2007. *Comprehensive assessment of water management in agriculture* (www.fao.org/nr/water/art/2007/scarcity.html)

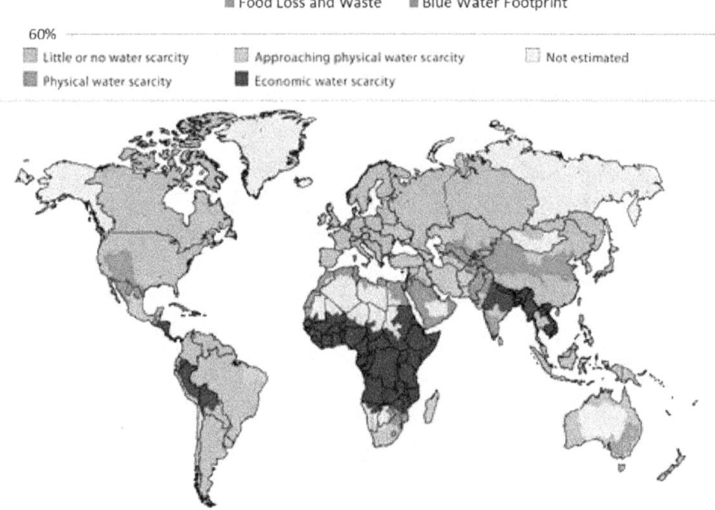


**1.5** Economic and physical water scarcity. Source: FAO, 2007

Globally, about 250 km$^3$ of irrigation water taken from the ground or surface water – also called blue water – is lost because of food that is produced but never consumed.[13] More water is lost and wasted than the amount of water used for irrigation in India, or in China.

Figure 1.6 shows the contribution of different commodities to food loss and waste and to the associated blue water footprint. Cereals represent the largest contributor to the blue water footprint of food loss and waste with 52%, although cereals only represent 26% of the weight of that loss. Wasted fruits are another large contributor, with 18% of the blue water footprint and 16% of total food loss and waste. Conversely, 19% of food loss and waste attributed to starchy roots only accounts for 2% of the water footprint, because their production generally does not require much irrigation.

Reducing food loss and waste will reduce stress on water resources. Other measures to save water include precision irrigation, better water management and the use of crops that require less water or are drought resistant.

**Is There Enough Land?**

Land is a limited natural resource with a number of competing uses.

---

[13] FAO, 2013. *Food wastage footprint – impacts on natural resources. Summary Report.* (www.fao.org/docrep/018/i3347e/i3347e.pdf)

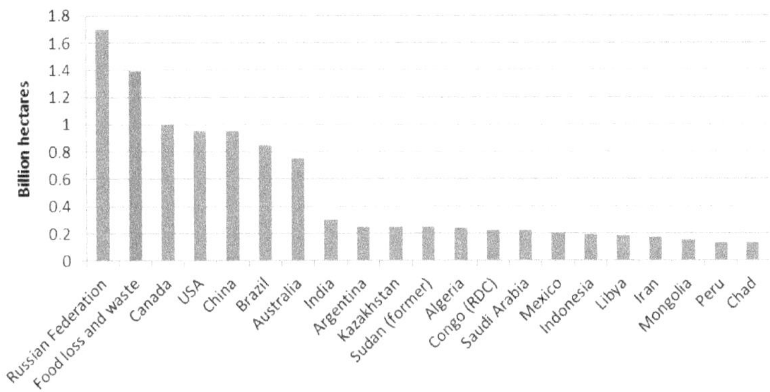


**1.7** Country areas: top 20 countries (FAOSTAT) vs. food loss and waste. Source: FAO, 2013

Today, only 37% of the planet's landmass outside of Antarctica is used to grow food, but future expansion potential is limited. Agricultural activity is not possible in many landscapes (ex. deserts, mountains) or is too costly. Land also may not be accessible because it is protected for conservation purposes or covered by forests.

At the global level, food lost and wasted in 2007 occupied almost 1.4 billion hectares (Figure 1.7), equal to about 28% of the world's agricultural land area, including cropland and grassland. If food loss and waste were a country, it would have the second largest land area in the world, just after the land area occupied by the Russian Federation.

Meat and milk are the most land-intensive commodities, representing 78% of the total land area occupied by lost and wasted food (Figure 1.8). However, meat and milk only represent 11% of total food loss and waste. Lost

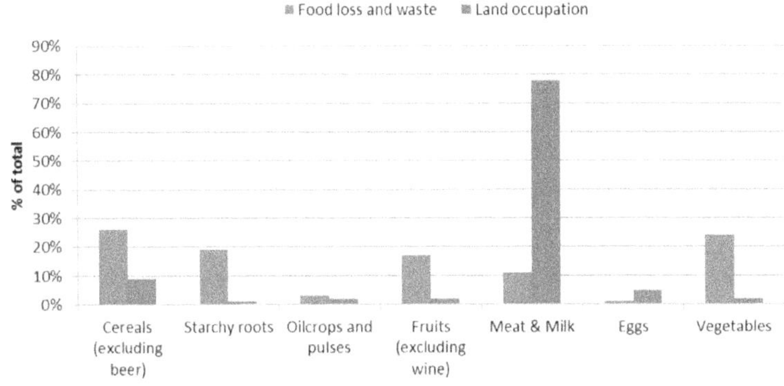


**1.8** Contribution of each commodity to food loss, waste and land occupation. Source: FAO, 2013

and wasted fruits and vegetables, while representing 16% of food lost and wasted, only accounts for 2% of the land area used for nothing.

Comparing yields of crop and animal products must be done with great caution, however. World averages hide the great variations that exist between different crops and different animals, as well as the productivity of a given crop in a given country or region.

Of course, the discussion of land surface area in agriculture is closely linked with the discussion of healthy soils. Soils constitute the foundation of vegetation and agriculture. They also host at least one quarter of the world's biodiversity. Soils play key roles in the carbon cycle and in water management, especially in improving resilience to floods and droughts. Yet, one-third of our soils have already degraded and, if current trends continue, the average surface of arable and productive land per person in 2050 will be reduced to a quarter of the land that was available for each person in 1960. It is time to give priority to the way we use land and to promote healthy soils. It is in this context that the UN named 2015 the International Year of Soils.

**Can We Reduce GHG Emissions?**

A number of recent studies provide scientific evidence linking climate change and greenhouse gas (GHG) emissions. Agriculture is affected by and contributes to climate change; it also provides solutions to mitigate, adapt and become resilient to climate change.

New FAO estimates show that GHG emissions from agriculture, including forestry and fisheries, have nearly doubled in the past fifty years and, in the absence of interventions to reduce emissions, could increase by an additional 30% by 2050.[14] In 2011, with 5.3 billion tonnes, emissions from agricultural activity accounted for more than one fifth (21%) of all emissions. [15] The largest source of GHG in agriculture is enteric fermentation from livestock farming (39%), distantly followed by application of synthetic fertilizers (14%).

According to a recent study conducted by FAO, the total carbon footprint of lost and wasted food, excluding land use, is estimated at 3.3 billion tonnes of $CO_2$ equivalent. If GHG emissions from food produced and never eaten were a

[14] FAO, 2014. *Agriculture, forestry and other land use emissions by sources and removals by sinks*. Working Paper Series (www.fao.org/docrep/019/i3671e/i3671e.pdf)
[15] This includes emissions within the farm gate and at the source. Does not include emissions from fertilizer manufacture, transport and refrigeration.

country, it would be the third largest GHG emitter after the U.S. and China (Figure 1.9).[16]

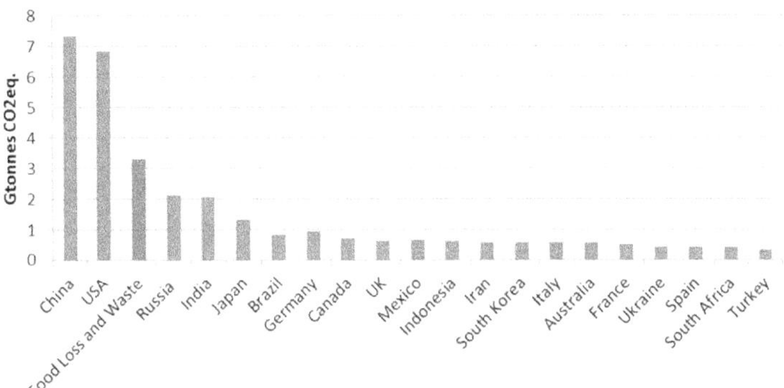


**1.9** Total greenhouse gas emissions in billions of tonnes: top 20 countries vs. food wastage. Source: FAO, 2013

Looking at the carbon print of the lost commodities, cereals contributed most, with 34% of GHG emissions associated with food loss and waste, followed by meat and vegetables with 21% each (Figure 1.10). It is interesting to compare cereals, which require nitrogen fertilizers and use of diesel for agricultural operations, with pulses, such as peas and beans. Pulses have the ability to fix nitrogen from air which means that they hardly need any nitrogen fertilizer. This shows that the choice of agricultural commodities matters when it comes to GHG emissions.

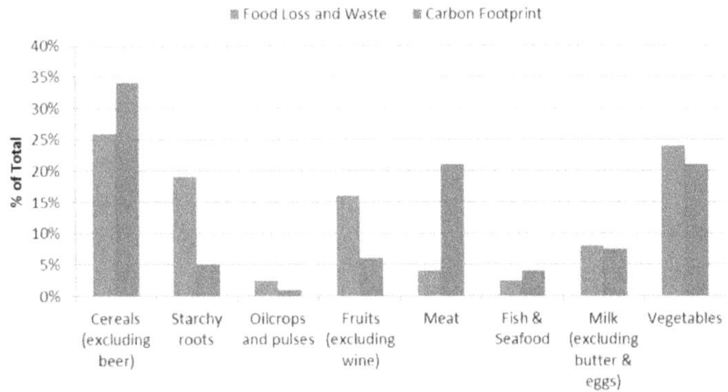


**1.10** Contribution of each commodity to food loss and waste and carbon footprint. Source: FAO, 2013

[16] FAO, 2013.

## CONCLUDING REMARKS

This article gives an overview of the relevance of food loss and waste in addressing the challenges of hunger and malnutrition in a context of limited natural resources. It shows the staggering toll that food loss and waste take on the environment, in particular with respect to land, water and GHG emissions.

Food loss and waste also poses a moral and ethical problem in a world where so many people go hungry. **Reducing food loss and waste must be an integral part of our common efforts to eradicate hunger from the planet.** FAO estimates that the food produced but never eaten would be enough to feed 2 billion people. If food losses and waste could be reduced by just half, the increase of food production needed to feed the world population by 2050 would be reduced from 60% to 25%. Food loss and waste also represents a huge economic loss of about 1 trillion USD. What a waste!

Food loss and waste is a complex issue. It involves many actors along the food chain: harvesting, transportation, storage. To reduce food loss and waste, the private sector is called upon to take the lead in many of these stages, but consumers, too, have a role to play by making intelligent choices. Governments can contribute through policies, legislation, programs; universities by promoting research activities and reviewing operations of their canteens; NGOs by linking retailers to food banks and raising awareness. Partnerships will be extremely important as we seek ways to reduce food loss and waste.

FAO is part of the global collective effort to reduce food loss and waste by providing a space for international discussion and consensus on the issue, by disseminating knowledge, data and analysis, by supporting countries in adopting strategies to reduce food loss and waste, and by providing a global platform to exchange information and lessons learned, such as the Save Food Network (www.fao.org/save-food/). FAO is also part of collaborative efforts to develop a Food Loss and Waste Protocol, which will help to better measure and monitor food loss and waste globally. FAO supports the UN Secretary-General's Zero Hunger Challenge launched in Rio in 2012, and discussions to include food loss and waste objectives into the Sustainable Development Goals to be adopted in 2015.

When we look at ways to eliminate hunger and malnutrition, we cannot afford to separate the food security of people, the economic perspective, and the environmental impact. As shown in this article on the issue of food loss and waste, **the people, the economy and the environment are inextricably intertwined and so must be the response.**

To conclude, let's focus on an important insight about hunger. We need to be reminded that **hunger is not a fatality**. It is about food production, certainly, but it is also about how society is organized. We can change this, but it requires vision, courage, political will. We are privileged in the sense that our generation has the knowledge and the technical means necessary to eliminate hunger from the planet during our lifetimes. **Again, it can be done and we must do it. Reducing food loss and waste is a first step that each one of us can take, starting today!**

# Chapter 2

## *Use of Natural Resources in the Global Food System*

**Robert Giegengack**

### ABSTRACT

While blatant food waste late in the supply chain is apparent to most of us, less apparent is the inefficiency of resource use in the processes whereby we produce, harvest, process, package, store, and deliver the food we eat.

Our food system extracts natural resources at rates many times faster than they are being replenished; such resources, thus, must be considered "finite" in human terms. We dispose of the waste products of our industrial and agricultural activities into natural systems at rates that exceed the capacity of those systems to neutralize or assimilate those wastes. Thus, we contaminate beyond use more of our renewable resources than we extract for use.

Of those natural resources essential to food production, water, oxygen, carbon dioxide, and nitrogen are abundant in the hydrosphere and atmosphere and, under "normal" circumstances, are fully renewable by natural processes. Solar energy required for photosynthesis, while not "renewable", is produced in abundance by the Sun, and will be available for the foreseeable future.

We are extracting other natural resources to support our food production at rates that are many times higher than the rates at which those resources are being naturally replenished. We can very likely sustain the current rate of food production as long as those resources last, but we must acknowledge that they are finite.

In 1900 the Earth supported 1.6 billion people, many of them not well. Futurists of that time estimated that the carrying capacity of Earth was not higher than 2.5 billion people. Today we feed 7.3 billion people, more calories/person than was the case in 1900, higher in the food chain, and on less land than was under cultivation in 1900. This has been achieved via development of a synthetic fertilizer industry, by selective development of high-yielding crops, by energy-intensive cultivation practices, by industrial-scale fish harvesting, and, most recently, by genetic manipulation of food plants and animals. We have not extended the global carrying capacity by reverting to traditional agricultural practices.

In the last 100 years, we have transformed our food industry on land from a system dependent on many small farmers practicing traditional forms of agriculture to a small number of multinational food-producing industries that utilize many forms of modern technology to extract the maximum amount of food from a given area of land.

Even with these advances, we do not feed 7.3 billion people well. However, nutrition deficiencies are not the consequence of inadequate global food production, but the consequence of distribution inefficiencies driven by economic and/or political factors.

We waste water: 72% of water "used" worldwide is applied directly to cropland, much of it via archaic technology. As the global population rises, and the demand for food also rises, water essential to irrigate crops to meet that demand must be sought from sources not previously exploited. We move water from where it is plentiful to places where we imagine it will be more useful. In the years since 1950, we have extracted groundwater at rates far greater than recharge at many aquifers worldwide to supplement surface water in irrigation.

We waste nutrients, even those that we know are in limited supply, by imprecise or excessive application, and by mismanagement of soil resources. Effluent from fertilized cropland has contaminated groundwater, streams, and vast areas of the ocean.

We waste our wild fisheries by extraction beyond their capacity to recover, by inadvertent destruction of non-target species that blunder into nets ("bycatch"), by wholesale destructive modification of sea-floor habitats by fishing equipment, and by contaminating the water on which marine resources depend. Exhaustion of marine fisheries was extensive before the first inventories were undertaken; thus, available baselines of fishery declines are not adequate to inform current management strategies.

We waste energy by pumping irrigation water against gravity, and in every stage of the food industry. Today, the U.S. food system invests >10 calories of energy for every food calorie delivered to an American household. Most of those invested calories, both in the USA and in other agricultural systems, come from fossil hydrocarbons. Surviving subsistence-agriculture societies deliver as much as 50 food calories for every fossil-fuel calorie invested. Many of those societies are now moving to more energy-intensive agricultural practices; thus, the percentage of the human population that feeds itself via "sustainable" agriculture is very small, and declining.

We have eliminated natural ecosystems across broad areas of land, replacing floral diversity with industrialized monoculture, and wild fauna with food animals. Fifty percent of the crops that we raise we use to feed food animals. Less than 1% of food we use from land-based sources is wild caught, but we still employ hunter-gatherer technology in pursuit of ~60% of marine food resources.

The systems that now produce food for 7.3 billion people can accommodate many more, as even newer technical advances are developed and implemented. But the most obvious and immediate strategy to feed the people we now have, and the people we expect, is to reform current practices to reduce waste in every stage of the food system.

## INTRODUCTION

In this volume, our authors explore the food waste that occurs in the long supply/consumption chain that extends from the farm to the table, to household disposal, and beyond.

I will summarize here the much larger waste of natural resources, many of them non-renewable, that occurs in processes before those resources reach the farms where our food products are grown.

Nothing in this chapter is new. The data from which I have drawn conclusions and recommendations have been available long enough to be considered part of the public domain of information. Thus, many of the references are to general articles, or to internet sources. But it may be useful to a discussion of the future of the global food industry to consider constraints imposed by availability of essential natural resources along with a review of the extent of the waste of produced food.

"Renewable resources" include water, atmospheric gases, energy*, and terrestrial soils.

"Non-renewable resources" include nutrients other than nitrogen, carbon, hydrogen, and oxygen.

*As a resource, energy is not renewable. But solar energy, essential to our global food industry, is produced in a quantity that vastly exceeds our needs, and will be available for the foreseeable future.

Fossil fuel, solar energy fixed via photosynthesis and stored in the rock column via convergence of natural processes, is now a very important input to our food industry. There is widespread disagreement as to the amount of extractable fossil fuel on Earth, but it is clearly being extracted and used at a rate that is much greater than the rate at which it is being replenished.

And, as we know, energy cannot be created or destroyed; it is conserved; but, as it is used, it becomes progressively less available.

Thus, energy is not renewable, but, on a human time scale, the Sun provides an inexhaustible supply of energy which, until recent centuries, represented the only energy input to the food industry. We will learn to revert to that source as other sources are exhausted or found to be environmentally too risky.

Major natural resources essential for production of human food, including solar radiation, energy from fossil fuels, water, soil, nutrients, and marine fisheries, are discussed below in terms of the nature, capacity, limitation, and implications regarding current human practice.

## SOLAR RADIATION

The Sun delivers 1,368 watts/$m^2$ (Solar Constant, across all wavelengths) to the outer Earth's atmosphere, which, when distributed over the Earth's spherical surface, provides an average of 342 watts/$m^2$ on a continuing basis. Actual solar irradiation varies by latitude and day of the year, influenced by the angle of the Earth as it orbits the sun. For example, the Equator receives ~3 times as much energy/$m^2$ as do the poles. Energy which reaches the Earth surface (insolation) is influenced by atmospheric attenuation, latitude, time of day, absorption, reflection, atmospheric dust, and cloud cover. Approximately 50% of that incoming energy is intercepted by the atmosphere and returned to space; thus, an average of 171 watts/$m^2$ reaches the Earth's solid/liquid surface. Of those 171 watts/$m^2$, photosynthesis uses ~2 watts/$m^2$ of light between 0.4 to 0.7 μm (the visible spectrum of light), termed photosynthetically active radiation.

Some of that solar radiation is used to produce photosynthate and is stored in plant and animal tissue as a form of chemical energy, primarily carbohydrate in plants and lipid in animals. Overall the efficiency of capture of total solar energy reaching plant surfaces into stored carbohydrate is only 1% to 3%, but can be as low as 0.2% and as high as 8% for sugar cane.

Not all of that chemical energy is recycled via Earth-surface processes; some is preserved, after death, as organic content of rocks, where, after burial, it is converted into coal, oil, and natural gas, fossil hydrocarbons that we choose to describe as "fossil fuels". The percentage of incident solar radiation so preserved is very small, but the process has been going on for at least 550 million years, so the inventory of fossil-fuel energy has become substantial. Today, however, we are extracting that fossil fuel and converting it into $H_2O$, $CO_2$, and energy at a rate many times faster than it is being replenished. Estimates of reserves of fossil fuel range widely, from decades to centuries, but few projections offer the prospect that much will be available in 500 years.

Thus, we are currently using fossil fuel at a rate that is at least *one million times* faster than it is being replenished. We may continue to use fossil fuels to drive our food industry, but we cannot rely on that source of energy as sustainable. Sooner or later, the continuing demand for energy that drives our food industry must be directed back to the Sun; it is worthy of note that the amount of energy delivered to Earth from the Sun is ~8,500 TIMES the amount of energy used by human civilization:

In 2012, human civilization used 524 Quads (Quadrillion btu) from all sources (U.S. Energy Information Agency, 2015). The U.S., representing ~4.5% of the global population, used 96 of those Quads, or 18%. To convert 524 Quads/yr to watts/m$^2$ of Earth's surface:

$\underline{524 \times 10^{15}\ \text{Quads/yr} \times 1055\ \text{joules/Quad}}$ $= 4 \times 10^{-2}$ watts $=$
0.04 watts/m$^2$

$31.5 \times 10^{6}\ \text{seconds/yr} \times 4\pi \times 36 \times 10^{12}\ \text{m}^2$

During 2012, the Sun delivered 342 watts/m$^2$ to the top of the atmosphere.

Thus, in 2012 the energy delivered by the Sun was *8500 TIMES* greater than the energy used by all of human civilization.

## ENERGY FROM FOSSIL FUELS

Until very recently, human societies produced all their food using fully renewable resources.

Fires for cooking and space heating were based on wood and other organic debris before coal was first used in China about 3500 BCE. Coal was later used in Greece and Roman Britain as a domestic fuel, but was not extracted on a commercial scale until coal became the fuel of choice in the Industrial Revolution that escalated in Europe and North America in the mid-18$^{th}$ century. Various engines, powered by wind or falling water prior to fossil hydrocarbons, were constructed to mill and process food products, but the primary energy input to food production remained human and animal labor. With the extraction of natural gas (1821, Fredonia, NY) and then petroleum (1859, Titusville, PA), liquid hydrocarbon fuels became available to power mobile internal-combustion engines that were used to drive tractors that, in the years after World War I, replaced animal and human labor in many agricultural economies.

Tied to agriculture, the human population remained rural, surrounding what would now be considered towns, until the population of Imperial Rome reached one million, briefly, in the 2nd century AD (estimates of rural populations have been difficult to make prior to the time of census records; these estimates are from Demographia, 2015). Four cities in China grew to one million inhabitants by 1000 AD, and Baghdad housed 1,000,000 people by 925 AD (but subsequently declined in population). London reached a population of 1,000,000 by 1825.

Mechanized transport late in the 19th century enabled food products to be brought unspoiled to the interiors of cities, and refrigeration made possible the storage of that food until it was distributed to end users and prepared for consumption. After 1900, the urban population grew rapidly. Today, more than half of the global population lives in 500 cities of 1 million inhabitants or larger; most of those cities are in the developing countries of Asia.

The amount of solar insolation incident on an urban landscape is not more or less than what falls on rural farmland. But the caloric demands of a densely populated city greatly exceed what might be grown on the area of land that a given city occupies.

In the U.S. food system today, ~1 acre (0.40 hectares) is required to raise the food for one person. Some agricultural practices feed slightly more than one person/acre; many are less efficient. One person/acre is a widely cited value.

New York City contained 8.2 million people in 2010; this population was distributed over a land area of 307 $mi^2$, or 795 $km^2$, or 196,000 acres. The population density of New York City was thus 103 people/hectare, or 42 people/acre, 42 times the population that could be supported if all of those 196,000 acres were converted to agriculture. Compared to many Asian cities, the population density of New York City is very low. New Delhi in 2010 had a population density of 255/hectare, or 100 people/acre; Dakka in Bangladesh contained 430 people/hectare, or 170 people/acre (Demographia, 2015; Wikipedia, 2015).

The projected population growth during the next 85 years – at least to 10 billion by 2060 and perhaps to 11 billion by 2100 – will be concentrated in existing cities; the UN estimates that the entire net growth will be accommodated in cities (Fischetti, 2014).

Urban agriculture may provide a fraction of the food needs of a low-density urban population, such as exist in North America and Western Europe. But urban agriculture cannot raise enough food to feed its urban population, let alone the increase in that population predicted for the next 50 years.

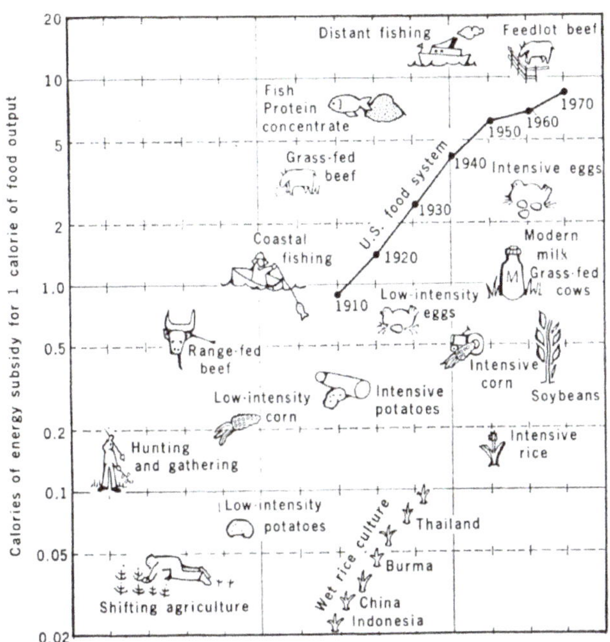


**2.1** Energy subsidies for various food crops. The energy history of the U.S. food system is shown for comparison (Steinhart and Steinhart, 1974, Fig. 5)

In the years after World War I, fossil-hydrocarbon fuel became essential to a modern food industry. In a much-cited paper, Steinhart and Steinhart (1974) charted the systematic growth of energy use in the U.S. food system, and compared that dependence to agricultural systems in other cultures. Steinhart and Steinhart showed (Figure 2.1) that energy use in the U.S. food system, expressed as calories of fossil fuel invested per calorie of food value produced, rose from <1 to 10 calories in the period 1910-1974. The same figure shows that traditional wet-rice culture, as then practiced in large areas of the world, yielded as many as 50 calories for each calorie of fossil fuel invested (and, of course, in some such cultures there was no investment of fossil fuel at all).

In the years since 1974, energy use in the U.S. food system has continued to rise, but at a slower pace than 1910-1970. Canning et al (2010) have shown that, from 1997 to 2002, energy use in the U.S. food system rose from 12.2% of total energy use to 14.4%. Most of the increase during those 5 years was in food processing.

Today we use fossil-hydrocarbon sources of energy, primarily petroleum, in many aspects of the food industry, from the fuel for agricultural machinery to the feedstock for synthetic fertilizers and pesticides/herbicides, to the production and distribution of seeds, to energy-intensive processing, to transportation, to refrigerated storage, to household storage, preparation, and disposal.

Today, with the universal availability of synthetic fertilizer and the mechanization of agricultural practices even in less-developed countries, the fossil-fuel subsidy is increasing worldwide, despite well-publicized efforts to return the food industry to "sustainability".

## WATER

Water is essential to every aspect of human life. Each of us consumes directly 2-4 liters of water per day, but we use vastly more in producing our food, and operating the complex system we call civilization.

Water is a fully renewable resource (see Figure 2.2). Water evaporates from the surface of land and ocean, and is transpired to the atmosphere by plants and animals. That water is carried by the circulating atmosphere, to fall back to Earth as rain or snow, to be taken up by plants and ingested by animals, to flow across the ground surface as flowing streams and masses of ice, to infiltrate into

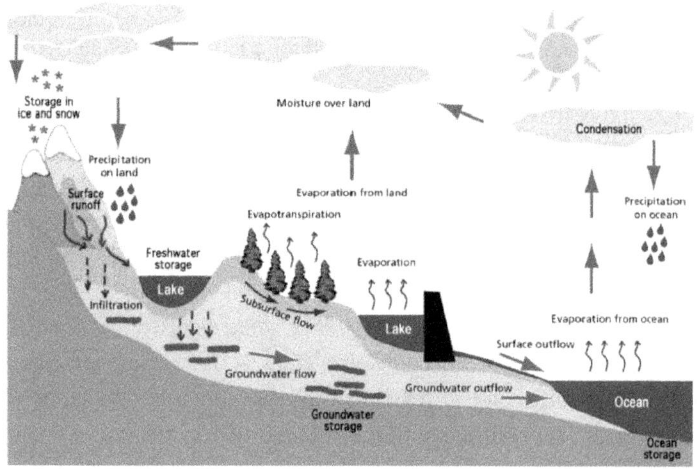


**2.2** The Hydrologic Cycle (USGS Water Science School; many other sources)

porous soil and rock, and to flow underground, eventually to the world ocean, from which water is continually removed by evaporation to re-cycle through these various reservoirs. The energy that drives this Hydrologic Cycle is the solar energy that evaporates water from the Earth's surface and moves that water around with circulating air masses; the acceleration of gravity draws that water back to the surface of Earth and, eventually, delivers it to the global ocean.

Today ~96.5% of the world's uncombined water resides in the global ocean, where it is available to marine organisms, but not usable directly for human ingestion. Salty lakes on land represent another 0.9% of the world's water. ~2.1% of the world's water is today retained in masses of ice on land, through which it moves very slowly; another ~0.5% lies beneath the surface of the land as groundwater, which is moving very slowly down gravitational gradients, eventually to the world ocean. A scant ~0.01% resides in lakes and flowing surface water, the atmosphere, and the biosphere (see Figure 2.3).

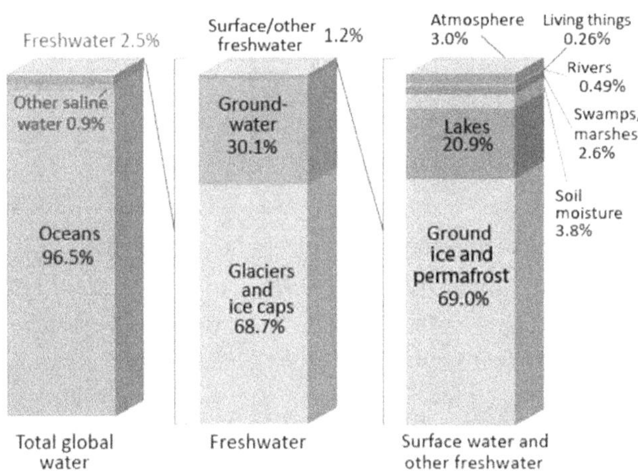


**2.3** Distribution of the Earth's water (Shiklomanov, 1993)

As scant as is this latter component, it has been enough to sustain a growing human population until the very recent past, when we learned to extract water by sinking wells into saturated rock and drawing that water to the ground surface, and, even more recently, to remove the dissolved salt from ocean water for direct human consumption and, eventually, to irrigate agriculture in regions not served by adequate direct precipitation.

As the human population has grown, we have vastly expanded our access to and use of water to the extent that availability of water has become a limiting factor to growth of human populations in many parts of the world.

Given today's configuration of international political boundaries, a majority of watersheds on Earth are "shared" by two or more sovereign states. This situation has led to repeated international hostilities in the Middle East, a region of scant rainfall, where large and growing human populations demand a growing supply of water for irrigation; and threatens to lead to similar political unrest among the rapidly growing economies of South Asia: India and China together represent 33% of the world's population, but control only 11% of the world's fresh water.

Even with our growing ability to find and extract water to support a growing human population, we still use directly barely 4% of the water that falls on the land (see Figure 2.4).

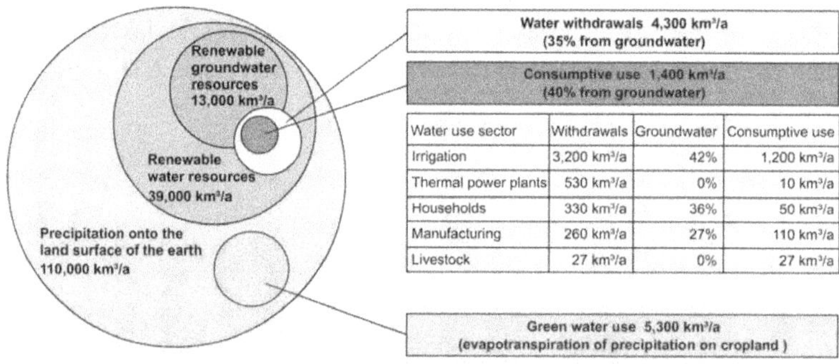


| Water use sector | Withdrawals | Groundwater | Consumptive use |
|---|---|---|---|
| Irrigation | 3,200 km³/a | 42% | 1,200 km³/a |
| Thermal power plants | 530 km³/a | 0% | 10 km³/a |
| Households | 330 km³/a | 36% | 50 km³/a |
| Manufacturing | 260 km³/a | 27% | 110 km³/a |
| Livestock | 27 km³/a | 0% | 27 km³/a |

**2.4** Human use of water (Wikimedia Commons, 2012)

Figure 2.4 summarizes the distribution of precipitation, 110,000 km$^3$/yr, that falls as rain or snow on the land area of Earth (largest green circle). Of that 110,000 km$^3$/yr, $\geq$55,000 km$^3$/yr is returned directly to the atmosphere in direct evaporation or via transpiration by plants. The smaller green circle, representing 5,300 km$^3$/yr, is water used and returned to the atmosphere by cropland. The blue circle, 39,000 km$^3$/yr, represents renewable water on the Earth's surface, of which 13,000 km$^3$/yr is resident below ground as groundwater. Humankind extracts 4,300 km$^3$/yr, of which 35% is groundwater, for all uses; 1,400 km$^3$/yr is converted to water vapor and returned to the atmosphere; that water use is described as "consumptive".

Seventy two percent (72%) of the water we use we deliver to food plants via many strategies of irrigation. That technology originated along the major rivers of the Middle East – the Nile and the Tigris-Euphrates – via engineered installations that retained and stored the annual floods from those two river systems, and then, as water levels fell seasonally, parceled that water out via engineered ditches to allow growing of food crops long after the time of year that they were supplied by the flood pulse or by direct rainfall. Gravity flow in these systems was supplemented by human- and/or animal-powered pumps. With the development of wind- and water-powered lifting devices and then, eventually, steam and internal-combustion engines, the regions of Earth on which food crops could be grown with artificial irrigation expanded; today we extract groundwater from aquifers across the world to deliver that water to surface crops. It has become apparent that, since World War II, the rate of extraction of groundwater in excess of recharge has depleted some aquifers (Wada and Bierkens, 2014; Richey et al, 2015) to the extent that the energy cost of extracting more water from them exceeds the agricultural benefit. Wada and Bierkens (2014, Fig. 6) conclude that the extraction of groundwater in excess of recharge, and the eventual transfer of that water, via irrigation, to the global ocean, contributes as much as 0.8 mm/yr to contemporary sea-level rise.

Clearly, there is enough fresh water available on Earth to sustain the direct needs of the human population; water crises occur because we use so much more to sustain our agricultural and industrial civilization, and because we dispose of the waste products of those activities directly into bodies of surface water at rates that exceed the capacity of those bodies of water to neutralize or assimilate those wastes. Thus, we contaminate beyond our use far more water than we extract to use. The infrastructure whereby we render much of the Earth's surface water unusable has evolved over 10,000 years of human civilization, during most of which time we were unaware that human disease was spread via water contaminated with human waste.

With modern understanding of the processes of the Hydrologic Cycle, and of the role of water in spreading human pathogens, we are now able to re-engineer our water systems to use that water more efficiently, to deliver it to meet the needs of a rapidly growing human population, and to protect an essential resource from avoidable contamination. Water remains a fully renewable resource; in recent millennia we have approached it as a finite natural resource.

## SOIL

With the invention of agriculture, human society evolved from small bands of foraging individuals to larger concentrations of people supported by food crops. Those crops were raised by a progressively smaller proportion of the population, freeing other members of society to develop specialized skills (crafts) and eventually, services. As agriculture continued to develop, farmers learned, first inadvertently and then quite systematically, to modify cultivated food plants by selection of the characteristics of the plants they grew from year to year, and yields grew.

The area of soil suitable for cultivation of human food is finite (about 11% of the Earth's land surface area is in permanent crops and pasture). However, the area of soil worldwide available to agriculture continues to diminish: excessive cultivation leads to accelerated erosion of agricultural soils. Erosion reduces the ion-exchange capacity of soil profiles, reduces the moisture-retaining capacity, and carries soil particles and adsorbed nutrient ions to the global ocean. Mismanagement of soil resources results in soil pollution and degradation. In large areas, soils are contaminated beyond their capacity to support food crops by accumulation of salts in arid regions where flood irrigation is practiced, and by accumulation of industrial effluents toxic to plant growth. Furthermore, large areas of land that previously had been dedicated to agriculture are lost each year by the expansion of cities and surrounding suburban regions (Eswaran et al, 2001; Montgomery, 2007; Crawford, 2012).

As the area of soil available worldwide continues to decline, humankind has learned to raise food crops via hydroponic technology at sites that would not support conventional agriculture, and to raise fish in artificial tanks that require neither soil nor even sunlight. These technologies offer promise to expand global food production to otherwise non-productive land, but will require inputs of energy.

## NUTRIENTS

Agronomists recognize 18 elemental nutrients that are essential to plant life, categorized as major, primary, secondary, and micro nutrients (Table 2.1). Most of the nutrients are available in minerals in near-surface rocks, and are made available to plants by chemical weathering of those minerals. That process is continuous, and, by most calculations, most mineral nutrients are replenished as fast as they are used in agricultural systems. Plants take up those nutrients through their root systems as ions dissolved in soil water. Plants extract carbon from the atmosphere as $CO_2$; hydrogen is taken up by plants as water, and split by photosynthesis to provide H and O. Plants also take up oxygen directly from the soil atmosphere.

C, H, and O are transferred among plant, atmospheric, and hydrospheric reservoirs, where all three are abundant. Primary, secondary, and micro-nutrients are fixed in plant tissues and returned to the soil upon the death of a plant, and by the loss of deciduous plant tissues. Plant tissues are ingested by animals and returned to the soil as excreta or dead tissue. As organic debris, known as soil litter, is broken down by soil microbes, the nutrient elements are released as dissolved ions, and either taken up directly by plants, or fixed on ion-exchange sites on mineral or organic particles in the soil, from which sites they may be extracted by plants as required. Because of the molecular configuration of weathered skeletons of silicate minerals, most of those ion-exchange sites are negatively charged. All primary, secondary, and micro-nutrients except N, P, and S predominantly form positively charged ions when dissolved; those ions may be stored in the soil adsorbed on ion-exchange sites.

**Table 2.1 Elemental nutrients essential to plant life (e.g.: Northeast Regional Certified crop Advisor, 2010)**

| TYPE | NUTRIENT | SOURCE |
|---|---|---|
| major non-mineral | Carbon, C | atmosphere |
| | Hydrogen, H | hydrosphere |
| | Oxygen, O | atmosphere, hydrosphere |
| Primary | Potassium, K | lithosphere |
| | Nitrogen, N | atmosphere, via biosphere |
| | Phosphorus, P | lithosphere |
| secondary | Calcium, Ca | lithosphere |
| | Magnesium, Mg | lithosphere |
| | Sulfur, S | lithosphere |
| micro *(not all are essential for all plants)* | Iron, Fe | lithosphere |
| | Manganese, Mn | lithosphere |
| | Chlorine, Cl | lithosphere |
| | Zinc, Zn | lithosphere |
| | Copper, Cu | lithosphere |
| | Boron, B | lithosphere |
| | Molybdenum, Mo | lithosphere |
| | Sodium, Na | lithosphere |
| | Silicon, Si | lithosphere |

Two essential nutrients, nitrogen and phosphorus, represent departures from the simplified analysis above.

**Nitrogen**

Nitrogen as an element was identified by Daniel Rutherford in 1772. Soon thereafter, the importance to agriculture of N as an essential nutrient was recognized, as was the fact that the atmosphere consists of ~78% elemental N (as $N_2$), which, known as non-reactive nitrogen, plants cannot directly utilize.

Under natural conditions, N becomes bioavailable (reactive N, or Nr) by lightning discharge, and by the activities of N-fixing micro-organisms, either resident in soil as free-living bacteria, or attached to some plants as symbionts.

Nitrogen forms are predominantly negatively charged ions that are not adsorbed on ion-exchange sites in the soil, and thus are carried by leaching water through the soil to groundwater reservoirs (aquifers), or by overland flow to surface streams, and, eventually, to the global ocean. Accumulation of excess N in groundwater, streams, and the ocean presents not only an avoidable waste of an essential plant nutrient, but leads to significant degradation of water resources.

Agriculture was developed in the absence of any understanding of modern chemistry. Trial-and-error practices led early farmers to preserve soil nutrients by leaving crop residue on fields, by supplementing that residue with human and animal manures, and by co-cultivating plants that harbor N-fixing symbionts, either by systems of alternate planting or by intercropping N-fixing plants with other crops. A comprehensive summary of humankind's quest for agricultural N is contained in the book *Enriching the Earth* (2000) by Vaclav Smil.

In natural terrestrial ecosystems, plant productivity is limited by the availability of nitrogen. Human efforts through many generations to extract plant-available N from the atmosphere were rewarded in 1911 by Fritz Haber's invention, in Germany, of a process to fix N as $NH_3$ in a high-temperature, high-pressure reaction catalyzed by Fe. By 1913, Carl Bosch had developed a procedure to produce Haber $NH_3$ on a commercial scale (e.g.: Smil, 2000). The process is now referred to as the Haber-Bosch process.

Haber-Bosch $NH_3$ was first used to manufacture munitions for the German military in World War I; after that War, Haber-Bosch $NH_3$ was applied directly to soils, or converted to plant-available nitrates that are more stable in soils. Today, 55% of N fixed in plants via agricultural activity worldwide is derived from the Haber-Bosch process (Smil, 2000), and 40% of the human population would not be alive if it were not for the Haber-Bosch invention.

Application of N to cropland in excess of plant needs, and accumulation of excess N in industrial-scale animal feedlots, produces N-enriched effluent. That effluent has contaminated valuable aquifers and has led to eutrophication of areas of the global ocean adjacent to the mouths of rivers that drain agricultural land.

**Phosphorus**

Among the 14 mineral nutrients, phosphorus is both needed in relatively large quantities, and appears to be in short supply (UNEP, 2011; Van Kauwenbergh, 2014). Supplemental P was first applied to agricultural fields on an industrial scale from the extensive supplies of bird droppings – guano – described by Alexander von Humboldt after his 1802 voyage along the west coast of South America.

The bird-guano deposits of South America, which supplied both N and P, were extracted and distributed globally during the 19$^{th}$ century. Competition for dwindling supplies of guano sparked the 1864-66 "Guano War" between Spain and her former colonies Peru and Chile. That conflict re-erupted later in the 19$^{th}$ century as a territorial dispute between Chile and her neighbors Peru and Bolivia (Sater, 2007), that briefly also engaged Argentina. Conflict over the combined N and P resources of seabird guano declined as the deposits were exhausted at the end of the 19$^{th}$ century. Today, most agricultural P comes from rock-phosphate deposits in Morocco, China. Russia, and the USA. Deposits of lower quality, and thus not of commercial value at current prices, are known from many other localities. Estimates of the capacity of those deposits to provide needed agricultural P vary widely – from decades to centuries, and the sizes of lower-grade sources of P are far more difficult to determine.

It is nonetheless clear that, among nutrients essential for plant growth, P is most limited in supply and as a natural resource nonrenewable. The nature of chemical forms of P in soils limits its bioavailability. Meanwhile, P losses through soil erosion and surface runoff, together with N, are the major culprits leading to water degradation worldwide and to many of the dead zones of the global ocean.

The manufacture and distribution of plant nutrients as fertilizer is an important part of modern agriculture, and uses energy, primarily from fossil fuels, to extract, process, and distribute fertilizers and, particularly, to convert atmospheric $N_2$ to $NH_3$ to be further processed to stable, but plant-available nitrogenous fertilizer. Production of $NH_3$ via the Haber-Bosch process alone uses ~2% of the energy used by humans.

## MARINE FISHERIES

Long after humans developed agricultural practices on land, we still extract most of the marine protein we eat using hunter-gatherer technology. Humans have been extracting fish and shellfish from fresh- and near-shore salt-water sources for many millennia, but it was not until the development of sea-going ships that fish resources could be sought on the high seas, far from coastal waters. Initially, those fish were dried or preserved in salt until the holds of fishing vessels were filled; with the invention of refrigeration, fish are now frozen at sea. Fisherpersons are thus able to range far and wide across the oceans of the world to deliver marine protein to populations far from the fishing grounds.

Although oceans cover ~70% of the Earth's surface, marine protein makes up a much smaller portion of the human diet than does animal protein raised on land. However, humans have been exploiting fish since long before accurate data of fish catch of any kind were recorded, so we don't know what the productivity of the world ocean might have been before humans began to take food from what was described in the 15th century as an inexhaustible supply.

The history of the groundfish (cod, haddock, yellowtail flounder, redfish) fishery in the North Atlantic is a good example of the rapid decline of an essential fishery in the face of uninformed over-exploitation (Kurlansky, 1998).

European fishermen, primarily from Spain, Portugal, and France, exploited the groundfish, primarily cod, on the Grand Banks and Georges Bank soon after John Cabot returned from his 1497 voyage of discovery to describe limitless supplies of fish in the Atlantic Ocean east of what is now Newfoundland, Nova Scotia, and New England. Viking and Portuguese fishermen exploited this resource before the "discovery" of the Americas by Columbus. Although word-of-mouth testimony attested to the decline of both size and abundance of cod through the 16th and 17th century, detailed monitoring of the condition of marine fisheries was not undertaken until early in the 20th century.

Thus, although the decline in productivity of groundfish was documented by the late 1800s, there are no reliable records of either the volume of fish collected or their mean size until evidence of the severe depletion of the cod fishery became apparent in the 20th century. Over-exploitation of cod, a key predator in the North Atlantic, has led to the proliferation of species on which the cod feed. Thus, species on which cod feed are now so abundant that they threaten the survival of cod eggs and juveniles. Depletion of cod in the North Atlantic may have fundamentally changed the structure of the food web in that fishery.

Many other fisheries around the world have suffered similar declines. Students of global marine fisheries (e.g. FAO 2015; MIT, 2015) now estimate that 52% of global fisheries are "fully exploited"; 17% are "overexploited"; 7% are "depleted"; 2% are "recovering"; 20% are "moderately exploited"; and only 3% are "underexploited."

The primary cause of that decline in fisheries production is unregulated overharvesting, aggravated by destruction of bottom habitat, ocean pollution, and bycatch, the inadvertent destruction of non-target marine species by non-discriminating fishing gear.

The state of ocean fisheries is now so dire that many students of marine ecology have asserted that some stocks have been depleted beyond their capacity to recover, and they have called for a wholesale moratorium on ocean fishing. Such a moratorium is difficult to enact and enforce, since so much of that activity is conducted in international waters.

## SUMMARY

During the last 100 years, and particularly since WW II, the global food industry has seized market opportunities as they emerged and, supported by low-cost energy and advancing technology, has extracted natural resources, both renewable and finite, at rates unprecedented in prior human history. As a consequence of this very recent trend, many key resources now appear threatened, and will be further strained to feed a human population that is predicted to approach 11 billion by 2100.

In 1900, the world contained ~1.6 billion people, and futurists of that time undertook to calculate the *carrying capacity* of Earth. Such calculations were based on measurements of available arable land, yields/acre of various crops at 1900 agricultural efficiencies, projected mix of plant vs animal sources, human caloric demands, etc. Such calculations led to estimates that the carrying capacity of Earth would not exceed 2.0 to 2.5 billion people.

In 2015, we feed 7.3 billion people (~10% of them not well) more calories/person than was the case in 1900; higher on the food chain (more meat and dairy products/person) than in 1900; and on less land than was under cultivation in 1900. We have achieved that capacity, considered impossible in 1900, through a number of advances:

1. We have learned to use solar energy stored as "fossil fuel" to:

    a) increase the efficiency of cultivation;

    b) manufacture and apply fertilizer, particularly nitrogenous fertilizer;

    c) manufacture and apply synthetic pesticides and herbicides;

    d) pump water against gravity, primarily to be used for irrigation;

    e) raise food animals intensively to add protein to the human diet;

f) extract fish from deep water far from land;

g) process raw foods to protect quality and to facilitate distribution;

h) maintain much of that food at low temperature;

i) manufacture and distribute antibiotics that not only protect food animals from pathogens that are efficiently distributed among animals kept in dense concentrations, but also enhance the rate at which those animals grow to harvestable size.

2. We have enhanced the yield of essential cereal grains, by selective breeding, to take advantage of higher levels of available nitrogen and mechanized cultivation, via the *Green Revolution* (e.g.: Nobelprize.org, 1970), initiated by Norman Borlaug in 1944.

3. In recent years, the food industry has benefited from the capacity to manipulate directly the genetic material of food plants, to enhance resistance to pests, to improve yields, and to lend desirable characteristics to various plants. Organisms so modified have come to be labeled Genetically Modified Organisms, or "GMOs".

The excesses of the modern global food industry, both real and perceived, have driven widespread reaction to the departure from traditional cultivation processes that that industry represents, and to the fear that technologies not tested by generations of experience offer the prospect of introducing into the food system unanticipated negative consequences of those technologies. Many social movements have arisen to restore the traditional agriculture that prevailed across the world as late as 1900 AD. These movements stress "natural", or "organic", practices as more conservative of resources, without acknowledging that even the "traditional agriculture" that fed 1.6 billion people in 1900 represented a dramatic departure from the natural conditions that prevailed across those continents before humans learned to raise food plants under controlled circumstances. Indeed, the conversion of vast areas of the world's continents from biologically diverse forests, grasslands, wetlands, etc., to productive farmland and pasture represents the most sweeping modification of "natural" conditions that humans have achieved. Because that conversion was largely accomplished before humankind began to document the systematic alteration of the environment that accompanied the rising success of *Homo sapiens*, many of us think of a rural agricultural setting as a "natural" environment.

Proponents of a return to traditional agriculture have not defined the concepts "natural" or "organic" in precise terms. The U.S. Department of

Agriculture undertakes to certify producers of food as "organic" if they meet specified requirements of land management, fertilizer/pesticide/herbicide use, animal husbandry, etc. (USDA, 2015a). The certification process is protracted and expensive; USDA (2015b) maintains *The National List of Allowed and Prohibited Substances*, which runs to many pages. The process allows for many exemptions.

Organic food production as a reaction to perceived excesses of the modern food industry emerged in the latter half of the 20th century, and represents a highly visible and growing component of the food industry today.

While it is clear that traditional agricultural technologies will deliver smaller burdens of synthetic pesticides, herbicides, antibodies, and growth-enhancing drugs to crops and to food animals, it is equally clear that the substantially lower yields/acre of such practices offer little hope of feeding 11 billion people. The technologies that today feed 7.3 billion people evolved in part in response to the need to feed an ever-growing human population, and cannot be abandoned without extreme human hardship. It seems likely that the modern organic alternative will remain a choice of Western populations who can afford to pay more for the perceived benefits of organic production. Globally speaking, regressing to traditional farming or organic methods as a strategy is unlikely to succeed in feeding a much larger population.

To provide food for 11 billion people by 2100, humankind must learn to produce and distribute food resources more efficiently than we do today, and must learn to curtail the avoidable waste that characterizes every step of that process. The following chapters in this volume will describe the success of efforts, in many arenas, to reduce waste in the later stages of the global food system.

Reducing the avoidable waste of natural resources that occurs before we actually plant and harvest our crops will not necessarily increase the amount of food that the Earth can produce to feed 11 billion people. But curtailing the wasteful exploitation of those resources will help ensure that whatever strategies are put in place to reduce food waste in the food-distribution system will be sustained through future generations when, presumably, the size of the human population, and its demand for natural resources, may start to decline.

To achieve this goal, humankind must, as a minimum:

1. Protect major renewable resources, air and water, from contamination with the waste products of our rapidly evolving technology;

2. Preserve the present inventory of arable land for food production;

3. Facilitate the transition to one or another form of abundant solar energy while fossil hydrocarbons can still be reserved for the future petrochemical and fertilizer industries;

4. Restructure contemporary irrigation technology to reduce losses through evaporation and infiltration;

5. Move food from regions where water is plentiful to locations where that food is needed, rather than move water from areas of abundance to grow food crops in water-stressed areas;

6. Apply fertilizer only as needed, and directly to the plants that will use it;

7. Use modern information technology to apply both water and nutrients with greater precision (and, thus, economy) than is used today;

8. Recycle plant nutrients by making more efficient use of agricultural waste products;

9. Develop hydroponic agriculture, which, while energy-intensive, allows the production of food at sites where soil resources are not available, and offers a productive strategy to recycle essential plant nutrients;

10. Greatly reduce, if not eliminate, extraction of fish protein from the world ocean until key fisheries have had an opportunity to recover, and monitor and regulate wild fish extraction as fisheries are gradually re-opened;

11. Continue to develop controlled aquaculture, both of fresh- and salt-water species. Effluent from aquaculture facilities can be applied directly to hydroponic agriculture;

12. Reduce the proportion of the human diet met by animal and dairy products in industrialized economies: eat "lower in the food chain";

13. Extend our concept of "food" to include resources not now extensively utilized: marine plankton and insects. Both resources are abundant, widely distributed, and nutritious, and lie lower in the food chain than much of the food we now use.

Taken together, these provisions may enable us to maintain, or perhaps even increase, the current agricultural productivity of Earth, while aggressive strategies to reduce food waste after production will enable the Earth to support 11 billion people.

## SPECULATIVE CONTRIBUTIONS TO OUR FOOD FUTURE

21st-century futurists undertake to look beyond strategies listed above to:

Aquaponics: the direct integration of hydroponic agriculture and aquaculture; and

Seasteading: the construction of fully sustainable human habitations on floating islands in international waters (www.seasteading.org).

Continued development of genetic engineering may enable plant scientists to improve the efficiency of photosynthesis by food plants.

Given the history of rapid change in the global food system, it seems likely that human ingenuity will develop other strategies, not imagined today, to feed a human population that is projected to continue to grow at least until 2100.

But, the realities of finite basic natural resources necessary to drive our convoluted food system must not be ignored. The fundamental laws of conservation of matter and energy will prevail. Human strategies to feed a larger population must be accommodated within those operational realities.

## REFERENCES

Canning, P., A. Charles, S. Huang, K.R. Polenske, and A. Waters. 2010. Energy use in the U.S. food system. USDA Econ. Res. Serv., ERR-94.

Crawford, J. 2012. What if the world's soil runs out? World Economic Forum, Dec. 14, 2012: http://world.time.com/2012/12/14/what-if-the-worlds-soil-runs-out/.

Demographia world urban areas, 2015, 11th annual edition: http://www.demographia.com/db-worldua.pdf.

Eswaran, H., R. Lal, and P.F. Reich. 2001. Land degradation: an overview. In: Bridges, E.M., I.D. Hannam, L.R. Oldeman, F.W.T. Penning de Vries, S.J. Scherr, and S. Sompatpanit (eds.). Responses to Land Degradation. Proc. 2nd. International Conference on Land Degradation and Desertification, Khon Kaen, Thailand. Oxford Press, New Delhi, India: Reprinted by USDA-SCS, in: http://www.nrcs.usda.gov/wps/portal/nrcs/detail/soils/use/?cid=nrcs142p2_054028.

FAO. 2015. Fisheries and Aquaculture Department. Rome. Updated 23 June 2015. http://www.fao.org/fishery/statistics/software/fishstatj/en

Fischetti, M. 2014. World Population Will Soar Higher Than Predicted. Sci. Am. http://www.scientificamerican.com/article/world-population-will-soar-higher-than-predicted/.

Kurlansky, M. 1998. Cod: A Biography of the Fish That Changed the World. Walker Publ., New York, NY.

MIT Mission Biodiversity, 2015 (fisheries): http://web.mit.edu/12.000/www/m2015/2015/fisheries.html.

Montgomery, D.R. 2007. Soil erosion and agricultural sustainability. Proc. Nat. Acad. Sci. 104(33):13268-13272. http://dx.doi.org/10.1073/pnas.0611508104.

Nobelprize.org. 1970. Norman Borlaug – Biographical. http://www.nobelprize.org/nobel_prizes/peace/laureates/1970/borlaug-bio.html.

Northeast Regional Certified Crop Advisor. 2010. Cornell University. http://nrcca.cals.cornell.edu/nutrient/CA1/CA010102.php.

Richey, A.S., B.F. Thomas, M.-H. Lo, J.T. Reager, J.S. Famiglietti, K. Voss, S. Swenson, and M. Bodell. 2015. Quantifying renewable groundwater stress with GRACE. Water Resour. Res. 51(7):5217-5238. http://dx.doi.org/10.1002/2015WR017349.

Sater, W.F. 2007. Andean Tragedy: Fighting the War of the Pacific, 1879–1884. University of Nebraska Press, Lincoln, NE.

Seasteading. www.seasteding.org.

Shiklomanov, I. 1993. World fresh water resources. in: P.H. Gleick (editor) Water in Crisis: A Guide to the World's Fresh Water Resources. Oxford University Press, New York, NY.

Smil, V. 2000. Enriching the Earth: Fritz Haber, Carl Bosch, and the Transformation of World Food Production. The MIT Press, Cambridge, MA.

Steinhart, J.S., and C.E. Steinhart. 1974. Energy use in the U.S. food system. Science 184(4134):307-316.

US Department of Agriculture (USDA) Organic Program. 2015a. http://www.ams.usda.gov/AMSv1.0/nop.

US Department of Agriculture (USDA). 2015b. The National List of Allowed and Prohibited Substances [for organic certification]. http://www.ecfr.gov/cgi-bin/text-idx?c=ecfr&SID=9874504b6f1025eb0e6b67cadf9d3b40&rgn=div6&view=text&node=7:3.1.1.9.32.7&idno=7.

UNEP Yearbook. 2010. Phosphorus and food production: http://www.unep.org/yearbook/2011/pdfs/phosphorus_and_food_production.pdf.

Van Kauwenbergh, S.J. 2014. Global Phosphate Rock Reserves and Resources, the Future of Phosphate Fertilizer: U.S. Department of Agriculture

2014 Agricultural Outlook Forum February 20-21, 2014. http://www.usda.gov/oce/forum/past_speeches/2014_Speeches/Kauwenbergh.pdf.

Wada, Y., and M.F.P. Bierkens. 2014. Sustainability of global water use: past reconstruction and future projections. Environ. Res. Lett. 9(10):104033. http://dx.doi.org/10.1088/1748-9326/9/10/104033.

Wikimedia Commons. 2014. Global Values of Water Resources and Water Use. https://commons.wikimedia.org/wiki/File:Global_Values_of_Water_Resources_and_Water_Use.jpg#filehistory.

Wikipedia. 2015. List of largest cities through history. https://en.wikipedia.org/wiki/List_of_largest_cities_throughout_history.

# Chapter 3

## *Food Loss in the United States at the Retail and Consumer Levels*

**Jean C. Buzby and Hodan F. Wells**

### ABSTRACT

The U.S. Department of Agriculture's Economic Research Service estimates the amount of food loss at the retail and consumer levels in the United States using their Loss-Adjusted Food Availability data series. In the United States, 31 percent – or 133 billion pounds – of the 430 billion pounds of the available food supply at the retail and consumer levels in 2010 went uneaten. The estimated value of this food loss was $161.6 billion using retail prices. This amount of food loss translates into 141 trillion calories in 2010. This chapter presents four key messages about food loss. First, quantities of food loss at the consumer level in the United States are larger than at the retail level for all food groups except added fats and oils. Second, the ranking of food loss varies depending on if measured by amount, value, or calories. Third, individual foods with the highest percent losses are not necessarily the foods with the most food loss. Fourth, measuring food loss is challenging and data intensive and as a result, data gaps exist for national estimates for individual commodities. In particular, data gaps exist at the farm level and between the farm and retail levels.

### INTRODUCTION

Definitions of food loss and waste vary worldwide, complicating the comparison of estimates across countries. The U.S. Department of Agriculture's Economic Research Service (USDA/ERS) defines *food loss* as the edible amount of food, postharvest, that is available for human consumption but is not consumed for any reason. These reasons include mold, pests, or inadequate climate control, cooking loss, natural shrinkage, and food waste. Some types of loss– such as spoilage of fresh strawberries and moisture loss in fresh leafy greens– occur at every stage of the farm to fork chain. Between the farm gate and retail stages, food loss can arise from problems during drying, milling, transporting, or processing that expose food to damage by insects, rodents, birds, molds, and bacteria. At the retail level, equipment malfunction of cold storage, over-ordering, and culling of blemished produce can result in food loss. Consumers also contribute to food loss when they cook more than they need and throw out the extras. Figure 3.1 provides an example of broccoli loss at different stages.

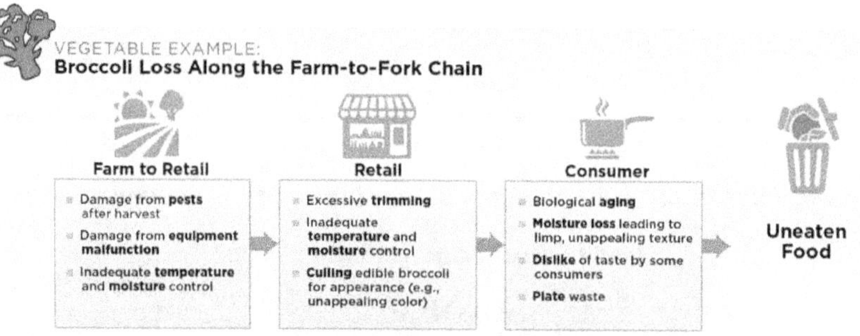


**3.1** Broccoli loss in the food supply chain

ERS' food loss estimates are only for the retail and consumer levels in the United States because of data limitations. Nationally-representative data on total food loss at the farm level and between the farm and retail levels are unavailable.

## WHAT ARE THE CONSIDERATIONS AND INCENTIVES CONCERNING FOOD LOSS?

Some loss is inevitable because food is inherently perishable, and spoiled or deteriorated food must be discarded to ensure the safety and wholesomeness of the food supply. For example, restaurant leftovers not taken home by patrons are appropriately discarded out of health considerations. Also, some meat, poultry, and other foods are recalled when there are health or safety concerns.

Individual tastes and preferences also come into play for consumers. For example, some people may not like to eat the crusts on their sandwiches or don't like or get around to using leftovers. Given the number of calories and overweight people in the United States, it would be detrimental for everyone to eat all the food that they are served or buy.

Economic factors may only provide limited incentives to reduce food loss, that is, some amount of loss may be *economically justifiable*. For example, it may not be worthwhile for a supermarket to pay for the labor and other costs to monitor and mark down foods as they approach the sell-by dates when considering the lower price they then might receive, and other factors.

Additionally, there are often tradeoffs between the advantages of using technologies that reduce loss and any disadvantages of using the technologies. For example, the chemical methyl bromide helps extend the shelf life of almonds but also acts as an ozone-depleting gas when released into the atmosphere.

One important point to understand is just because food goes uneaten, that does not mean there has to be a 100 percent loss in value or that it is a total waste.

If the food is edible and safe to eat, it could be donated to feed those in need. However, logistical challenges of getting wholesome food to the hungry exist, such as the dispersion of uneaten food among millions of households, food processing plants, and food-service locations, and the time and expense needed to deliver food to a new destination, such as to a food bank. Additionally, uneaten food can be diverted to other economic uses such as, energy creation and composting or even feeding livestock, zoo animals, pets, and other animals. However, there are tradeoffs and limits to how much food loss the United States could realistically prevent, recover for human consumption, or divert to another economic use.

Advances in food packaging, handling, and tracking technologies show promise in reducing food loss. For example, special plastic films – which allow produce to breathe – continue to be developed and improved. Other examples of innovations that reduce the creation of food loss include nanoclays used in beer bottles, nanosilver used in food storage containers and refrigerator compartments, and fruit-and vegetable-based products that inhibit spoilage of fresh cut produce (Buzby, 2010; Golan and Buzby, 2015).

**ERS' Food Availability Data System**

ERS' core Food Availability data series is the foundation for two other data series in the Food Availability Data System (FADS), the Loss-Adjusted Food Availability data (LAFA) compiled by ERS and the Nutrient Availability data computed by USDA's Centers for Nutrition Policy and Promotion (CNPP) in what they call their Nutrient Content of the U.S. Food Supply series. In short, the FADS provides estimates of the supply of food and nutrients available for consumption in the United States.

For a given year, the supply of each commodity in the core Food Availability data series is the sum of production, imports, and beginning inventories, and from this amount, ERS then subtracts out exports, farm and industrial uses, and ending stocks (Figure 3.2). Another USDA agency – the National Agricultural Statistics Service (NASS) – and the Bureau of the Census collect data on these components directly from producers and distributors using techniques that vary by commodity. These data are not collected from surveys of individual consumers, and thus provide an independent basis for examining food consumption trends. Per capita availability estimates are calculated by dividing the total annual domestic availability for a commodity divided by the U.S. population for that year.

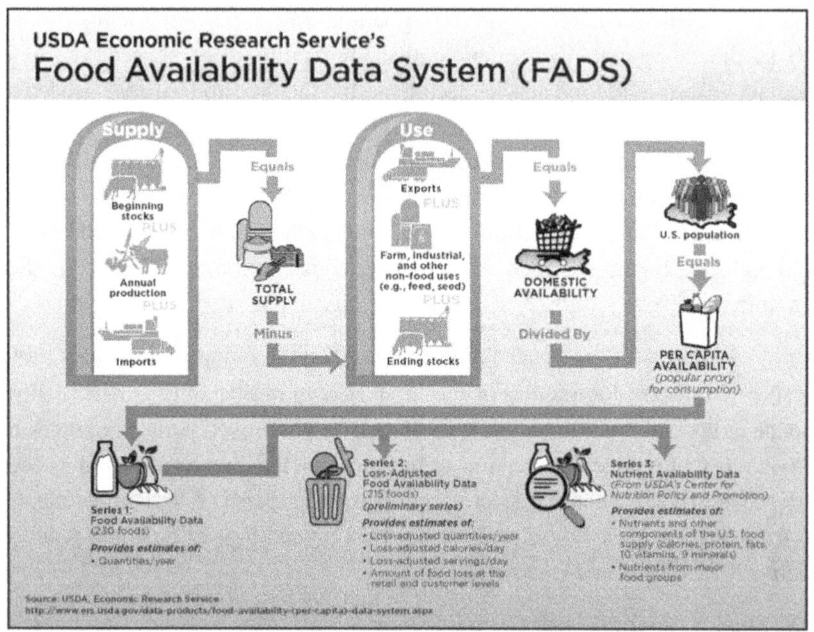


**3.2** The USDA-ERS Food Availability Data System

Like the Food Availability estimates, the LAFA estimates serve as popular proxies for actual consumption for roughly 215 commodities (e.g., fresh spinach, beef, and eggs) in the United States. Both of these series provide per capita estimates of food availability for individual commodities and food groups and where appropriate, in total. Here, we focus on the LAFA series because the underlying loss estimates are used to estimate food loss in the United States. Details on both the core Food Availability data series and the LAFA series can be found on the ERS website (ERS, 2015a, 2015b).

The LAFA data series was developed by ERS in the late 1990s because the core Food Availability data series overstates the amount of food ingested. In short, the LAFA series takes into account the substantial quantities of food that goes uneaten because of spoilage, moisture loss, plate waste and other reasons from farm to fork. The primary goal of this series is to more closely approximate actual intake. The series is considered to be preliminary as there are initiatives underway to update and improve the data series.

Unlike the core FA data series which has data back to 1909 for many commodities, the LAFA data series begins in 1970 and both series have data now available through 2012. Both the FA and LAFA estimates are useful for studying food consumption trends, among other uses. The LAFA data series is constructed by taking the Food Availability data for each commodity and taking into account three types of losses to more closely estimate actual intake:

1) The Loss from primary or farm weight to the retail weight (for limited commodities, such as fresh apples, canned peaches, and pork).

2) The Loss at the retail level, such as in supermarkets, supercenters, convenience stores, mom-and-pop grocery stores, and other retail outlets (but not including restaurants and other foodservice outlets). Retail losses include dented cans, unpurchased holiday foods, spoilage, and the culling of blemished or misshaped foods.

3) The Loss at the consumer level includes losses for food consumed at-home and away-from-home and includes "cooking loss and uneaten food" from the edible share, such as extra tomato sauce poured down the drain, plate waste, such as broccoli served to children who dislike the taste, and fresh strawberries that are thrown out because they turned blue on the kitchen counter. The "non-edible share" of certain commodities, such as fresh fruits, vegetables, and eggs are also removed at the consumer level (e.g., broccoli stalks, peach pits, and apple cores). Data on the non-edible share are from the *National Nutrient Database for Standard Reference*, compiled by USDA's Agricultural Research Service (U.S. Department of Agriculture, 2007).

The end result is that the Loss-Adjusted Food Availability data more closely approximates actual food intake than the Food Availability series. In addition to loss-adjusted estimates of per capita consumption, the data are presented in two forms: the calories available per capita per day, and the daily allowance or servings available per capita per day. Importantly, the data can be used to compare with dietary recommendations, such as from the *Dietary Guidelines for Americans*, because they can be aggregated to the main food commodity groups. Additionally, the LAFA data series is used to estimate food loss at the retail and consumer levels.

## SO HOW BIG IS THE PROBLEM?

In the United States in 2010, 31 percent of the available food supply was lost at the retail and consumer levels (Buzby et al., 2014). Of this amount, retail-level losses tallied 10 percentage points and consumer level losses totaled an additional 21 percentage points. Losses on-farm and between the farm and retailer were not estimated due to data limitations for some of the food groups. Had these losses been included, total postharvest loss in the United States would be well over 31 percent.

In terms of the amount of food loss, about 133 billion pounds of the available food supply went uneaten at the retail and consumer levels in the United States in 2010 (Table 3.1). Of this amount, retail level losses accounted for 43 billion pounds and consumer level losses totaled almost 90 billion pounds. On a per capita basis, this amount of food loss is roughly 429 pounds per year at both levels of which 290 pounds went uneaten at the consumer level.

**Table 3.1 Estimated Total Food Loss in the United States, 2010. Source: Buzby, Wells and Hyman, 2014**

| Commodity | Losses from Food Supply* | | |
|---|---|---|---|
| | Retail | Level | Total |
| | | *Billion pounds* | |
| Dairy products | 9.3 | 16.2 | 25.4 |
| Vegetables | 7.0 | 18.2 | 25.2 |
| Grain products | 7.2 | 11.3 | 18.5 |
| Fruit | 6.0 | 12.5 | 18.4 |
| Added sugar and sweeteners | 4.5 | 12.3 | 16.7 |
| Meat, poultry, and fish | 2.7 | 12.7 | 15.3 |
| Added fats and oils | 5.4 | 4.5 | 9.9 |
| Eggs | 0.7 | 2.1 | 2.8 |
| Tree nuts and peanuts | 0.2 | 0.3 | 0.5 |
| Total | 43.0 | 89.9 | 132.9 |

*Totals may not add due to rounding.

Figure 3.3 shows the amount of food loss for each food group by the size of the bar and also the breakdown between the retail loss in yellow and the consumer-level loss in blue. In short, the share of each depends on the food group. For example, out of the total amount of loss of grain products, 39 percent of loss occurred at retail level and 61 percent at the consumer level. Added fats and oils was the only food group where a larger portion of loss occurred at the retail level than at the consumer level. In terms of pounds of food loss, dairy products had the largest loss at the retail level, while vegetables had the greatest loss at the consumer level.

The estimated total value of food loss at the retail and consumer levels in the United States was around $162 billion as measured using average retail prices for each commodity. This amount of food loss translates to an estimated $522 per capita at both levels and of this amount, $371 was at the consumer level.

ERS also estimates the amount of food loss at the retail and consumer level at over 141 trillion calories.

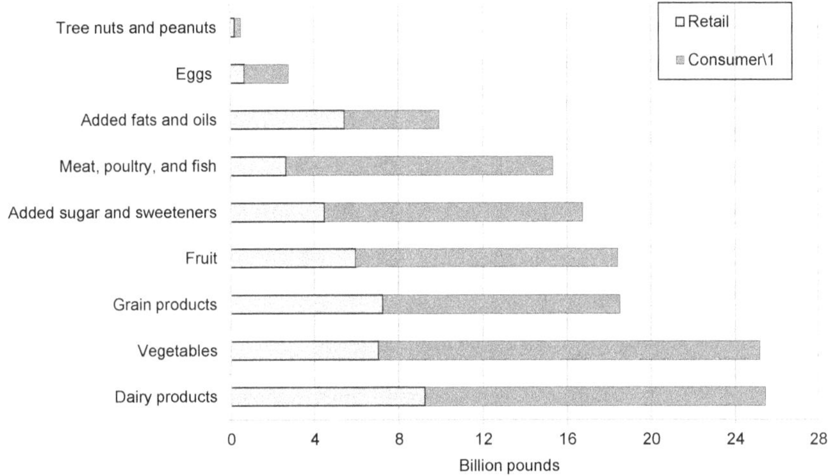


**3.3** Quantity losses at the consumer level are larger than retail level losses for all categories except added fats and oils. Source: Buzby, Wells, and Aulakh, 2014

In the United States, the food groups with the highest share of food loss in 2010 depended on the type of measurement (Figure 3.4). In terms of amount or weight of food loss, three food groups made up slightly over half (52 percent) of the food loss at the retail and consumer levels. They are dairy products, vegetables, and grain products. On a total value basis, the meat, poultry, and fish group comprises almost a third (30 percent) of the total (versus 12 percent by weight). This is because foods in this group tend to cost more per pound than many other foods. Vegetables and dairy products come in second and third in terms of share of total value, as measured using retail prices. The top three food groups in terms of shares of total calories uneaten are noticeably different – shares for added fats and oils, added sugars and sweeteners, and grains are much higher in terms of calories, reflecting these foods' caloric density per pound.

In addition to relative weight, prices, and caloric content, two major reasons why food loss estimates vary by food group is because there is variation in loss rates among individual commodities and some foods are more important to consumption to Americans than others (Table 3.2). Looking more closely at fruit loss for example, we see that the three fruits with the highest loss rates (fresh grapefruit, fresh tangerines, and fresh papaya) have loss rates in the range of 60-64 percent. These three fruits are different than the top fruits in terms of amounts (fresh oranges, orange juice, fresh bananas) or in terms of value of food loss (fresh apples, fresh grapes, and fresh strawberries). Buzby et al. (2011) provide more detail on the range of loss rates, amounts, and values for fruit and vegetables and the loss estimates currently used in the LAFA data series can be accessed on the ERS website for each commodity in all food groups (ERS, 2015c).

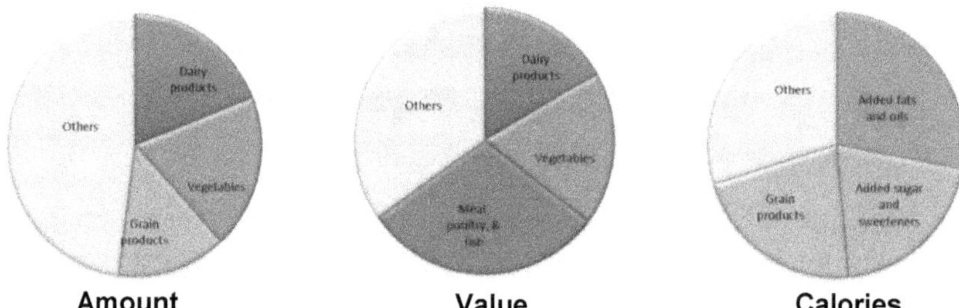


**3.4** The top three food groups in terms of annual food loss vary depending on if measured by amount, value or calories. Source: Buzby, Wells, and Aulakh, 2014

**Table 3.2 U.S. Fresh Fruit Loss in 2010. Source: ERS Loss-Adjusted Food Availability data, September 2014**

| Food | Retail and Consumer Level Losses | | |
|---|---|---|---|
| | *Million pounds* | *Million dollars* | *Percent loss* |
| **Top 3 foods by percent loss** | | | |
| Fresh grapefruit | 497 | $320 | 60 |
| Fresh tangerines | 683 | $855 | 62 |
| Fresh papaya | 219 | $541 | 64 |
| **Top 3 foods by losses in dollars:** | | | |
| Fresh apples | 1,236 | $1,224 | 27 |
| Fresh grapes | 879 | $1,451 | 38 |
| Fresh strawberries | 858 | $1,806 | 41 |
| **Top 3 foods by losses in pounds:** | | | |
| Fresh oranges | 1,264 | $863 | 43 |
| Orange juice | 1,684 | $974 | 15 |
| Fresh bananas | 2,092 | $964 | 26 |

## SUMMARY

There are four key messages presented here. First, quantities of food loss at the consumer level in the United States are larger than at the retail level for all food groups except added fats and oils. Second, the ranking of food loss varies depending on if measured by amount, value, or calories. Third, individual foods with the highest percent losses are not necessarily the foods with the most food loss. Fourth, measuring food loss is challenging and data intensive and as a result, data gaps exist for national estimates for individual commodities. In particular, data gaps exist at the farm level and between the farm and retail levels.

## REFERENCES

Buzby, J.C. 2010. Nanotechnology for food applications: More questions than answers. J. Consum. Aff. 44(3):528-545. http://dx.doi.org/10.1111/j.1745-6606.2010.01182.x.

Buzby, J.C., J. Hyman, H. Stewart, and H.F. Wells. 2011. The value of retail- and consumer-level fruit and vegetable losses in the United States. J. Consum. Aff. 45(3):492-515. http://dx.doi.org/10.1111/j.1745-6606.2011.01214.x.

Buzby, J.C., H.F. Wells, and J. Aulakh. 2014. Food Loss—Questions About the Amount and Causes Still Remain. U.S. Department of Agriculture, Economic Research Service, Amber Waves, Feature, June 2. http://ers.usda.gov/amber-waves/2014-june/food-loss—questions-about-the-amount-and-causes-still-remain.aspx.

Buzby, J.C., H.F. Wells, and J. Hyman. 2014. The Estimated Amount, Value, and Calories of Postharvest Food Losses at the Retail and Consumer Levels in the United States. Economic Research Service, U.S. Department of Agriculture, Economic Information Bulletin No. (EIB-121) 39 pp. http://www.ers.usda.gov/publications/eib-economic-information-bulletin/eib121.aspx.

Economic Research Service (ERS). 2015a. Food Availability (LAFA) data documentation. ERS, U.S. Department of Agriculture. http://ers.usda.gov/data-products/food-availability-(per-capita)-data-system/food-availability-documentation.aspx.

Economic Research Service (ERS). 2015b. Loss-Adjusted Food Availability (LAFA) data documentation. ERS, U.S. Department of Agriculture. http://ers.usda.gov/data-products/food-availability-(per-capita)-data-system/loss-adjusted-food-availability-documentation.aspx.

Economic Research Service (ERS). 2015c. Loss-Adjusted Food Availability (LAFA) data. ERS, U.S. Department of Agriculture. http://ers.usda.gov/data-products/food-availability-(per-capita)-data-system.aspx.

Golan, E., and J.C. Buzby. 2015. Innovating to meet the challenge of food waste. Food Technol. 69(1):21-25.

The views expressed here are those of the authors and cannot be attributed to the Economic Research Service or the U.S. Department of Agriculture.

# *Chapter 4*

## *Wasted Food, Wasted Resources*

### *Land, Irrigation Water, and Nutrients Associated with Food Wastage in the U.S.*

**John D. Toth and Zhengxia Dou**

## ABSTRACT

In 2012, the U.S. food system had 518 billion lbs (235 million metric tonnes; MMT) of food leaving primary production (farms) and entering the domestic supply chain for humans. Approximately 45% (234 billion lbs; 106 MMT) was lost or wasted before reaching a human stomach. The total food loss/waste was partitioned into 79 billion lbs (36 MMT), or 34% during processing/handling/manufacturing; 45 billion lbs (20 MMT), 19%, in retail; and 110 billion lbs (50 MMT), 47%, in the consumer sector (edible 90.3 billion lbs and inedible 19.7 billion lbs; 41 and 9 MMT), respectively. Cropland embedded in the edible food loss/waste (retail plus consumer-level edible food loss) was 42 million acres (16 million ha), accounting for 27% of the total cultivated cropland for domestically-consumed human food production in the nation. Similarly, 25% of the annual irrigation water and 26% of fertilizer nutrients were wasted as embedded in the edible food loss/wastage, amounting to 4.5 trillion gallons (17 billion $m^3$) and 8.6 billion lbs (3.9 MMT), respectively. In an era of growing challenges to sustainably feed the world with dwindling resources, the magnitude of food and resource wastage is scandalously unacceptable, food waste reduction is paramount.

## INTRODUCTION

The many aspects of the burden that food wastage places on society have become the focus of attention from popular writers such as Tristram Stuart (2009) and Jonathan Bloom (2010) and various organizations e.g. the U.S. Department of Agriculture Economic Research Service (USDA-ERS; Buzby et al., 2014), the Natural Resources Defense Council (NRDC; Gunders, 2012), the Waste and Resources Action Programme in the U.K. (Quested and Johnson, 2009), and the U.N.'s Food and Agriculture Organization (FAO; Gustavsson et al., 2011).

There is a general notion that wasting food also means wasting resources, from preparing the land base and growing the crops to raising the animals, through the harvest, processing, transport and finally consumption. However, research on quantitative assessment of resources or environmental impacts associated with food wastage has been lacking except for a few recent reports. At the global scale, Kummu et al. (2012) examined food waste and its effect on productive land area, crop nutrient use and water inputs; an FAO study (2013) addressed greenhouse gas emissions and biodiversity impact in addition to land, water, and nutrients associated with food wastage. According to the two studies, the North America-Oceania region (NAO; including the U.S., Canada, Australia and New Zealand) had roughly 30 to 35% of resource inputs (in terms of land, fertilizer and water) associated with consumer-level food waste; comparable values for Europe were 25 to 31%. Assessing the nitrogen burden associated with food wastage, Grizzetti et al. (2013) concluded that an average of 900 g (2 lbs) per capita per year fertilizer nitrogen inputs were embedded in food waste globally, but for the European Union nations the amount is as high as 2300 g (5.1 lbs). Liu et al. (2013) examined the extent of food waste in China, and the land area and irrigation inputs associated with that waste. They estimated that 21% of land and 20% of water inputs were embedded in food waste in China, with food loss (2010 data) of 19% for the entire food chain and 7% for the consumer sector (substantially lower than the U.S. or many developed countries). For the U.S. at the national level, the only relevant study (to our knowledge) was on energy embedded in food waste by Cuéllar and Webber (2011). These researchers reported that over 2,000 trillion BTU ($2.1 \times 10^{18}$ J) of energy was associated in the edible food waste at the national level (2007 data), which was 25% of total energy use in the entire food system (from agricultural production through consumption), and approximately 2% of national energy use for all purposes.

The current chapter fills an information gap by quantitatively assessing three types of resources embedded in wasted food in the U.S., namely, land, irrigation water, and fertilizer (nitrogen, phosphorus and potassium). Total food loss/waste in the U.S. supply chain is partitioned into three streams associated with the processing/handling/manufacturing (processing in short), retail, and consumption sectors, respectively. By linking with the intrinsic nature of different waste streams (edible and potentially avoidable; inedible and thus unavoidable), we distinguish the magnitude of resource expense that represents "true" wastage as well as the amounts of resources that are an integral part of growing food and thus unavoidable. Our data demonstrate that potential "savings" of the resources through food waste reduction are immense.

## DATA SOURCES AND CALCULATIONS

Data on U.S. food supply for domestic human consumption (i.e. excluding all other uses) and calculations for food loss and waste were derived from the USDA-ERS Loss-Adjusted Food Availability Tables (LAFA; USDA-ERS, 2015a) for the year 2012. Listed in LAFA are over 200 commodities, including fresh, canned, frozen and dried vegetables and fruits, fruit juices, milk and processed dairy products, eggs, red meat, poultry and seafood, nuts, grain products, fats and oils, and added sweeteners. The primary weight (in lbs $capita^{-1}$ $yr^{-1}$ in LAFA) was converted into quantity of food supply by multiplying by the 2012 U.S. population of 313.9 million. (The primary weights correspond to foodstuffs leaving the farm gate for fruits, vegetables, dairy, eggs, cereals, nuts and oil crops, carcass weights for meat and poultry, and boneless fillets for fish.)

Included in the LAFA tables were per capita food amounts entering the retail and consumer sectors of the food chain, and percent losses between the sectors. Consumer-level losses were subdivided into unavoidable losses (i.e. inedible parts such as trimmings and peelings for meal preparation) and edible-avoidable loss/waste (i.e. edible parts, including cooking loss, plate waste and food discarded for various reasons). Waste factors were derived from the losses incurred during processing, at retail, and at consumption sectors. To create values for amounts of waste by sector and commodity, the food supply was multiplied by the corresponding waste factor; thus for losses in processing, the weight of food entering processing was multiplied by the processing waste factor, the food entering retail times the retail waste factor yielded retail-sector waste, and consumer-level losses were consumer weight multiplied by the proportion of waste that was in the edible or inedible fractions. To facilitate presentation and discussion, the food commodities are grouped into 9 categories (Table 4.1).

Crop acreages for the (domestically-produced) food supply were obtained from the USDA National Agricultural Statistics Service 2012 Agricultural Census Survey Program, Crops Sector (USDA-NASS, 2015a). Unlike the LAFA tables from which we selected 151 food commodities, the crop acreage (and fertilizer data as well) include 117 crop-commodities, therefore certain aggregation was made to align the two datasets for calculation. Supplemental information needed for partitioning of multiple-use crops such as grains and oilseeds into food, feed, and oil crush were obtained from the USDA-ERS Feed Grains (USDA-ERS, 2015b), Oil Crops (USDA-ERS, 2015c) and Wheat (USDA-ERS, 2015d) Yearbooks.

Crop acreages associated with the food loss/waste streams at processing, retail, and consumer levels were calculated using weighted waste factors derived from those for the food supply. For animal product commodities (red meat, poultry, eggs, dairy, animal fats), crop acreages were that portion of total production for grain, forage and hay crops used as animal feed; the weighted waste factors were based on relative contribution to total animal products from beef and dairy cattle, swine and poultry, and the proportion of feed crop consumption by each species (Eshel et al., 2015). Harvested acreage was available for all 117 commodities, while planted acreage was generally lacking for orchard fruit and nuts, and perennial forage crops. Food supply acreage, fertilizer and irrigation use are based on resources used for domestic human consumption, excluding resource inputs embedded in exported commodities.

Data on fertilizer nutrients were from the 2012 Agricultural Census Survey Program, Environmental Sector (USDA-NASS, 2015b), which provides average annual application rate in lbs $ac^{-1}$ of nitrogen, phosphate ($P_2O_5$), and potash ($K_2O$) for the same 117 crop-commodities used in the crop acreage section. Total applications per crop-commodity were calculated by multiplying annual application rate by planted acres for each crop. Waste factors were the same as for crop acreage, and losses calculated for each nutrient were aggregated into fertilizer use totals lost/wasted in the processing, retail, and consumer sectors.

Data on irrigation water were derived from the USDA-NASS Census of Agriculture 2013 Farm and Ranch Irrigation Survey, Tables 35 and 36 (USDA-NASS, 2013), which provide irrigated acres for 21 crop-commodities and average annual irrigation rate in acre ft $ac^{-1}$ $yr^{-1}$. Total amount of irrigation was calculated from irrigation rate multiplied by the fraction of total acres receiving irrigation, and converted into gallons of irrigation water. To reconcile the different extent of details in commodity grouping (151 commodities for food supply and food loss/waste but 21 crop-commodities for irrigation), certain arbitrary aggregation and grouping are necessary. All small grains were grouped together, oilseed crop data was only for soybeans, fruit and nuts included only those borne on orchard trees, plus vineyards, and vegetables were partitioned into lettuce, sweet corn, tomatoes, dried beans, potatoes, and "other vegetable crops." Weighted waste factors were calculated for the 21 crop-commodity groups, and irrigation water embedded in food loss/waste calculated similarly to the other resource groupings. Partitioning of water use between groundwater and surface water sources for crop production on a by-state basis was derived from Maupin et al. (2014) in Estimated Use of Water in the United States in 2010.

## FOOD SUPPLY AND FOOD LOSS/WASTE STREAMS

Our results show the total U.S. food supply in 2012 to be 518 billion lbs. This is the amount of foodstuffs by weight that leaves the farm or site of primary production, entering the processing-retail-consumption chain for American consumers. Note that the food supply of 430 billion lbs widely known through the publication of Buzby et al. (2014) refers to the quantity of food entering the *retail* sector (as their purpose was to delineate food loss/waste in the retail and consumer sectors). By commodity class, the total food supply consists of 124 billion lbs. of vegetables, or 24% of the total; dairy products (93 billion lbs; 18%), meat, poultry and seafood (83 billion lbs; 16%), fruit (76 billion lbs; 15%), grain products (61 billion lbs; 12%), added sweeteners (40 billion lbs, 8%), fats and oils (26 billion lbs, 5%), eggs (10 billion lbs, 2%) and nuts (4 billion lbs, 0.7%).

Of the 518 billion lbs food supply, 284 billion lbs were consumed while the remaining 234 billion lbs (45% of the total) were lost or wasted along the supply chain. The food loss/waste streams were: 79.2 billion lbs in processing, handling, and manufacturing, 44.5 billion lbs at retail, and 110.0 billion lbs at the consumer level (including 19.7 billion lbs inedible and 90.3 billion lbs edible), respectively (Figure 4.1). For the food loss/waste at retail as well as the edible food loss at consumer levels, our results are essentially the same as Buzby et al's (2014; Chapter 2), as they should because the calculations were based on the same set of data and parameters. The 79.1 billion lbs food loss for the processing, handling, and manufacturing sector is the first estimate derived from an authoritative dataset. A previous report estimated the food loss for the sector to be roughly 44 billion lbs (BSR, 2013), which was derived from 13 survey entries then extrapolated to the entire sector.

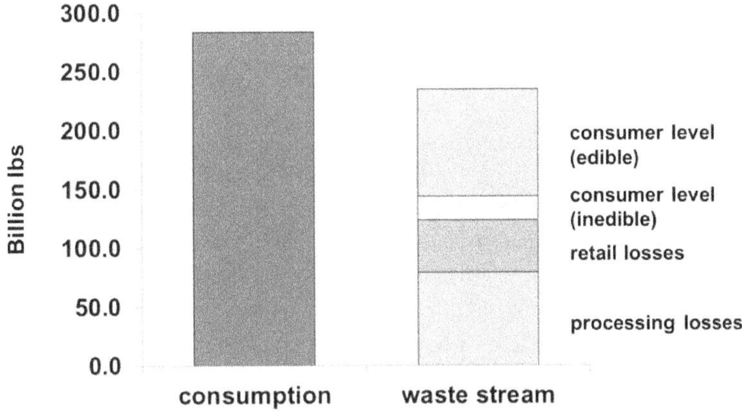


**4.1** Partitioning of the food supply in the U.S. in 2012, consumption and distribution of losses and waste across the food supply chain. Food supply was 518 billion lbs

Individual commodities with the highest edible food loss/waste (retail and consumer levels) by weight are: fluid milk plus yogurt (17.0 billion lbs), wheat products (12.5 billion), sugar (8.6 billion) and corn-based sweeteners (8.0 billion), red meat (8.4 billion) and poultry (4.7 billion), and cooking oil (5.5 billion). For fruits and vegetables, among the largest contributors are fresh onions and canned tomatoes (2.8 billion lbs each), fresh (2.3 billion) and frozen potato products (1.6 billion), and oranges for table use and juice (1.4 billion each).

Table 4.1 lists the nine commodity categories ranked in their respective quantities of food lost/wasted, percent contribution to the total wastage (234 billion lbs), and percent of each category's amount of supply at the beginning of the chain.

It is important to point out that food loss/waste generated along the supply chain has different "intrinsic nature." The waste stream in the processing sector is primarily food processing residues (e.g. wheat middlings, citrus pulp, etc.) that cannot be made into consumer food products (refer to Chapter 15 for a detailed discussion). Such food residues exiting the supply chain can be considered unavoidable. Meanwhile, the processing waste stream is largely concentrated at a limited number of sites (as in contrast to the retail and consumer waste streams), making its recovery and re-use feasible and profitable (BSR, 2014; FWRA, 2014). On the other hand, food loss/waste at the retail sector consists of consumer food products that are pulled out of the supply chain for various reasons (Buzby et al., 2014), whereas the 90.3 billion lbs of edible food loss at the consumer level are basically food discarded by consumers as affected by complex social, cultural, and economic factors (see discussions in Chapters 5, 7, and 9). Processing sector food loss is an analog to point sources; food waste streams at the retail and consumer sectors are non-point sources (think of millions of households, restaurants, and grocery stores) with the form and content extremely variable.

The intrinsic nature of food waste streams determines their end point. According to the Analysis of Food Waste conducted by BSR (2013; 2014) based on interviews with several dozen food supply chain professionals, 98% of the food residues at the processing/manufacturing sector were recovered or recycled while only 2% ended up in landfills. In contrast, more than half of retail sector food waste ended up in landfills, whereas at the consumer level landfilling is the ultimate destination for 84% of the restaurant food waste and 95% of household waste.

**Table 4.1 Losses and waste by food commodity in the food chain for the U.S. in 2012**

| Commodity Class | Total waste/losses | Waste/losses % of total waste | Waste/losses % of commod. supply |
|---|---|---|---|
| | billion lbs | | |
| Vegetables | 73.0 | 31.2 | 58.9 |
| Fruit | 41.8 | 17.9 | 55.0 |
| Meat poultry fish | 41.5 | 17.7 | 49.8 |
| Dairy products | 27.1 | 11.6 | 29.1 |
| Grain products | 19.1 | 8.2 | 31.2 |
| Sweeteners | 16.6 | 7.1 | 41.1 |
| Fats and oils | 10.0 | 4.3 | 38.5 |
| Eggs | 4.3 | 1.8 | 42.2 |
| Nuts | 0.6 | 0.3 | 16.2 |

## CROPLAND EMBEDDED IN WASTED FOOD

A total of 156 million acres (harvested[1]) were associated with food production for U.S. domestic human consumption in 2012, including the portion of feed grains as well as forages for animals and subsequently animal products entering the food supply chain. The total acreage was partitioned into 87 million acres for animal feeds, 28 million acres oilseed crops, 22 million acres grain crops, 10 million acres as sweetener crops (sugar cane, sugar beet and corn), and 5.6, 1.9, and 1.2 million acres for vegetables, fruits, and nuts, respectively.

[1] Approximately 11 million acres were planted but not harvested in 2012 (USDA-NASS, 2015a), which were excluded in our calculation.

Crop acreage embedded in wasted food amounted to 65 million acres, accounting for 42% of the 156 million cultivated total. Considering the food waste streams, consumer level edible food loss/waste is the largest component of land wastage (26 million acres), followed by processing wastage (24 million), the retail (14 million), and the inedible food loss at consumption (0.9 million). Relevant percentage distribution of the total crop acreage embedded in food consumed vs. food loss/waste streams is in Figure 4.2.

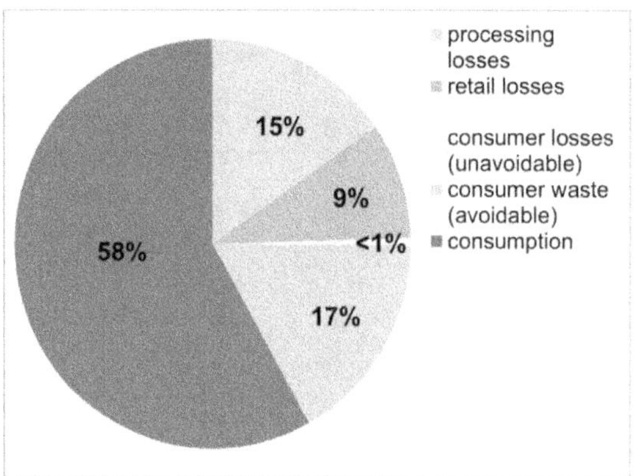


**4.2** Percent of total crop acreage embedded in consumed food and losses by sector of the food chain. Total harvested area was 156 million acres

Further examination of the results revealed that of the 65 million acres embedded in wasted food, 40.0 million acres (61%) were associated with feed crops and the loss/waste of animal products; the remainder consisted of 10.2 million acres with oilseed crops, 7.0 million acres grain crops, 4.2 million acres sweetener crops, 2.5 million acres vegetables, 1.2 million acres fruit and 0.2 million acres nuts. Interestingly, animal products contribute 35% by weight of the food supply, 38% of food consumed, and 30% of calories in American diets (Buzby et al., 2014), but producing these animal products required 56% (87 out of 156 million acres) of the crop acreage whereas 46% (40.0 out of 87 million acres) of these acres were wasted in vain because of food loss/waste. It must be noted that some of the acres used for producing roughages are land that is not suitable for other crops. Also, the importance of animal production systems in utilizing enormous amount of food residues (which may be sent to landfills otherwise) and the critical role of nutrient-rich animal foods in the human diet should not be overlooked.

## FERTILIZER NUTRIENTS EMBEDDED IN WASTED FOOD

About 33 billion lbs of nitrogen, phosphate and potash fertilizers were used to produce the food and feed crops for domestic human consumption, with 42% as nitrogen, 23% $P_2O_5$ and 35% $K_2O$ by weight. Well over half (19 billion lbs) of the fertilizer inputs were associated with feed crops for animal production; the remaining included oilseed crops 4.1 billion lbs, grain crops 4.2 billion lbs, sweetener crops 3.1 billion lbs, and vegetables, fruit and nuts 2.4 billion lbs together.

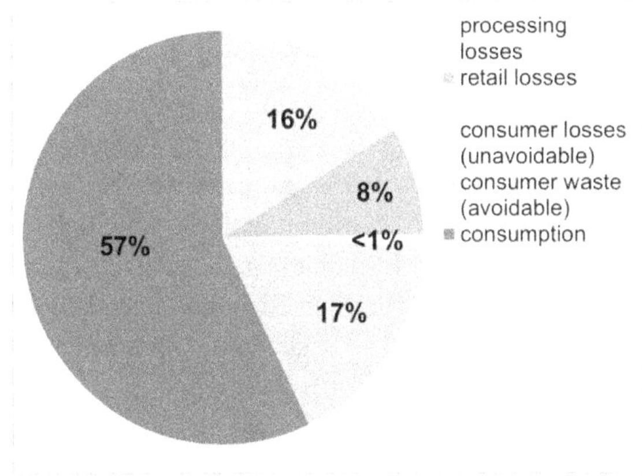


**4.3** Total N-P-K fertilizer applications embedded in food losses by sector of the food chain. Values are the percent of the 33 billion lbs of fertilizer either in primary consumption or in losses and waste

A total of 14.3 billion lbs fertilizer were associated with wasted food (43% of total applications). This can be attributed to 5.8 billion lbs as edible consumer waste, 5.4 billion lbs in the processing and 2.8 billion in the retail sectors, and 300 million lbs as inedible food loss at the consumer level. The relevant percentage distribution of fertilizers into consumed vs. wasted food is shown in Figure 4.3.

For comparison, the global-scale study by Kummu et al. (2012) estimated total fertilizer use excluding that for animal feed crops to be 24 billion lbs (10.9 MMT) and fertilizer embedded in food losses to be 7.3 billion lbs for North America and Oceania. Another study (Grizzetti et al., 2013) reported nitrogen fertilizer embedded in consumer food waste to be 900 g (2.0 lbs) per capita per year averaged globally, but 2300 g (5.1 lbs) per capita per year for

European Union nations. Our result would be equivalent to 7.7 lbs nitrogen fertilizer per capita for the edible food loss/waste in the U.S. consumer sector.

## IRRIGATION WATER EMBEDDED IN WASTED FOOD

Approximately 13% of the cropland received irrigation in 2012, with the water amount totaling 18 trillion gallons. About 80% of vegetable crops, 94% of orchard fruit and nut crops, and 43% of forage alfalfa were irrigated, along with all rice crop acreage (USDA-NASS, 2013; 2015a). Feed crops used for animals received 11.7 trillion gallons (64% of the total), along with grain crops 3.2 trillion gallons, orchard fruit and nuts 900 billion gallons, and vegetables 1.9 trillion and oilseed crops (here soybeans alone) 700 billion gallons. Aquaculture and livestock rearing also used water drawn from ground and surface sources, 3.4 and 1.1 trillion gallons, respectively. Alfalfa hay was the leading crop-commodity receiving irrigation, 4.8 trillion gallons. Aggregated tree crops and vineyards had 900 billion gallons applied, rice 2.5 trillion gallons, non-alfalfa hay 2.5 trillion and corn grain for animal feed 1.6 trillion gallons. Pastures used by grazing animals, which otherwise were not considered in this chapter, received 1.4 trillion gallons; the pasture area irrigated was a small fraction (<3%) of the enormous 101 million acres used for food animal production.

According to Mekonnen and Hoekstra (2011), the blue water footprint (i.e. irrigation use) for all crop production in the U.S. was 95.9 $Gm^3$ (25.3 trillion gallons) per year for the period of 1996-2005, greater than our result of 18.4 trillion gallons for 2012. Our calculations exclude water used to produce commodities for non-food and feed uses (biofuels, seed) and crops destined for export.

As with the crop acreage and fertilizer analyses, irrigation water embedded in processing losses and edible consumer-level waste were the largest fractions across food chain sectors, 3.7 trillion gallons for processing and 3.2 trillion gallons for consumer waste. The relevant distribution of irrigation water among consumed food and food waste streams is shown in Figure 4.4.

There was a greater proportion of irrigation water embedded in retail- and edible food loss/waste at the consumer level from vegetables and orchard fruit and nuts relative to their harvested area, likely a result of the relatively high demand for irrigation required by these high-value commodity crops.

Irrigation water usage from ground and surface water sources varied a great deal by crop type and by state. For groundwater, the food groups and states with the greatest water volume embedded in edible retail and consumer food waste were cereal crops in Arkansas (170 billion gallons), and animal feed and

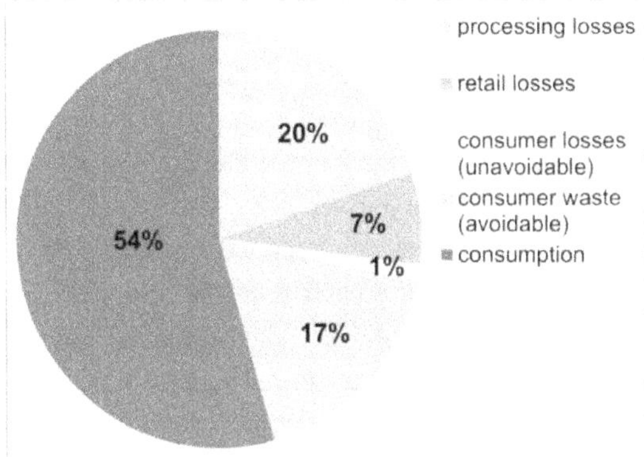


**4.4** Percent of total irrigation water that was embedded in consumed food or food losses by sector of the food chain. Irrigation applications were 18 trillion gallons

vegetable crops in California (160 and 100 billion gallons, respectively). For surface water, the largest volumes were 260, 240 and 170 billion gallons for animal feed, cereal crops and vegetables, all in California. Also, irrigation water pumped from sensitive or over-exploited aquifers that is subsequently lost in retail- and consumer-level food waste was 500 billion gallons for the Central Valley aquifer in California and 590 billion gallons for the High Plains aquifer in Colorado, Kansas, Nebraska and Texas. In areas of Kansas and Texas, the High Plains aquifer levels have dropped by over 40 meters (130 ft) since the early 20th Century; in parts of California's Central Valley the water table drop was over 120 m (Scanlon et al., 2012; McGuire, 2013). Water use embedded in food waste exacerbates already-threatened water resources and sustainability in these high-production but sensitive regions.

## GENERAL DISCUSSION AND CONCLUSIONS

Our data show that at the national level, 46% (234 billion lbs) of the total annual food supply for domestic human consumption is not consumed by Americans, but exited the supply chain as: (i) processing/ handling/manufacturing food residues, 79 billion lbs, which is primarily unavoidable but readily recoverable for non-human uses; (ii) retail loss/waste, 45 billion lbs, as consumer food products, which is basically avoidable, some can be recovered for human consumption, some can be re-purposed for other beneficial uses; and (iii) consumer food loss/waste including inedible 20 billion lbs and edible 90 billion lbs, the latter primarily avoidable but difficult to recover and re-purpose once discarded (except for some extent of composting).

To put the massive amounts of food wastage (avoidable, i.e. retail and consumer edible portion only) and the associated resources into perspective, some comparisons would be illustrative. At the U.S. national level, the edible/avoidable food loss/wastage amounts to 135 billion lbs annually, enough to feed 50 million people. The 41 million acres of land embedded in the edible/avoidable food loss/waste is approximately half of the total area of the U.S. National Park System. The amount of fertilizer nutrients embedded in the edible/avoidable food loss/waste, 8.6 billion lbs, is 50% greater than the total fertilizer usage in Sub-Saharan Africa (FAO, 2015). The 4.5 trillion gallons of irrigation water lost in avoidable food waste is equivalent to the area of the City of Philadelphia covered by 160 feet of water. For a typical family of four, edible/avoidable food loss/waste averages 1,700 lbs annually, relevant resources embedded would be 1/2 acre of land, 110 lbs fertilizer, and 60,000 gallons water.

In addition to the food loss/waste streams along the supply chain, a presumably considerable amount of food is left in the field never harvested because of overplanting, low market prices, lack of harvest labor, and culling prior to processing due to cosmetic issues with crop appearance. An NRDC study (2012) interviewed 16 growers and packing houses in the prime fresh produce area in central California and found that for tree fruit, 15% of the otherwise-edible crop was unharvested, 5% of head lettuce and 13% of broccoli. Packing culls accounted for a further 13% of tree fruit and 3% of lettuce. The authors acknowledge that the small sample size precludes general conclusions on the true extent of pre-harvest food loss.

The largest share of resources (cropland, fertilizer nutrients, and irrigation water) embedded in wasted food is associated with the production of feed crops for livestock animals – very nearly half of the total resources consumed. This opens to discussion the much broader issue of the environmental, economic and ethical aspects of American consumer (other developed countries as well) preference for a meat- and animal product-rich diet. The implication is profound at the global scale considering the rapidly growing middle class populations in developing countries and their desire for more meat and dairy products. Putting the "should vs. should not" debate aside, at the least reducing the wastage of animal food products, along with the reduction of other wastage, is paramount if we are to build a sustainable food system.

The data we used for the calculations are well-sourced, authoritative, and quantifiable to the extent possible. There are other dimensions to the drain on resources (that are beyond the scope of this study) associated with food loss/waste. Examples are the labor and energy costs for producing and provisioning food;

energy entailed in fertilizer manufacture and water pumping and dispensing in irrigation; decreases in biodiversity due to cropland cultivation, greenhouse gas emissions associated with agricultural production processes as well as the degradation of wasted food in landfills, etc. Food is a basic necessity of man, food wasting is ultimately the consequence of man's ignorant food behavior. The sustainable future of food security hinges upon many variables, among which changing our food behavior and substantially reducing food wastage is paramount and indispensable.

## REFERENCES

Bloom, J. 2010. American Wasteland: How America Throws Away Nearly Half its Food (And What we can do About it). Da Capo Press, Cambridge, MA.

Business for Social Responsibility (BSR). 2013. Analysis of U.S. food waste among food manufacturers, retailers, and restaurants. http://www.foodwastealliance.org/wp-content/uploads/2013/06/FWRA_BSR_Tier2_FINAL.pdf.

Business for Social Responsibility (BSR). 2014. Analysis of U.S. food waste among food manufacturers, retailers, and restaurants. http://www.foodwastealliance.org/wp-content/uploads/2014/11/FWRA_BSR_Tier3_FINAL.pdf.

Buzby, J.C., H.F. Wells, and J. Hyman. 2014. The estimated amount, value and calories of postharvest food losses at the retail and consumer levels in the United States. Economic Information Bulletin Number 121. Economic Research Service/USDA.

Cuéllar, A.D., and M.E. Webber. 2010. Wasted food, wasted energy: The embedded energy in food waste in the United States. Environ. Sci. Technol. 44:6464-6469. http://dx.doi.org/10.1021/es100310d.

Eshel, G., A. Shepon, T. Makov, and R. Milo. 2015. Partitioning United States' feed consumption among livestock categories for improved environmental cost assessments. J. Agric. Sci. 153(3):432-445. http://dx.doi.org/10.1017/S0021589614000690.

Food and Agriculture Organization of the United Nations (FAO). 2015. World Fertilizer Trends and Outlook to 2018. Food and Agriculture Association of the United Nations, Rome, Italy.

Food Waste Reduction Alliance (FWRA). 2014. Best Practices & Emerging Solutions Toolkit. FWRA, Washington, DC.

Grizzetti, B., U. Pretato, L. Lassaletta, G. Billen, and J. Garnier. 2013. The contribution of food waste to global and European nitrogen pollution.

Environ. Sci. Policy 33:186-195. http://dx.doi.org/10.1016/j.envsci.2013.05.013.

Gunders, D. 2012. Wasted: How America is Losing up to 40 Percent of its Food from Farm to Fork to Landfill. Natural Resources Defense Council Issue Paper IP:12-06-B.

Gustavsson, J., C. Cederberg, U. Sonesson, R. van Otterdijk, and A. Meybeck. 2011. Global food losses and food waste: Extent, causes and prevention. Food and Agriculture Organization of the United Nations, Rome.

Kummu, M., H. de Moel, M. Porkka, S. Siebert, O. Varis, and P.J. Ward. 2012. Lost food, wasted resources: Global food supply chain losses and their impacts on freshwater, cropland, and fertilizer use. Sci. Tot. Environ. 438:477-489. http://dx.doi.org/10.1016/j.scitotenv.2012.08.092.

Liu, J., J. Lundqvist, J. Weinberg, and J. Gustafsson. 2013. Food losses and waste in China and their implication for water and land. Environ. Sci. Technol. 47:10137-10144. http://dx.doi.org/10.1021/es401426b.

Maupin, M.A., J.F. Kenny, S.S. Hutson, J.K. Lovelace, N.L. Barber, and K.S. Linsey. 2014. Estimated use of water in the United States in 2010: U.S. Geological Survey Circular 1405.

Mekonnen, M.M. and A.Y. Hoekstra. 2011. The green, blue and grey water footprint of crops and derived crop products. Hydrol. Earth Syst. Sci. 15:1577-1600. http://dx.doi.org/10.5194/hess-15-1577-2011.

McGuire, V.L. 2013. Water-level and storage changes in the High Plains aquifer, predevelopment to 2011 and 2009-11. U.S. Geological Survey Scientific Investigations Report 2012-5291.

Natural Resources Defense Council (NRDC). 2012. Left-Out – An Investigation of the Causes and Quantities of Crop Shrink.

Quested, T. and H. Johnson. 2009. Household food and drink waste in the UK, Final report. Waste & Resources Action Programme (WRAP), Banbury, UK.

Scanlon, B.R., C.C. Faunt, L. Longuevergne, R.C. Reedy, W.M. Alley, V.L. McGuire, and P.B. McMahon. 2012. Groundwater depletion and sustainability of irrigation in the US High Plains and Central Valley. Proc. Nat. Acad. Sci. 109(24): 9320-9325. http://dx.doi.org/10.1073/pnas.1200311109.

Stuart, T. 2009. Waste: Uncovering the Global Food Scandal. W.W. Norton and Co., Inc., New York, NY.

U.S. Department of Agriculture Economic Research Service (USDA-ERS). 2015a. Feed Grains: Yearbook Tables.

U.S. Department of Agriculture Economic Research Service (USDA-ERS). 2015b. Food Availability (Per Capita) Data System – Loss-Adjusted Food Availability.

U.S. Department of Agriculture Economic Research Service (USDA-ERS). 2015c. Oil Crops Yearbook.

U.S. Department of Agriculture Economic Research Service (USDA-ERS). 2015d. Wheat Data.

U.S. Department of Agriculture National Agricultural Statistics Service (USDA-NASS). 2013. Census of Agriculture – 2013 Farm and Ranch Irrigation Survey.

U.S. Department of Agriculture National Agricultural Statistics Service (USDA-NASS). 2015a. 2012 Census of Agriculture – Survey Program – Crops Sector.

U.S. Department of Agriculture National Agricultural Statistics Service (USDA-NASS). 2015b. 2012 Census of Agriculture – Survey Program – Environmental Sector.

# Chapter 5

## *Food Waste and Food Security in a Globalizing World*

### *The Social Factor*

**Brian Spooner**

Some [foods] should be eaten before fully ripe, such as capers, asparagus, sucking pigs, and pigeons...; others, at the moment of perfection, such as melons, most fruit, mutton, and beef...; others, when they start to decompose, such as medlars, woodcocks, and especially pheasants, others finally, after the methods of art have removed their deleterious qualities, such as the potato and the cassava root.

Brillat-Savarin (1854)

**ABSTRACT**

Humans are social beings. Food has always been a medium of social interaction. Our interest in food comes not only from our biological need to satisfy hunger. Eating is a social package, part of the cultural order we have developed for managing the relationships on which we depend as we navigate the continuing social change of everyday life. The package has always involved a certain amount of waste. But the way we relate to food now is different from a century ago, a millennium ago, in the Ancient World, or in prehistory. As population grows, more people interact, and society becomes more complex, social life continuously changes in quality. As it changes, the package changes, and the way food gets wasted, as well as the amount of waste, also changes.

Until recently the rate of social change was slow enough to be manageable. But under globalization it has accelerated, diluting the relationships on which our food habits were built, and causing our social packages to unravel. There is no longer any cultural control over the amount of food that goes to waste. As we become more aware of the global limits on our ability to produce food, we have become anxious about waste. But if population growth levels off towards the end of this century, as is predicted, and globalization completes its course, the rate of change will slow down, new types of relationships will form, and waste will once again become culturally manageable. Although it will never be possible to eliminate waste, history suggests it may be brought back under control.

## THE PROBLEM

The quotation from Brillat-Savarin, whose book is a foundational text for the historical study of the Western approach to food, illustrates a number of factors that condition our behavior in ways that lead to waste. Most importantly, it shows that our attitude to food starts not with nutrition, but with taste, which is gained by experience, and reinforced by social norms. Where hunger is the biological need to eat, taste is cultural, and is learned (socially) in the company of others. We have always eaten with particular groups of other people as part of our social identity, our place in society. Taste is therefore not just psychological, but can change with social context, and with class. Brillat-Savarin presents the cultural rules that ordered the social eating of the French upper classes of the 19th century, rules that were emulated in much of Western civilization, providing a code for restaurants, which have become one of the major generators of waste in the modern world.

Eating, like all human behavior, whether thoughtful or spur-of-the-moment, is conditioned by three sets of variables: biological (derived from inherited genetic programming), cultural (acquired by experience and habituation), and social (our need for company and identity). For any given individual the biological variable may not change. The cultural is cumulative, and is likely to change with experience, because cultural difference, variation and change are embedded in social processes. The social, on the other hand, depends on context, and can easily change according to how we choose to belong.

Food was the first medium of social interaction, with a history and prehistory that goes back over two hundred thousand years. Whether or not we are hungry, we tend to eat when it is socially appropriate, and we often over-eat. Much of what we expect to be able to eat on demand is nutritionally unimportant, or even inadvisable. Eating has always been a social activity, and while there may be particular cultures of waste in the modern world, waste is usually a function of social variables. If our social situation changes, anything we do that is wasteful is also likely to change. There is also much that we do not eat that would be nutritionally satisfying, if we could bring ourselves to eat it, such as insects (see Chapter 2), apart from many animals and plants. We have always been surrounded by plant and animal products that are edible. But we have always been selective about which of them we consider as food and choose to eat, and as our social lives have become more complex, we have become more selective. We are also generally inflexible about how we choose to eat what we have selected. What do we eat it with? Is it breakfast food, lunch food or dinner food? or just a snack? Can we eat it just as it is? Should it be raw or cooked? How should it be prepared? Alone or with other ingredients? Boiled or broiled, steamed, roasted, or grilled?

or baked? Hot or cold? And food is a social marker. We like to seal a new social relationship with a meal, to give it symbolic value. Our first choice used to be exchange of personnel, usually in the form of women as spouses. But that was not always possible, and the next choice is to link our relationships to something external. Food has been the most common choice. We still seal important agreements with a common meal.

While some practices are indisputably wasteful, others may be a matter of opinion. Apart from being a problem of human behavior, food waste is also a problem of human perception. In order to explain the current scourge of food waste, therefore, we will define food as edible substances that are culturally acceptable, and waste as the failure to match the biological need with the cultural acceptability and the social function. We will now explore how the mismatch evolved historically and where it seems to be heading.

## WHY IS WASTE A HOT TOPIC NOW?

As society has changed, our perception of waste has changed. To understand why it has become a hot topic in the 21st century, when little attention was paid to it earlier, it is necessary to see it in the context of three related variables – variables that (like waste) also change with perception, and though not new, are also receiving more attention today than earlier: population growth, ecological degradation and social change.

Whatever the definition, it would be difficult to argue that food waste is a new problem. But since in the past we were always able to increase production when we needed to increase supply, we paid little attention to waste. Concern about the possibility of feeding increasing numbers of people began with the publication of Malthus' essay on population two centuries ago (1798). The modern concern with population growth intensified in the late 1960s (cf. Ehrlich, 1968). But it was soon allayed in the early 1970s when the Green Revolution transformed India from a food-deficit economy to a food exporter. Its modern growth since the final years of the last century is related to our rising global awareness (cf. Livi-Bacci, 1991), which for the first time suggests overall limits to our ability to produce food. Concern with these limits is exacerbated by increasing evidence of degradation in the renewable natural resources which define them, even though our rate of technological innovation has been increasing.

Ecological degradation had already been noticed in the Ancient World, by none other than Plato, who remarked on the denudation of Cape Sounion, south east of Athens, as a result of over-grazing by goats. But modern concern began with increasing dust storms in the 1950s and the desertification of semi-arid areas either side of the northern arid zone following the Sahelian drought of 1968-1973.

Since then (as with population growth) it has intensified as our environmental awareness has become global, leading to a series of international conferences organized by the UN. As population continues to grow, it is likely that increased human activity will exacerbate degradation.

Both these problems are trumped by accelerating social change. There has never been a society without change. Social life everywhere is a continuing struggle between built-in demographic change and our cultural effort to create and maintain order. We get the order and security we need in our lives through the social relationships we develop with each other, which make us interdependent, give us a sense of security, provide the collaboration we need for dealing with the uncertainty in the natural world around us, and allow us to learn collectively. But our relationships are never entirely secure. Our social lives are always subject to change – if only because our life cycles are staggered, and births and deaths are irregular and unpredictable. As a result of population growth over the past century the rate of social change now exceeds anything known in the past. The faster the change, the greater the disruption, and the greater the potential for waste. But food has continued to be involved in the familiar patterns of social life, irrespective of change and disruption, and in the past population growth has always spurred agricultural innovation (cf. Boserup, 1964; Spooner, 1972).

## THE HISTORY OF CHANGE

From the beginning people collaborated in small groups as they moved around to gather plant foods and to hunt. Beginning some twelve thousand years ago climate change created opportunities for them to settle and gather in larger numbers. No longer mobile, their fertility rate increased and continuing climate change created the need for them to produce food. The preparation of land for cultivation required investment of labor, which led to agricultural landownership becoming the main criterion for social differentiation and identity as communities grew in size. With the beginning of food production and the gradual institutionalization of agricultural arrangements beginning in the Fertile Crescent of the Middle East some twelve thousand years ago, followed in subsequent millennia in Egypt, China, and the Indus Valley, their social lives changed significantly. A little later (because of climate change again) increasing numbers made it possible for them to organize labor for investment in the development of irrigation engineering. A little later, since agriculture limited their hunting ability, they began to bring animals into their villages and domesticated them, first for meat and later for secondary products, especially milk.

Since all these activities required organization and investment by larger and larger numbers of people, their communities grew in size and became more socially complex, and landownership soon became the main criterion for social differentiation and identity. The increasing social complexity continued to be mediated and symbolized in terms of food: its production and distribution on the one hand, and preparation, consumption on the other. Much of what the landed elite left over was passed on down to the less privileged and consumed by the poor, who had no land and little to waste. This process could still be observed in the middle of the 20th century in societies that had not yet industrialized. As the numbers of people living together continued to increase, the social arenas within which people routinely interacted with each other expanded, the rules of social behavior became more elaborate, and food-related behavior became culturally organized in more detail. Food that did not fit the rules was likely to be wasted. As the value of agricultural land increased, society became more stratified.

World population has continued to grow at an accelerating rate. It finally reached a billion in 1830, two billion in 1930, and over seven billion today, though now it is beginning to slow. At the same time the proportion of the world's population living in cities has risen even faster: from just 13% at the beginning of the 20th century, to 29% in 1950, to 54% percent in 2015. As population has shifted away from agricultural areas, ownership of agricultural land has lost its social role, the rug has been pulled out from under the old landed aristocracy, wealth is in the cities, and more people with different cultural backgrounds from different parts of the world are coming into contact with each other. More people are becoming familiar with new ways of doing things, increasing the rate of social change.

This accelerating change began in the West. In the 14th century new opportunities opened up as a result of the catastrophic social disruption caused by the Black Death, which killed rich and poor indiscriminately, and destroyed vested interests. Pursuit of these opportunities led to further disruptions, of which the most important were the Reformation and religious wars starting in the 16th century, the rise of science and engineering and the English Revolution in the 17th century, and the French Revolution and the Industrial Revolution in the 18th century (cf. Herlihy, 1997). The population of London had reached one million by 1700 – the first city to reach that size since Rome at the height of the Roman Empire. When Voltaire visited London in 1726 (after the English Revolution, but before the French) he was impressed by its coffee shops, where people of all social classes met and talked. In the West population continued to grow. But there were no comparable disruptions in other parts of the world, and social change that would generate similar growth elsewhere was inhibited by the administrative straightjacket of Western colonialism until the middle of the 20th century. When colonial administrations were withdrawn following World War II, the rates of growth and change in China, India and Africa began to accelerate to unprecedented degrees with obvious modern implications for the production, distribution, consumption and waste of food.

Population growth is never simply an increase in numbers. Urban populations have grown faster than overall world population ever since the emergence of the first urban clusters some eight thousand years ago. As the number of people grows, densities increase, and more people come together and interact with each other, changing the quality of social relationships. In the past the historical growth of cities was limited by carrying capacity, by the problem of local food supply before the mechanization of transportation. By the end of the 19th century advancing technology was providing transportation that removed those limits. In today's global food economy local carrying capacity no longer limits urban growth. Our food economy is now global, and the problem is reversed. Global urbanization is now introducing unprecedented difficulties in the supply and distribution of food within cities. By separating an increasing percentage of the global population from the resources they rely on, we have created a new problem that may be more difficult to solve. The change in geographical distribution of world population from widespread rural settlement to dense concentration in expanding conurbations, has opened up the social arenas within which people routinely interact, changing their culture of interaction and the quality of their relationships, raising their level of awareness, increasing collective learning and the rate of innovation (Christian, 2011:306-331), and driving the process of globalization in its final, crucial phase (Spooner, 2015). Is the increase in food wastage at a time when under-nutrition is also increasing, related to increasing social separation from food production? This process began with the Industrial Revolution which not only separated workers from food production and created a new food economy, but changed their diet, with new opportunities for waste.

It is easy to see that increasing urbanization and the separation of urban life from areas of agricultural production lead to changes in each of the stages of our social relationship with food, from production to consumption, which relax controls on the amount likely to go to waste. When production, supply, preparation and consumption are managed within a single family, or even a small community of families, the potential for waste is limited. Although we have always wasted food, as the community expands and the production, supply, preparation and consumption become separate functions performed by different groups, the limits disappear, and the potential for waste increases. Waste is seen to be a major problem now because we have only recently come to see all these variables – the conflict between population growth, ecological degradation and urbanization – in global terms.

The difference between everyday social life today and in 1950 is greatest in the Western world, but the rest of the world is catching up. When few adults were unmarried, and few people lived on their own, food was a family affair, and to eat alone was unusual. When we lived in small groups, our relationships with each other were personal and our use of food in the mediation of those relationships was attentive. Now we live in arenas of social interaction that have grown beyond face-to-face, to mobile, to social media, beyond local to global, and our social interaction continues to intensify. As our relationships have grown in numbers they have become diluted. The increase in the quantity of our social interaction has changed the quality. Where we used to have special types of relationships with family, friends, colleagues, neighbors, now we live in a social ocean, in which we cannot keep up with all our relationships however hard we try, with or without social media. Families are not what they used to be. Traditional social distinctions such as sex, sexuality, gender, and even age are disappearing, and food no longer has much to do with it. To make things worse the more we eat outside the home, the less control we have over the amounts of food we are served, or have available to us when we feel like eating. Now in the 21st century, although the use of food for social engagements is always an option, eating alone is common. The rate of change throughout most of human history has been slow enough for us to be able to manage it fairly easily in our everyday lives. But as the overall global population continues to increase, and more and more people come into daily contact with each other, the way we think about food is changing faster than at any time in the past. Food no longer fits into our social habits the way it always has. In the past, appetite, taste and nutrition, and the social and cultural meaning of eating, were all taken care of in the family. Food came in a complex social package, a package that had evolved over time, becoming gradually more complex as human societies expanded from the Paleolithic, through the Neolithic, into the Ancient, Mediaeval and Modern worlds. Each society had a different package. Now that globalization is blurring the boundaries between cultural traditions, the packages are unraveling.

This process of the unraveling of social packages and the cultural rules of behavior that developed in the history of civilization has been recognized with the term "informalization" (Wouters, 2007). In recent decades informalization has increased waste, because it has changed our social context too fast for us to manage the change. It has de-socialized, and de-personalized most of what we do that has to do with our relationship to food. Until the middle of the last century the problem of managing interaction among increasing numbers of people was managed by elaborating the rules of civilized behavior, and those rules were especially noticeable in relation to food. For example, upper-class table manners became more obsessive in the first half of the 20th century. Since the middle of the 20th century we have been outgrowing the rules. For a while new situations were managed tweaking the rules. But as more and more people from different cultural backgrounds came into contact with each other in the second half of the century, the rules were disregarded. Eating began to lose its value as medium of interaction and as a seal for agreement. The change is by no means complete. But the increasing complexity of our social lives, resulting from increasing numbers of people and expanding arenas of social interaction, can no longer be accommodated by further elaboration of formal ways of doing things. We have broken through formality, and everything in our daily social lives is becoming more and more informal. Nevertheless, each ingredient of the old package will continue for a while to influence our food-related behavior. We still need to analyze the package carefully, in order to understand how the past continues to condition the present. But while the rate of change continues to increase, it will be difficult to put the package back together.

## FOOD AND CIVILIZATION

Although it is not the sort of issue that interests most archaeologists of the period from the early Paleolithic into the early millennia of food production, their prehistoric research has produced considerable of evidence of food waste. But since we were small in numbers (no more than a million worldwide in 10,000 BCE), and were dealing with small quantities of food, the amounts wasted were insignificant in relation to what was available. The most wasteful use of food may well have been in the commitment of animals to sacrifice. But large amounts of food have been found in deposits that appear to represent feasts of meat and associated fats just tossed away. There is a well-known example of funerary feasting (Kansa et al., 2004). An extreme example of sacrifice with no associated eating from Classical times is also documented (Hojlund, 1983). And there are examples of foragers who mismanaged their affairs so badly that they ended up wasting fabulous amounts of food. This was at one time considered to be the hallmark of the kind of "affluent" hunter-gatherer who drove the large fauna of the Ice Age in North America to extinction. The most famous case is documented in Wheat, Malde, and Leopold (1972), where a hundred out of three hundred bison killed weren't even cut up, but just lay at the bottom of the pile and rotted. Plant food is generally less evocative of quantity and intention than the animal bone data. But there are also important cases of deposits where large amounts of pottery show a general conspicuous discard event that would be considered wasteful on one level that must have been involved with food. An interesting series of ethnographic examples of the food experience of foraging populations that survived into the 20th century is given in Lee and DeVore (1968), and examples of the production and social uses of food in small tropical communities of slash and burn horticulturalists that survived into the modern world may be seen in Chagnon's description of the Yanomamo either side of the border between Brazil and Venezuela (Chagnon, 2013).

The long prehistoric and historical period after the beginning of agriculture saw the development of cities, states and interurban trade east and west through the arid zone of the northern hemisphere. The resulting expansion of arenas of social interaction was facilitated by writing which was developed first in trade and later in bureaucratic administration for the extension of empires. The last three millennia BCE saw the growth of empires from the Sumerians in southern Mesopotamia to the Persians and the Romans extending from central Asia to the western Mediterranean and Europe. From the 7th century CE down to the Colonial Period Eurasia was divided among major civilizations – Buddhist in the east, Islamic in the middle and Christian in the west – that provided an even larger organizational framework than empires. The larger they became, the more elaborate their institutions. We know the institutionalization of food selection, preparation and consumption in each of them as cuisines. Each cuisine was a complex cultural repertoire of recipes and rules (see Brillat-Savarin, 1854 and Laudan, 2013), based on a hierarchy of staples, starting with a grain or other food that could easily be produced and distributed in large quantities (wheat, rice, maize, taro, etc.). These became the basis for the later development of the elite cuisines of the modern world (Chinese, French, Indian, Japanese, Persian, Thai, etc.). Each generates different patterns of waste.

As industrialization advanced in the 18th century, a new sort of diet developed in the working class areas of Western cities. As more people moved out of food-producing areas into industrial communities, food was selected, prepared, and eaten in new ways. Towards the end of the 19th century some foods began to be pre-prepared and packaged – the beginning of food processing. By the turn of the 20th century many foods were already being mass produced, packaged in pre-determined quantities, branded and advertised. Food processing had begun. In some ways it may have reduced waste since it had a longer shelf-life, but in others it may have increased it by facilitating hoarding. The spread of domestic refrigerators in the 1950s changed the approach to freshness and spoilage. And by the end of the 1950s "Fast Food" was changing the food habits of an increasing proportion of industrialized populations. The value of table manners, which represented the peak of cuisine-related behavior, was already fading among a growing number of Western populations. English table manners were derided by Americans because the American experience unraveled the package before Europe.

## FUTURE FOOD SECURITY

Although the primary concern is whether food production can keep up with the needs of a global population that not only continues to grow but also to cluster in large conurbations away from the main areas of agricultural food production, the biggest problem today, for which no solution is in sight, is not production but distribution. We currently produce more food than we need, but a large proportion of world population is under-nourished, and that proportion is growing. So global food security is currently a problem not of production, either ecological or technological, but of social inequality, economic and political (cf. Pottier, 1999). In recent decades as we have become more concerned about global limits to production, we have come to understand that large numbers of people in developed as well as developing countries suffer from inadequate supply, and are undernourished. Food continues to be the most crucial factor of global inequality. As world population continues to grow, and to cluster in cities, away from major areas of food production, the problem of ensuring adequate nourishment for all will become more difficult. Future food security for a global population that is predicted to plateau later this century in the region of ten billion in a very different geographical distribution from the past, is the central problem for our global community. It cannot be solved simply by increased production.

We have given a brief characterization of the relationship between food and society from the emergence of our species down to the present – a story of qualitative as well as quantitative change (more people, greater densities, more complex distribution patterns, greater social complexity, more waste) – in which the past continues to condition the present and suggest what might be expected in the future. The constant factor is social: eating has continued to be a social activity, though conducted in somewhat different ways in different cultural communities and different historical periods, and institutionalized differently in different cuisines. But the accelerating rate of social change since the middle of the last century is now disrupting all these conventions. We still make cultural selections of what we consider to be food, when and how we want to eat it, and with whom, but we are now all aware of other cultural ways of doing it. For each of us there is no longer just one socially appropriate way. Cultural boundaries between food tastes and eating habits are fading, and we are more likely to be faced with unfamiliar items. Since we are all interacting with larger numbers of people, our idea of a social relationship has changed, and with it the social use of food. Many people no longer see a necessary relationship between eating and socializing. Since we are not always eating with particular others, food now has less to do with our sense of identity. We eat more packaged or served foods, with little or no control over the quantities set before us. Furthermore, while we are all changing, some of us change faster than others. We are further ahead on our trajectory. Rates of change vary not only (as in the past) in different parts of the world, but (now the borders between cultural traditions are fading) among different people and different groups who come in contact with each other. Many of us still subscribe to the cuisines and table manners of the past, but we are aware that they are no longer typical of most of the eating around us. Restaurants do not want to get a reputation for serving less than their rivals. While misuse is mainly in the form of discarding or allowing to spoil, another important form of misuse is in the form of over-consumption or (as in the case of restaurants) presenting for over-consumption in excessive amounts. As their industry expands, so do the predetermined portions they serve, irrespective of individual appetites. We pay the same for a restaurant meal whether we eat it all or not. The further separation of the functions of production, supply, distribution, preparation and consumption increases the potential for waste.

Until the middle of the 20th century food waste patterns appear to have remained roughly the same, though the amount of food wasted was growing along with the number of people eating. Since then the amount of food wasted per head of population appears to have risen faster, first in the West, and then gradually here and there in other parts of the world. Although hard figures will never be available, the amount of food wasted per head of population is probably now greater than at any previous time in world history, and growing. We live in an individualistic society, which sees psychology as the primary factor underlying individual behavior, and have been told (by Margaret Thatcher in 1987) "there's no such thing as society." But this is not a problem of individual behavior or psychology. It is a social problem. Since the primary mechanism that maintains and develops human society of any size is social interaction, when patterns of interaction change there are consequences. As population has grown, society has become more complex and social inequality has increased. Inequality has always grown with the size of society, and it began with the distribution of food. Food may continue to be the most significant marker of inequality in a globalized world, because the larger the society the more difficult it is either to manage inequality or to pass surplus on from the over- to the under-privileged. When our arenas of interaction expand, our patterns of interaction become more intensive, generating more social relationships, most of which get less attention than before, with the result that the use of food to mediate them is less controlled. This is a problem that has emerged at the current stage of world history not because of our diet, our nutritional needs, or our carelessness, but because of the qualitative change in the history of our relationship to food as a medium of social interaction, and the change in the way we are interacting with each other under globalization. The current scourge is one of the problems of accelerating globalization, which changes our relationships with others and therefore also with the way we relate to the material world through those relationships, especially with food. But whatever we are experiencing now is likely to change again when the rate of change slows, as is predicted, in the second half of the current century.

Everything that happens in the coming decades of this century will be conditioned by our historical experience, that has brought us – globally – to the patterns of food waste we are observing today. We need detailed knowledge of this past in order to understand how these patterns might be changed. This knowledge will not enable us to predict with any certainty what will happen in the future, in terms either of our abilities, our needs or our interests. What it will do is show us the various trajectories of change that have brought us to where we are today, from the time we became a fully global species, no more than fifteen thousand years ago, down to the present, as we become a fully globalized community. Although much could be done by policy and regulation to discourage waste, like much else in the modern world that we would like to change (including crime) the problem of waste is more complex than at first appears, and top-down efforts to change problematic behavior tend to be ineffective. But careful attention to the problem will enable us to make the most of the opportunities to reduce waste as the rate of change slows and we develop a new cultural understanding of our social uses of food on a global level.

## NOTES

I am grateful to Dr. Katherine M. Moore for discussion and for references relating to archaeological evidence for food waste.

## REFERENCES

Boserup, E. 1964. The Conditions of Agricultural Growth. Aldine Press, Chicago, IL.

Brillat-Savarin, J.A. 1854. The Physiology of Taste, or Transcendental Gastronomy, illustrated by anecdotes of distinguished artists and statesmen of both continents, translated from the last Paris edition by Fayette Robinson. Lindsay and Blakiston, Philadelphia, PA.

Chagnon, N.A. 2013. Yanomamo, Legacy Sixth Edition. Wadsworth Publ., Belmont, CA.

Christian, D. 2011. Maps of Time. University of California Press, Los Angeles, CA.

Ehrlich, P.A. 1968. The Population Bomb. Buccaneer Books, Cutchogue, NY.

Herlihy, D. 1997. The Black Death and the Transformation of the West. Harvard University Press, Cambridge, MA.

Hojlund, F. 1983. The Maussolleion sacrifice. Am. J. Archaeol. 87(2):145-152. http://dx.doi.org/10.2307/504931.

Kansa, S.W. and S. Campbell. 2004. Feasting with the dead?. A ritual bone deposit at Domuztepe, south eastern Turkey (c. 5550 cal BC). pp. 2-13 in: S.J. O'Day, S.W. van Neer, and A. Ervynck, (eds.), Behaviour Behind Bones: The Zooarchaeology of Ritual, Religion, Status, and Identity. Oxbow Books, Oxford, UK.

Laudan, R. 2013. Cuisine and Empire, Cooking in World History. University of California Press, Berkeley, CA.

Lee, R.B. and I. DeVore (eds.). 1968. Man the Hunter. Aldine Press, Chicago, IL.

Livi-Bacci, M. 1991. Population and Nutrition, An Essay on European Demographic History. Cambridge University Press, Cambridge, UK.

Malthus, R. 1798. An Essay on the Principle of Population. republished as First Essay on Population, 1798. Macmillan Publ., London, UK.

Pottier, J. 1999. Anthropology of Food, the Social Dynamics of Food Security. Polity Press, Cambridge, UK.

Spooner, B. 1972. Population Growth: Anthropological Implications. MIT Press, Cambridge, MA.

Spooner, B. 2015. Globalization via World Urbanization. pp. 1-21 in: Globalization: The Crucial Phase, edited by Brian Spooner, (editor), Penn Museum Publications, University of Pennsylvania Press, Philadelphia, PA.

Wheat, J.B., H.E. Malde, and E.B. Leopold. 1972. The Olsen-Chubbuck site: a Paleo-Indian bison kill. Memoirs Soc. Am. Archaeol. 26:1-180.

Wouters, C. 2007. Informalization: Manners and Emotions Since 1890. Sage Publ., London, UK.

# *Part Two*

# *Consumer food behavior*

# *Chapter 6*

## *Wasted Food: U.S. Consumers' Reported Awareness, Attitudes, and Behaviors*[1]

**Roni A. Neff, Marie L. Spiker, and Patricia L. Truant**

**ABSTRACT**

The U.S. wastes 31 to 40 percent of its post-harvest food supply, with a substantial portion of this waste occurring at the consumer level. Globally, interventions to address wasted food have proliferated, but efforts are in their infancy in the U.S. To inform these efforts and provide baseline data to track change, we performed a survey of U.S. consumer awareness, attitudes and behaviors related to wasted food. The survey was administered online to members of a nationally representative panel (N=1010), and post-survey weights were applied. The survey found widespread (self-reported) awareness of wasted food as an issue, efforts to reduce it, and knowledge about how to do so, plus moderately frequent performance of waste-reducing behaviors. Three-quarters of respondents said they discard less food than the average American. The leading motivations for waste reduction were saving money and setting an example for children, with environmental concerns ranked last. The most common reasons given for discarding food were concern about foodborne illness and a desire to eat only the freshest food. In some cases there were modest differences based on age, parental status, and income, but no differences were found by race, education, rural/urban residence or other demographic factors. Respondents recommended ways retailers and restaurants could help reduce waste. This is the first nationally representative consumer survey focused on wasted food in the U.S. It provides insight into U.S. consumers' perceptions related to wasted food, and comparisons to existing literature. The findings suggest approaches including recognizing that many consumers perceive themselves as being already-knowledgeable and engaged, framing messages to focus on budgets, and modifying existing messages about food freshness and aesthetics. This research also suggests opportunities to shift retail and restaurant practice, and identifies critical research gaps.

---

[1] A version of this chapter was published previously in the journal PLoS ONE: Neff, R.A., M.L. Spiker, and P.L. Truant. 2015. Wasted food: U.S. consumers' reported awareness, attitudes, and behaviors. PLoS ONE 10(6):e0127881. http://dx.doi.org/10.1371/journal.pone.0127881.

## INTRODUCTION

In the U.S., 31 to 40 percent of the U.S. post-harvest food supply goes to waste (Hall et al., 2009; Buzby et al., 2014). A substantial portion of this waste occurs at the consumer level, driven not only by consumer behaviors but also by practices at the processing, retail and restaurant levels and by broader social and economic factors (Evans, 2011; Quested et al., 2013b). This waste has immense consequences.

The lost nutritional value of post-harvest waste in the U.S. represents an estimated 1,249 calories per capita per day, with the greatest amount by weight coming from fruits and vegetables (Buzby et al., 2014). Waste impacts public, food industry and household budgets; food lost from harvest to consumer in 2010 cost $161.6 billion; losses at the consumer level averaged $371 per capita, or 9.2% of average food spending (Buzby et al., 2014). Addressing wasted food puts that food and/or money back into circulation, potentially contributing to improved nutrition and, among those with lower incomes, improved food security. More broadly, reducing waste could help offset the 60% increase in food the United Nations projects we will need from 2009 to 2050 (Alexandratos and Bruinsma, 2012). Because wasting food means wasting all the food's "embodied" social and environmental impacts, this loss contributes extensive water, air and soil contamination (Eshel et al., 2014) and harm to workers (Lo and Jacobson, 2011). Wasted food in North America/Oceania also accounts for an estimated 35% of freshwater consumption, 31% of cropland, and 30% of fertilizer usage (Kummu et al., 2012); as well as 2% of U.S. greenhouse gas emissions (Venkat, 2012); and 21% of post-recycling municipal solid waste (US EPA, 2014). The avoidable use of limited resources and additional environmental impacts from wasted food contribute to the challenge of providing a sustainable and affordable food supply for the future.

While well-supported efforts to understand and address wasted food have proliferated around the world (WRAP, 2014; European Commission, 2015; FAO, 2016), in the U.S. these efforts are nascent, piecemeal, and primarily entrepreneurial, though there are some federal and state interventions (Johnson, 2013; Pascaud, 2014). Intensive, multifaceted efforts supported by research can work; for example, following a range of activities, the UK achieved a 21% reduction in avoidable consumer food waste in five years (Quested et al., 2013a).

As prevention efforts ramp up in the U.S., there is a need for evidence to inform the approaches taken, as well as baseline data to assist in tracking progress. We performed a nationally representative consumer survey aimed at addressing research questions including:

- How aware are Americans of wasted food generally, and of food they waste?
- What attitudes shape their decisions about purchasing and discarding food?
- What would motivate them to waste less?
- To what extent do they perform behaviors known to increase or decrease waste?
- What retail and restaurant industry actions to reduce consumer-level waste of food are supported by consumers?

Surveys in several countries have addressed these topics, but due to differences in culture and society, food system, infrastructure, policy, and geography, we cannot presume how these findings might translate to the U.S. context. This survey identifies points of similarity and difference with that work, and highlights areas in need of additional in-depth research.

We define "wasted food" per the U.S. Department of Agriculture (USDA) as "reductions in *edible* food mass anywhere along the food chain." (USDA, 2015). (We prefer the term, "wasted food" to "food waste," emphasizing that the item is essentially food rather than essentially waste.) For the purposes of the survey, in most cases, we used terms such as "discarding food" rather than "wasting food," to reduce implied judgment and thus reduce bias in responses.

## MATERIALS AND METHODS

### Survey Development

We designed a survey instrument to examine consumer awareness, knowledge, attitudes, and behaviors related to wasted food. To enable comparison, many of the questions were replicated from other surveys; in some cases the number of response options was reduced, or questions were edited for U.S. language usage or clarity (Brook Lyndhurst, 2007; Sustainable America, 2013; Sonntag, 2014). Additional questions were added to expand upon topics of interest. The survey does not aim to quantify the amount of food consumers actually waste (surveys are inappropriate tools for measuring waste) but does ask qualitative questions about waste quantity. One section requests information about respondent performance of a set of behaviors we characterize as waste-promoting and waste reducing. We based these classifications on evidence summarized in literature from the UK and Australia (Baker et al., 2009; Sustainable Victoria, 2011; Quested et al., 2013b). Most of the demographic information was gathered in advance from panel participants by the GfK

Knowledgeworks firm. The survey was administered online, enabling randomization of response option order for questions with multiple options. The survey instrument was reviewed by more than 20 colleagues, including those engaged in research and communications on wasted food, and survey design experts from our network and the firm, GfK Knowledgeworks, which administered the survey. The firm prepared a user-friendly interface reflecting principles of online survey design. The questionnaire is provided at the PLoS ONE website: http://dx.doi.org/10.1371/journal.pone.0127881.s006.

**Survey Sampling and Implementation**

GfK/Knowledgeworks maintains a nationally representative online panel with members randomly recruited using probability-based geographic criteria (GfK, 2012). A key benefit of this approach is that it "covers" 97 of U.S. households in sampling nomenclature, regardless of whether they have cell phones, landlines or neither. To improve representation in its panel, GfK oversamples census blocks with high concentrations of African American and Hispanic residents, and provides Internet access and devices to those lacking them. The firm collects extensive demographic and background data on participants, supporting its ability to create survey samples that are representative of the U.S. non-institutionalized population. For sample selection purposes, the firm applies an adjustment based on the updated national demographic distribution for nine demographic variables. Following survey administration, the firm also supplies a set of post-stratification weights for use in analysis, based on benchmark distributions of seven demographic variables. GfK runs a modest incentive program including raffles or sweepstakes for cash and prizes; this program encourages participation in general, and is not linked to specific surveys.

The survey was formally piloted for two days with a random sample of respondents, and further modifications were made. The survey was administered from April 16 to 20, 2014, to a nationally representative sample of 1,998 non-institutionalized adults ages 18 and above. Reminders were sent on day 3. The response rate was 51%, yielding a sample size of 1,010 respondents.

**Analysis**

Results were analyzed in Stata (version 13.1). We used chi-square tests of independence to test for associations with demographic variables, with statistical significance determined by $p<0.05$. Demographic variables reported here include gender, age ("older respondents" referring to those age 65 and above, and "younger respondents" referring to those under age 65), parental status ("parents" referring to respondents with a child age 18 or under in the household), education (less than high school, high school, some college, or completed college), and household income quintiles (less than $29,999; $30,000 to $59,999; $60,000 to $84,999; $85,000 to $124,999; and $125,000 or more). Other demographic variables, such as household size and employment status, were not significantly associated with outcomes of interest reported here. In chi-square analyses, some categorical dependent variables were analyzed as binary variables so that results would be more readily interpretable. A future multivariate analysis will describe non-demographic predictors of awareness, attitudes and behaviors related to wasted food. Survey response data can be accessed at the PLoS ONE website: http://dx.doi.org/10.1371/journal.pone.0127881.s007.

**Ethics Statement**

This study was reviewed by the Johns Hopkins Bloomberg School of Public Health Institutional Review Board (IRB), which determined it non-human subjects research. All subjects had previously consented to participate in GfK Knowledgeworks surveys. Due to the non-sensitive nature of the research and subject anonymity, the IRB determined there was not a need for additional consent procedures.

## RESULTS

Table 6.1 describes the characteristics of the unweighted sample of 1,010 respondents. Survey weights were applied to further improve the sample's representativeness, and are used in all analyses. Table 6.2 displays results of the chi-square analyses.

**Awareness and Knowledge**

The survey assessed respondents' reported level of awareness and knowledge about wasted food. In the past year, 42% indicated they had seen or heard information about wasted food and 16% had sought information about reducing it. In describing their knowledge about how to reduce the amount of food they discard, 24% described themselves as "very knowledgeable" and 38% described themselves as "fairly knowledgeable." Age and parental status were significantly associated with self-reported knowledge: 30% of older respondents reported being "very knowledgeable," compared to only 23% of younger respondents. Among non-parents, 27% felt "very knowledgeable," compared to only 13% of parents (Table 6.2). Figure 6.S1 describes additional information respondents would like to help reduce food discards.

**Table 6.1 Respondent demographics, unweighted+**

| | Survey % | U.S. % |
|---|---|---|
| **Gender** (1) | | |
| Male | 50 | 49 |
| Female | 50 | 51 |
| **Age** (1) | | |
| 18-24 | 7 | 14[a] |
| 25-44 | 28 | 35[a] |
| 45-64 | 40 | 35[a] |
| 65 and older | 25 | 18[a] |
| **Education** (2) | | |
| Less than high school | 11 | 12[a] |
| High school | 32 | 30[a] |
| Some college | 25 | 19[a] |
| College graduate | 33 | 39[a] |
| **Race** (3) | | |
| White, non-Hispanic | 75 | 63 |
| Black, non-Hispanic | 8 | 13 |
| Other, non-Hispanic | 4 | 9 |
| Hispanic | 10 | 17 |
| **Household Income** (3) | | |
| <$19,999 | 12 | 19 |
| $20,000 - $39,999 | 20 | 20 |
| $40,000 - $59,999 | 17 | 17 |
| $60,000 - $99,0999 | 23 | 22 |
| > $100,000 | 29 | 22 |
| **Live with children <18** (2) | 21 | 32[b] |
| **Self or parent is immigrant** (1) | 17 | 13[c] (self; NA for parents) |

**+**Due to rounding, some categories do not sum to 100 percent.
[a] percentage is based on population age 18, not total population.
[b] refers to percentage of households with members under age 18.
[c] refers to percentage of foreign-born individuals.
Sources for U.S. data: (U.S. Census Bureau, 2013; 2014; 2015).

**Table 6.2 Survey results (selected) and chi-square tests [a]**

| | ALL | Age | | Gender | | Children <18 in household | | Household income quintile [b] | | | | | Highest educational attainment | | | |
|---|---|---|---|---|---|---|---|---|---|---|---|---|---|---|---|---|
| | | <65 | ≥65 | Female | Male | Yes | No | Q1 | Q2 | Q3 | Q4 | Q5 | <High school | High school | Some coll. | Coll. grad |
| **Perceived <u>knowledge</u> about reducing waste of food** | | | | | | | | | | | | | | | | |
| Very knowledgeable | 24 | 23* | 30* | 24 | 24 | 13* | 27* | 26 | 27 | 22 | 20 | 23 | 25 | 22 | 27 | 22 |
| Not very, somewhat, or fairly | 76 | 77* | 70* | 76 | 76 | 87* | 73* | 74 | 73 | 78 | 80 | 77 | 75 | 78 | 73 | 78 |
| **Estimated <u>U.S. waste of food</u>** | | | | | | | | | | | | | | | | |
| 5 or 10 | 10 | 10 | 11 | 10* | 10* | 13 | 9 | 15* | 7* | 7* | 9* | 12* | 17* | 7* | 8* | 12* |
| 20 | 33 | 33 | 35 | 32* | 35* | 29 | 35 | 26* | 27* | 45* | 36* | 41* | 24* | 32* | 35* | 36* |
| 40 | 45 | 45 | 43 | 42* | 47* | 43 | 45 | 37* | 52* | 43* | 47* | 41* | 39* | 47* | 43* | 46* |
| 60 | 12 | 12 | 11 | 16* | 8* | 15 | 11 | 23* | 14* | 5* | 5* | 5* | 19* | 14* | 14* | 6* |
| **Estimated <u>household food discards</u>** | | | | | | | | | | | | | | | | |
| 0 | 13 | 12* | 20* | 12 | 14 | 6* | 15* | 15 | 14 | 17 | 9 | 13 | 18 | 11 | 13 | 13 |

| | ALL | Age | | Gender | | Children <18 in household | | Household income quintile [b] | | | | | Highest educational attainment | | | |
|---|---|---|---|---|---|---|---|---|---|---|---|---|---|---|---|---|
| | | <65 | ≥65 | Female | Male | Yes | No | Q1 | Q2 | Q3 | Q4 | Q5 | <High school | High school | Some coll. | Coll. grad |
| 10 | 56 | 55* | 59* | 53 | 59 | 56* | 56* | 51 | 58 | 50 | 62 | 56 | 51 | 56 | 54 | 59 |
| 20 | 21 | 22* | 17* | 24 | 19 | 26* | 20* | 24 | 16 | 21 | 22 | 21 | 16 | 22 | 25 | 19 |
| 30 | 10 | 11* | 5* | 11 | 8 | 12* | 9* | 10 | 12 | 11 | 2 | 10 | 15 | 11 | 8 | 9 |
| **Estimated household waste of food compared to the <u>average American</u>** | | | | | | | | | | | | | | | | |
| More | 3 | 3* | 2* | 4* | 2* | 7* | 2* | 4* | 2* | 3* | 2* | 6* | 4 | 2 | 3 | 4 |
| The same | 24 | 26* | 15* | 27* | 20* | 27* | 23* | 28* | 21* | 16* | 20* | 35* | 30 | 24 | 25 | 20 |
| Less | 73 | 71* | 84* | 69* | 78* | 66* | 75* | 68* | 77* | 80* | 77* | 59* | 66 | 74 | 73 | 76 |
| **Acceptance of <u>brown banana</u>** | | | | | | | | | | | | | | | | |
| 0-24 | 35 | 35 | 36 | 40* | 30* | 31 | 37 | 47* | 33* | 34* | 27* | 34* | 46 | 38 | 29 | 34 |
| 25-49 | 32 | 32 | 33 | 30* | 35* | 37 | 31 | 26* | 36* | 30* | 40* | 28* | 29 | 31 | 35 | 33 |
| 50-74 | 19 | 19 | 17 | 18* | 19* | 22 | 18 | 13* | 17* | 25* | 22* | 19* | 12 | 19 | 20 | 20 |
| 75-100 | 14 | 14 | 14 | 12* | 16* | 10 | 15 | 14* | 14* | 11* | 11* | 19* | 13 | 12 | 16 | 13 |

| | ALL | Age | | Gender | | Children <18 in household | | Household income quintile [b] | | | | | Highest educational attainment | | | |
|---|---|---|---|---|---|---|---|---|---|---|---|---|---|---|---|---|
| | | <65 | ≥65 | Female | Male | Yes | No | Q1 | Q2 | Q3 | Q4 | Q5 | <High school | High school | Some coll. | Coll. grad |
| **Perceived current effort to minimize waste of food** | | | | | | | | | | | | | | | | |
| A lot of current effort | 35 | 33* | 47* | 36 | 34 | 38* | 28* | 43 | 37 | 30 | 32 | 35 | 37 | 35 | 39 | 32 |
| None, a little, or a medium amount | 65 | 67* | 53* | 64 | 66 | 62* | 72* | 57 | 63 | 70 | 68 | 69 | 63 | 65 | 61 | 68 |

* $p<0.05$.
[a] For each chi-square test, the percentages shown represent column proportions.
[b] Household Income Quintiles: Q1: less than $29,000; Q2: $30,000 to $59,999; Q3: $60,000 to $84,999; Q4: $85,000 to $124,999; Q5: $125,000 or more.

As another gauge of awareness, respondents were asked to estimate the percentage of food in the U.S. that is discarded or otherwise not eaten by humans. Media stories about wasted food commonly share Hall et al.'s (2009) estimate that 40% of the nation's post-harvest food supply is wasted. As shown in Table 6.2, 45% of respondents provided this 40% figure, with most others providing lower estimates. Estimates of average household food waste in the U.S. were significantly associated with gender, household income quintile, and education.

Respondents were also asked to estimate the total percentage of food they themselves discard. While these estimates should not be taken literally, they are useful for gaining insight into how Americans perceive their waste levels and for comparing with evidence-based averages and perceptions about national waste. Table 6.2 shows that respondents overwhelmingly reported discarding low percentages of food they purchase, with 13% reporting that they did not discard any food, and 56% indicating they discarded 10% of purchased food. Only 10% said they discarded 30% or more. Age and parental status were significantly, but non-linearly associated with self-reported estimates of household discards (Table 6.2).

When asked to compare the amount of food they discard to that of others, 73% of respondents reported that they discard less than the average American household, and only 3% reported that they discard more. These comparisons were significantly associated with demographic factors including age, gender, parental status, and household income quintile, as shown in Table 6.2.

Respondents were asked to estimate qualitatively how much of their household's food discarding could be avoided; 29% reported that "a fair amount" or "a lot" was avoidable. The only significant demographic association was with age; 10% of younger respondents reported that "a lot" of discarding was avoidable, versus only 3% of older respondents.

Qualitative ratings were also requested regarding the amount of food discarded in each of six possible categories. Respondents perceived themselves as throwing out the highest amount of fruits and vegetables, followed by homemade meals, bread, meat, milk, and packaged foods. Strikingly, 37% saw themselves as throwing out either "none" or "hardly any" food in all six categories.

**Attitudes and Motivations**

Respondents were asked how much it bothered them to throw out food because it was not eaten, according to a 3-point Likert-type scale including "does not bother me at all," "bothers me a little," and "bothers me a lot." Fifty-two percent reported that discarding food bothered them "a lot", while 9% reported, "not at all." In comparison, more respondents were bothered "a lot" by letting the faucet drip (72%) or leaving the lights on (57%). Fewer were bothered "a lot" by discarding rarely-used clothes (48%) or books (33%).

Respondents were also asked to rate a set of potential motivations for reducing food discards according to a 4-point Likert-type scale of importance, ranging from "not at all important" to "very important." As shown in Figure 6.1, the most important motivation was saving money, with setting an example for children coming in second among parents. Notably, 22 percent of respondents said that the environmental concerns of greenhouse gas emissions, energy and water were "not at all important" motivations.

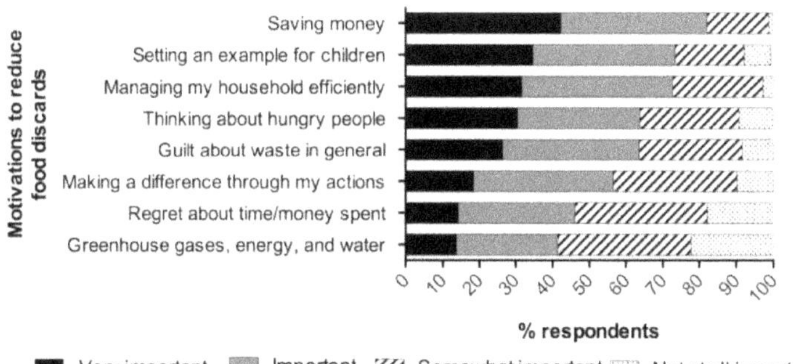


**6.1 Reported Importance of Motivations to Reduce Food Discards.** Responses to four-point Likert-type questions about eight possible motivations for reducing the amount of food discarded. Percentages indicate the proportion of respondents who chose each response, adjusted to 100

To understand respondents' reasons for discarding food, they were asked to indicate their agreement or disagreement with a set of statements. Responses suggest the most common reasons were concern about foodborne illness and desire to eat only the freshest food (Figure 6.2). Among people who reported composting, 41 indicated that because they compost, discarding food does not bother them.

## Behaviors

Respondents were asked to indicate the frequency with which they performed a variety of food shopping and food preparation behaviors, presented in random order. Figure 6.3 presents reported performance of shopping behaviors, and Figure 6.4 presents food preparation behaviors. We characterize these behaviors as either waste-reducing (italicized) or waste-promoting. The figures

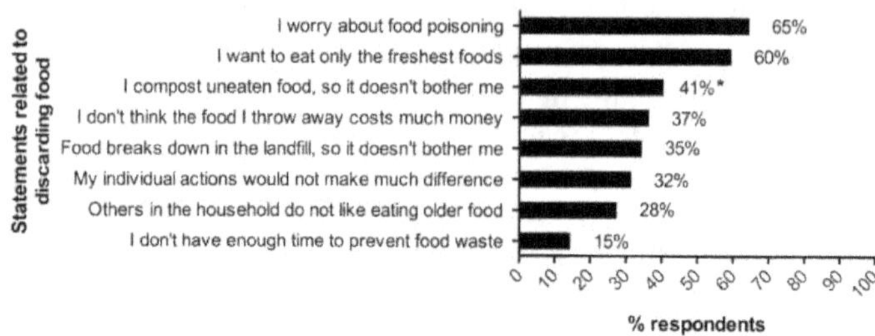


**6.2 Agreement with Statements Related to Discarding Food.** Responses regarding eight possible reasons for discarding food. Percentages indicate the proportion of respondents who chose each response

* Restricted to respondents reporting in a separate question that they compost at least some of their food; percentages for all other motivations reflect the entire sample.

depict substantial portions of respondents performing waste-reducing behaviors always or often, and waste-promoting behaviors rarely or never – though there remains considerable space for improvement.

Several questions were asked to gain insight into consumer decision-making about discarding food. First, respondents selected which of five indicators they used when deciding when to throw out milk. Most common was "use my senses," reported by 72% of respondents; 39% reported using the "use by" date; 22% the "sell by" date [an indicator used by stores, leaving a buffer of time when products still remain high quality]; 18% reported thinking about how long the milk had been opened; and 12% reported thinking about how the milk had been stored (e.g., how long it had been left out). The mean number of indicators used was 1.5 (median 1). In a separate question assessing use of date labels overall, 91% of respondents said they pay attention to date labels. Another aspect of consumer decision-making is willingness to use food that is safe but older. Consumers were asked to move a slider to indicate the maximum percentage brown at which they would accept bananas. The mean response was 40% brown (IQR: 25, 64). As shown in Table 6.2, this decision was significantly associated with gender and household income quintile.

The next set of questions addressed respondents' current level of effort to reduce the amount of food discarded, and their interest in and perceived difficulty of doing so. Over 1/3 said they exerted "a lot" of effort, with the figure rising to nearly 3/4 when the second of four categories, "a medium amount," was added. Reported effort in reducing food discards was significantly associated with age and parental status (see Table 6.2). As for next steps, 23% of respondents reported being very interested in taking action (or additional action) to reduce the amount of food discarded, 65% reported being fairly or somewhat interested, and only 12% reported being not at all interested. When asked how difficult it would be for their household to significantly reduce the amount of food discarded, 43% said easy or very easy, while 16% said difficult or very difficult. Perceived level of difficulty was significantly associated with parental status; 45% of non-parents reported that it would be "very easy" to reduce their household's discards, compared to only 35% of parents.

**6.3 Reported Frequency of Shopping Behaviors.** Responses to five-point Likert-type questions about the frequency of performing nine behaviors related to food shopping. Percentages indicate the proportion of respondents who chose each response. Behaviors classified as "food waste reducing" are italicized; behaviors classified as "food waste promoting" are non-italicized

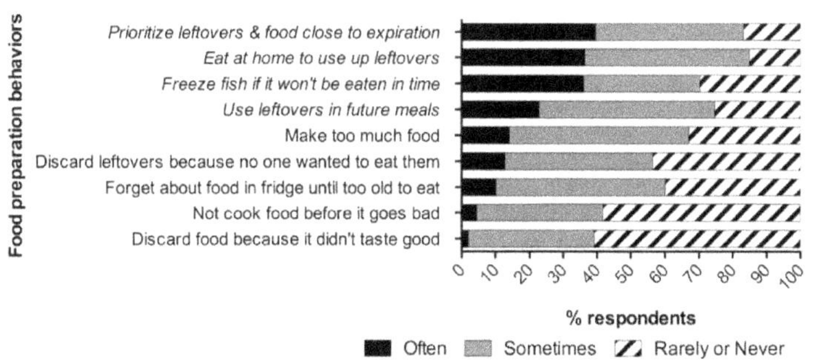


**6.4 Reported Frequency of Food Preparation Behaviors.** Responses to three-point Likert-type questions about the frequency of performing nine behaviors related to food preparation. Percentages indicate the proportion of respondents who chose each response. Behaviors classified as "food waste reducing" are italicized; behaviors classified as "food waste promoting" are non-italicized

**Recommendations for Retail and Restaurants**

Respondents were provided with a list of possible changes retailers could make to help reduce household discarding of food (Figures 6.S2 and 6.S3). The most popular changes respondents said they "would like to see" were more resealable packages (57%), more variety in product sizes (50%), "buy one, get one later" sales (48%), and discounting foods that are over-ripe or near expiration (48%). Respondents were also asked which products they wished were sold in smaller packages. Top responses included baked goods, bagged salad, bread, and meat (43, 41, 39 and 29%, respectively). Frequent write-in responses included milk, fruit, vegetables and canned goods.

Although restaurants could make many changes to reduce discarding of food, some of these changes may be perceived as unacceptable to patrons. Accordingly, respondents were queried about which of a list of changes they would consider acceptable. The leading items were donating excess food (73%), serving smaller portions (61%), taking time to make items to order rather than serving only ready-made items (37%), and providing smaller salad bar plates (30%). Respondents were less accepting of reduced menu variety (15%) and eliminating salad bar trays in favor of plates (8%) (Figures 6.S4 and 6.S5).

## DISCUSSION

This survey paints a portrait of Americans' awareness, perceived knowledge, attitudes and behaviors regarding wasted food as of mid-2014, as prevention efforts are beginning to build. Overall, the survey suggests substantial receptivity to waste prevention activities, with a high self-reported baseline level of awareness, knowledge, and positive attitudes; and moderate rates of desired behaviors. There were important commonalities between our findings and those from international surveys, suggesting that there is some benefit to utilizing insights from international research on other aspects of food waste prevention while awaiting further evidence from the U.S. context. That said, there are substantial cross-national differences in culture and society, food system, infrastructure, policy, and geography; shopping patterns differ in terms of the amount of food purchased in a single trip, the number of days between shopping trips, and the amount of food stored in the household. Accordingly, even similar survey responses may mask meaningful differences in determinants of waste that would lead to differing prevention strategies.

We found that Americans perceive themselves as wasting little, with nearly three-quarters reporting that they discard less food than the average American. International surveys also found high percentages of the public self-reporting very low amounts of wasted food, although the percentages in the U.S.

seem to be even higher. For example, compared to our finding of 69% reporting discarding 0-10% of food, in the UK in 2007 as food waste activities were developing, 43% of consumers said they discarded "hardly any" or no food (Brook Lyndhurst, 2007), and in New South Wales, Australia, 63% said they threw out "very little" food (NSW, 2010). We also found – as did WRAP in the UK – that a majority of respondents reported investing a great deal or a fair amount of effort in minimizing the food discarded (73% in our study; 67% in WRAP's) (Brook Lyndhurst, 2007). WRAP found that younger age groups reported less effort, while we noted no differences by age group.

Based on what is known about wasted food in the U.S., it is clear that respondents as a group are substantially underreporting their waste levels, and they may also be overreporting their effort levels. It is well-known that surveys are an ineffective tool for assessing waste levels (Quested et al., 2013b), and thus our questions regarding the amount of food wasted were asked only with the intent of understanding respondent perceptions. In-depth research using trash sorting and diaries is needed to gain insights into actual waste, while qualitative and other methods will be more effective for understanding the underlying social phenomena. It is possible that in comparing their own waste to national figures, respondents commonly used judgment heuristics involving anchoring their estimates to known figures. For example, the estimate that 40% of all food is wasted would lead most people to say that they waste less than that (Tversky and Kahneman, 1974; Hall et al., 2009; Quested et al., 2013b). Regarding effort, we asked only a general question about the amount of effort invested, aimed at drawing a broad-brush picture of respondent perceptions; it would be valuable for follow-up research to ask detailed questions about effort. This survey does not enable assessing the extent to which the above findings reflect a lack of awareness, aspirational reporting, cognitive dissonance, social desirability bias, or biases due to judgment heuristics.

Regardless of the accuracy of these estimates, given the volume of food wasted in the U.S. and lack of attention to the issue thus far, it would be easy for interventionists to assume they should "start from zero." These self-reports send a message that communications may resonate more if they are framed to recognize people's view of themselves as engaged and knowledgeable.

Among reported motivations for reducing food discards, saving money topped the list, with setting an example for children second (a motivation that did not appear in other surveys reviewed), followed by other concerns related either to efficiency or guilt. These findings were similar to those found in other studies (Brook Lyndhurst, 2007; Quested et al., 2013b; Stefan et al., 2013; Parizeau et al., 2015). Environmental concerns ranked last, with just over 10% of respondents rating them as "very important." Surveys in other countries have also found environmental concerns to rank behind others – though perhaps not with quite as low priority as we found. For example, 20% of respondents indicated a priority on environmental concerns in both UK and the U.S. Sustainable America survey; and in Canada, when asked to characterize the problem of food waste, "an environmental problem" (68%) came in third behind "social" (83%) and "economic" (72%) problem (Brook Lyndhurst, 2007; Quested et al., 2013b; Stefan et al., 2013; Sustainable America, 2013; Parizeau et al., 2015). Many interventionists and food waste prevention organizations are driven by environmental concerns, however, our finding suggests that their work might have more resonance if it focused on budget and other factors that more strongly motivate American consumers. That said, we note that it is possible that if consumers were aware of the actual cost of the consumer portion of food that is wasted (an average of $371 per capita per year according to the USDA's estimate), it might not be sufficient to motivate most non-low income consumers.

It is an open question why respondents placed so little priority on environmental concerns; possibilities include a lack of knowledge, a desire to avoid thinking about the environmental consequences of wasting food, and a weighting of "altruistic" environmental factors as less motivational than the more personally relevant concerns of saving money and aligning one's behaviors and moral attitudes (particularly if the environmental impacts are not viewed as significant). Other research has found that Americans have low levels of environmental knowledge (with considerable variation across topics) (Robelia and Murphy, 2012), and also that despite reported concerns about the environment, levels of action are low (Kollmuss and Ageyman, 2002). Knowledge and concern are commonly found to be important but insufficient precursors to pro-environmental behaviors (Bamberg and Moser, 2007; Polonsky et al., 2012). The relationship may be even more complex in the case of behaviors that can alternatively be motivated by non-environmental concerns. These findings suggest support for all of these possibilities, and a likely need both for education about environmental impacts of wasted food, and for more sophisticated strategies aimed at addressing this discordance.

Based on our findings, there are several critical areas for further work. First, concern about foodborne illness was the most common reason given for discarding food, yet, for most foods, contamination and spending too long at the wrong temperature are key to risk, rather than primarily food age – although certainly those factors intertwine (USDA-FSIS, 2015). There is thus a great need for clearer food safety guidance, for different foods, presented in the joint context of waste prevention and food safety. There is also a need for improved understanding of how Americans make decisions about when to discard foods, including their level of knowledge and their rational thought processes, as well as their implicit, unconscious and habitual attitudes and behaviors. Additionally, there is a need for a nationally harmonized and well-communicated expiration date label system to reduce consumer confusion and resultant unnecessary wastage (Brook Lyndhurst, 2007; Broad Leib and Gunders, 2013; Hanssen and Møller, 2013; Sustainable America, 2013; Newsome et al., 2014).

Second, a majority of respondents reported that they only wanted to eat the freshest food. While humans may have a natural preference for freshness, this concept has also been heavily promoted by health advocates, cooking shows, local food supporters and others as a strategy for making produce more palatable, to add economic value to fresh and local foods, and to enhance both enjoyment and status (Freidberg, 2010). Food waste prevention must incorporate efforts to expand the acceptability of still-good produce and other foods that are older and/or less aesthetically pleasing, and those nearing their expiration dates. Such items can be sold at a discount. There is a substantial space where food may appear less attractive but remains healthy, and when well-prepared, equally palatable.

To improve our ability to intervene and target appropriate interventions, there is a need for research drilling deeper into every one of these reasons for waste. For example, in the case of food safety, *which* foods are consumers discarding for food safety reasons, and how much of that legitimately *should* be discarded? What are the key reasons for food remaining uneaten long enough to become unsafe? What drives incorrect perceptions of food safety?

Communications research can provide additional depth of understanding regarding what messages motivate actual behavior change, and what messages influence people to prioritize food safety, even as they increase waste prevention. There is a related need to probe more deeply into consumers' self-reported relatively high level of knowledge about wasted food; understanding the extent to which they are actually knowledgeable can help inform whether interventions ought to assume people are knowledgeable, versus merely catering to their self-perceived high level of knowledge by avoiding messages that treat them as new to the issue.

Other areas indicated for additional communication include: the cost of wasted food, because that speaks to people's concerns; the environmental impact, because of the possible low awareness; and promoting food-related behaviors that may take place far in advance of the waste and that may not be conceptually associated with waste in peoples' minds (Quested et al., 2013b). Further, the concept that composting food still represents waste (albeit a preferable alternative to sending uneaten food to landfills) should be highlighted, as a significant portion of composters (41%) reported that discarding food does not bother them *because* they compost (studies elsewhere have also highlighted similar concerns, although commonly at lower levels than we found) (Brook Lyndhurst, 2007; Koivupuro et al., 2012; Hanssen and Møller, 2013; Parizeau et al., 2015).

In addressing consumer-level waste of food, it is easy to assume a personal responsibility frame, with most efforts focused on education and communication (Evans, 2011). Extensive evidence from public health and related fields shows limitations of the individual responsibility frame (Marmot, 2005), and highlights that shame and blame can be counterproductive (Marmot, 2005). Consumer education and behavior change interventions must be complemented by approaches making use of entrepreneurial, policy, economic, behavioral economic, and other tools. Indeed, our findings suggest consumer interest in and acceptability of several changes that could be made by the food industry at little cost or even at profit, including smaller packages, donating food, smaller portions, discounts for less aesthetic foods, and making food to order. Behavioral economics approaches would additionally be valuable, to address shopping patterns and their influence on the kinds and amounts of food purchased. There is a need for additional research to understand the tradeoffs for consumers in increased unit price of smaller packages versus cost of food that would otherwise be wasted. While more environmentally-friendly packaging is always desirable, some evidence suggests that for many foods and packaging types, the environmental impacts of extra packaging needed to create smaller sizes may be less than that of the food that would be wasted, although again, additional research would be valuable (Williams and Wikström, 2011; Silvenius et al., 2014).

Overall, this survey found few differences in reported waste or other factors based on race, education, or rural/urban status. Modest differences by age and parental status suggest that older people report less waste of food, and parents of children under 18 report feeling less knowledgeable about how to reduce food discards in their households. In the latter case, it is interesting to note one recent study finding that while adults consumed 89-92% of what they served themselves, children consumed only an average of 59%, suggesting that eating less of one's food may be a normal child behavior – albeit one frustrating to many parents (Wansink and Johnson, 2015). We identified some differences based on income that reached statistical significance, however, there were not clear linear relationships leading to readily interpretable messages. International surveys have varied considerably regarding which demographic predictors rose to the top. For example, a survey in the UK identified age, status as a parent, income and gender (Brook Lyndhurst, 2007) as predictors of wasted food, while one in Finland found household size and type, gender of grocery shopper, valuation of low prices, and views of potential to reduce food waste among the top predictors (Koivupuro et al., 2012). Unlike some other studies, ours did not observe a linear relationship between income and reported level of food wasted; additional inquiry is needed in the U.S. and internationally to gain insight into that relationship.

## STRENGTHS AND LIMITATIONS

As the first national consumer survey focused on wasted food (one prior survey included a set of questions on the topic), this study provides important insights that can guide intervention and serve as a baseline for activities to address such waste in the U.S. Effort was made to improve national representation through the use of stratification and clustering to develop the survey firm's database of respondents and application of post-survey weights, though we do note that the unweighted survey sample underrepresented those in the lowest income quintile and minorities. An additional strength is that questions used in other surveys were incorporated for comparability. The chief limitation is that surveys cannot provide information about actual behaviors or attitudes, only reported ones. The survey used anonymous administration and randomized question order to reduce bias (Nederhof, 1985). Survey administration was slightly expedited, due to anticipated food waste communication activities on Earth Day; those who respond to a survey soon after being contacted may differ in unmeasured ways from those responding later.

## CONCLUSIONS

Consumer waste of food in the U.S. represents a powerful quintuple threat; reducing it may improve food security, nutrition, budgets, environment and public health. Evidence from the UK suggests that multifaceted interventions combined with research can lead to substantial reductions in a short time. This survey, the first known nationally-representative consumer survey focused on wasted food in the U.S., provides essential data to assist those developing interventions using education, changes in business practices, and policy, as well as baseline data against which to measure effectiveness. It suggests approaches including avoiding treating consumers as if they are completely new to the issue of wasted food, framing messages focused on budgets, and providing smaller portions. In addition, there is a need to sharpen our messages about food safety and freshness in order to expand acceptability of those foods that may be less attractive or nearing their expiration dates but are unlikely to be hazardous, while avoiding waste prevention messaging that could increase food safety hazards. The survey also suggests opportunities to shift business practices to support consumer-desired changes that prevent waste. It highlights critical research gaps, particularly in better understanding the actual behaviors underlying the survey responses provided. Finally, it provides reason for optimism about waste prevention in the U.S. – because respondents are concerned about wasted food, and are interested in taking further action.

## ACKNOWLEDGEMENTS

For manuscript review, we thank the three anonymous reviewers, as well as Dave Love and Brent Kim of the Johns Hopkins Center for a Livable Future (CLF), Claudia Giordano, Department of Agricultural and Food Sciences, University of Bologna, and Dana Gunders, NRDC. We thank Sameer Siddiqi, Research Assistant at CLF for very helpful research assistance. We thank the following for in-depth feedback on the survey instrument: Viki Sonntag, Ecopraxis; Dana Gunders, NRDC; Corinne Shefner-Rogers, Consultant; and Dave Love, Anne Palmer, and Shawn McKenzie (CLF). Appreciation also goes to Colleen Barry, Johns Hopkins Bloomberg School of Public Health, for useful guidance. Lastly we thank Carolyn Chu of GfK for assistance in managing, reviewing and administering the survey.

## REFERENCES

Alexandratos, N. and J. Bruinsma. 2012. World agriculture towards 2030/2050: The 2012 revision. Working paper No. 12-03. FAO, Rome, Italy.

Baker, D., J. Fear, and R. Denniss. 2009. What a waste: An analysis of household expenditure on food. The Australia Institute, Canberra City, Australia.

Bamberg, S., and G. Möser. 2007. Twenty years after Hines, Hungerford, and Tomera: A new meta-analysis of psycho-social determinants of pro-environmental behaviour. J. Environ. Psychol. 27(1):14-25. http://dx.doi.org/10.1016/j.jenvp.2006.12.002

Broad Leib, E., and D. Gunders. 2013. The Dating Game: How Confusing Food Date Labels Lead to Food Waste in America. Natural Resources Defense Council Report R:13-09-A.

Brook Lyndhurst. 2007. Food Behaviour Consumer Research: Quantitative Phase. Waste & Resources Action Programme (WRAP), Banbury, UK.

Buzby, J.C., H.F. Wells, and J. Hyman. 2014. The estimated amount, value and calories of postharvest food losses at the retail and consumer levels in the United States. Economic Information Bulletin Number 121. Economic Research Service/USDA.

Eshel, G., A. Shepon, T. Makov, and R. Milo. 2014. Land, irrigation water, greenhouse gas, and reactive nitrogen burdens of meat, eggs, and dairy production in the United States. Proc. Nat. Acad. Sci. 111(33):11996-12001. http://dx.doi.org/10.1073/pnas.1402183111.

European Commission. 2015. Food Waste. http://ec.europa.eu/food/safety/food_waste/index_en.htm.

Evans, D. 2011. Blaming the consumer – once again: the social and material contexts of everyday food waste practices in some English households. Crit. Public Health 21(4):429-440. http://dx.doi.org/10.1080/09581596.2011.608797.

Food and Agriculture Organization of the United Nations (FAO). 2016. Save Food: Global Initiative on Food Loss and Waste Reduction. FAO, Rome, Italy.

Freidberg, S. 2010. Fresh: A Perishable History. Harvard University Press, Cambridge, MA.

GfK. 2012. Knowledge Panel Design Summary. http://www.knowledgenetworks.com/ganp/docs/KnowledgePanel28R29-Design-Summary.pdf.

Hall, K.D., J. Guo, M. Dore, and C.C. Chow. 2009. The progressive increase of food waste in America and its environmental impact. PLoS ONE 4(11):e7940. http://dx.doi.org/10.1371/journal.pone.0007940.

Hanssen, O.J., and H. Møller. 2013. Food Wastage in Norway 2013 – Status and Trends 2009-2013. FAO and Østfoldforskning Sustainable Innovation.

Johnston, M.W. 2013.Getting the public tuned in to food waste reduction. BioCycle 54(11):17.

Koivupuro, H.-K., H. Hartikainen, K. Silvennoinen, J.-M. Katajujuuri, N. Heikintalo, A. Reinikainen, and L. Jalkanen. 2012. Influence of socio-demographical, behavioural and attitudinal factors on the amount of avoidable food waste generated in Finnish households. Int. J. Cons. Stud. 36:183-191. http://dx.doi.org/10.1111/j.1470-6431.2011.01080.x.

Kollmuss, A., and J. Agyeman. 2002. Mind the gap: Why do people act environmentally and what are the barriers to pro-environmental behavior? Environ. Educ. Res. 8(3):239-60. http://dx.doi.org/10.1080/13504620220145401.

Kummu, M., H. de Moel, M. Porkka, S. Siebert, O. Varis, and P.J. Ward. 2012. Lost food, wasted resources: Global food supply chain losses and their impacts on freshwater, cropland, and fertilizer use. Sci. Tot. Environ. 438:477-489. http://dx.doi.org/10.1016/j.scitotenv.2012.08.092.

Lo, J., and A. Jacobson. 2011. Human rights from field to fork: Improving labor conditions for food-sector workers by organizing across boundaries. Race/Ethnicity 5(1):61-82.

Marmot, M. 2005. Social determinants of health inequalities. Lancet 365(9464):1099-1104. http://dx.doi.org/10.1016/S0140-6736(05)71146-6.

Nederhof, A.J. 1985. Methods of coping with social desirability bias: A review. Eur. J. Soc. Psychol. 15(3):263-280. http://dx.doi.org/10.1002/ejsp.2420150303.

Newsome, R., C.G. Balestrini, M.D. Baum, J. Corby, W. Fisher, K. Goodburn, T.P. Labuza, G. Prince, H.S. Thesmar, and F. Yiannis. 2014. Applications and perceptions of date labeling of food. Comp. Rev. Food Sci. F. 13(4):745-769. http://dx.doi.org/10.1111/1541-4337.12086.

NSW. 2010. Food Waste Avoidance Benchmark Study 2009. New South Wales Government, Australia.

Parizeau, K., M. von Massow, and R. Martin. 2015. Household-level dynamics of food waste production and related beliefs, attitudes, and behaviours in Guelph, Ontario. Waste Manage. 35:207-217. http://dx.doi.org/10.1016/j.wasman.2014.09.019.

Pascaud, L. 2014. A food waste reduction movement gathers steam. *Forbes*, July 24. http://www.forbes.com/sites/techonomy/2014/07/24/a-food-waste-reduction-movement-gathers-steam/.

Polonsky, M.J., A. Vocino, S.L. Grau, R. Garma, and A.S. Ferdous. 2012. The impact of general and carbon-related environmental knowledge on attitudes and behaviour of US consumers. J. Marketing Manage. 28(3-4):238-263. http://dx.doi.org/10.1080/0267257X.2012.659279.

Quested, T., R. Ingle, and A. Parry. 2013a. Household Food and Drink Waste in the UK. Final report. Waste & Resources Action Programme (WRAP), Banbury, UK.

Quested, T.E., E. Marsh, D. Stunell, and A.D. Parry. 2013b. Spaghetti soup: The complex world of food waste behaviours. Resour. Conserv. Recy. 79:43-51. http://dx.doi.org/10.1016/j.resconrec.2013.04.011.

Robelia, B., and T. Murphy. 2012. What do people know about key environmental issues? A review of environmental knowledge surveys. Environ. Educ. Res. 18:299-321. http://dx.doi.org/10.1080/13504622.2011.618288.

Segrè, A., and M. Pessato, translated by C. Giordano. 2014. Waste Watcher – Knowledge for Expo Rapporto 2014 – Executive Summary.

Silvenius, F., K. Gronman, J. Katajajauuri, R. Soukka, H. Koivupuro, and Y. Virtanen. 2014. The role of household food waste in comparing environmental impacts of packaging alternatives. Packaging Technol. Sci. 27:277-292. http://dx.doi.org/10.1002/pts.2032.

Sonntag, V. 2014. Food Too Good to Waste Pre-engagement Survey (unpublished).

Stefan, V., E. van Herpen, A.A. Tudoran, and L. Lähteenmäki. 2013. Avoiding food waste by Romanian consumers: The importance of planning and shopping routines. Food Qual. Prefer. 28(1):375-381. 10.1016/j.foodqual.2012.11.001.

Sustainability Victoria. 2011. Food Waste Avoidance Studies 2010. State Government, Victoria, Australia.

Sustainable America. 2013. Food/Fuel Public Poll: Key Findings. http://www.sustainableamerica.org/downloads/presentations/SustainableAmericaFinalDeck.pdf.

Tversky, A., and D. Kahneman. 1974. Judgment under uncertainty: Heuristics and biases. Science 185(4157):1124-1131. http://dx.doi.org/10.1126/science.185.4157.1124.

U.S. Census Bureau. 2013. Current Population Survey, 2012 Annual Social and Economic Supplement.

U.S. Census Bureau. 2014. Current Population Survey, 2014 Annual Social and Economic Supplement.

U.S. Census Bureau. 2015. American Community Survey, 2013 Population Estimates.

U.S. Department of Agriculture. 2015. Food Waste Challenge: Frequently Asked Questions. http://www.usda.gov/oce/foodwaste/faqs.htm.

US EPA. 2014. Municipal Solid Waste Generation, Recycling, and Disposal in the United States: Tables and Figures for 2012. http://www3.epa.gov/epawaste/nonhaz/municipal/pubs/2012_msw_dat_tbls.pdf.

USDA Food Safety and Inspection Service (USDA-FSIS). 2015. Food Safety Education – Food Product Dating.

Venkat, K. 2012. The climate change and economic impacts of food waste in the United States. Int. J. Food Syst. Dynam. 2(4):431-446.

Wansink, B., and K.A. Johnson. 2015. Adults only: Why don't children belong to the clean plate club? Int. J. Obesity 39:375. http://dx.doi.org/10.1038/ijo.2014.205.

Waste & Resources Action Programme (WRAP). 2014. Food Waste Reduction. WRAP, Banbury, UK. http://www.wrap.org.uk/food-waste-reduction.

Williams, H., and F. Wikström. 2011. Environmental impact of packaging and food losses in a life cycle perspective: A comparative analysis of five food items. J. Cleaner Prod. 19(1):43-48. http://dx.doi.org/10.1016/j.jclepro.2010.08.008.

## SUPPORTING INFORMATION

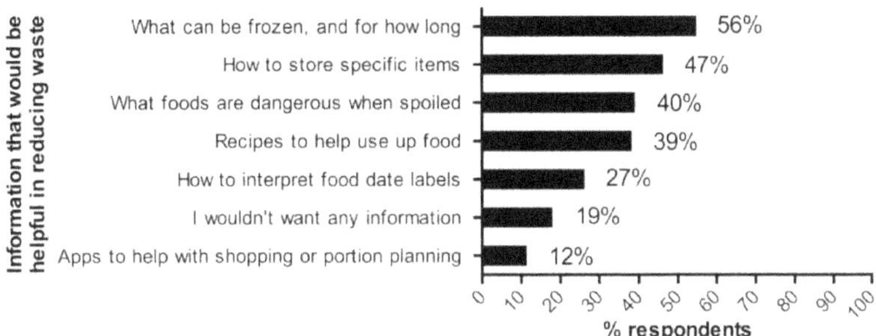


**6.S1** Desired information to assist in reducing food discards

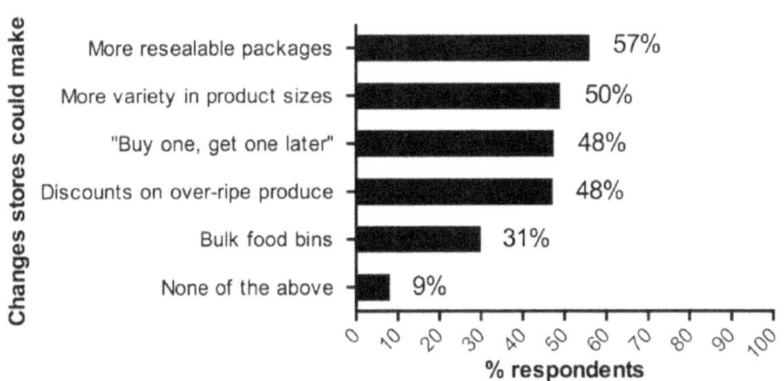


**6.S2** Changes retailers could make to reduce food discards

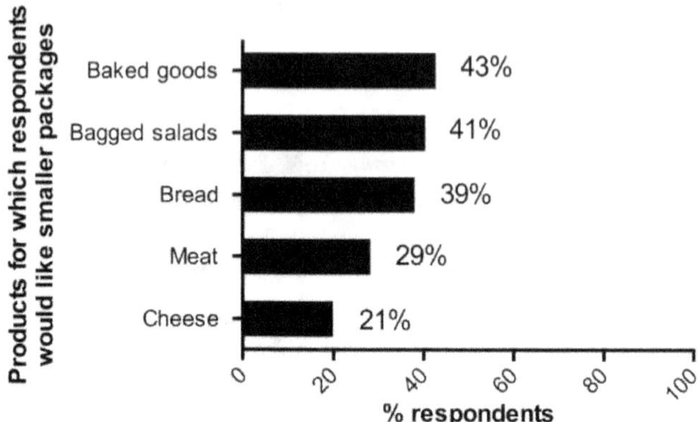


**6.S3** Products for which respondents want smaller packages

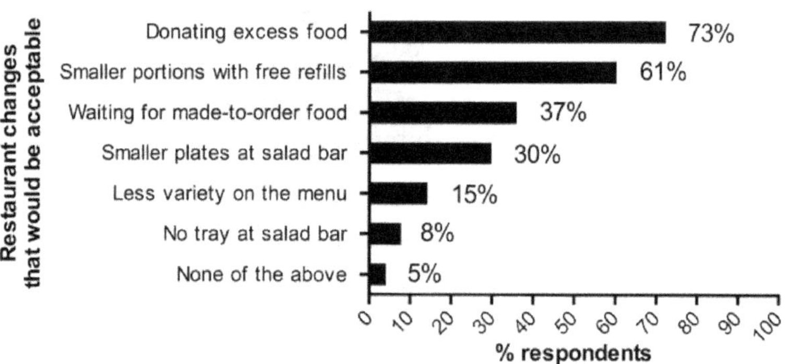


**6.S4** Potentially unaccepted restaurant changes that respondents consider acceptable

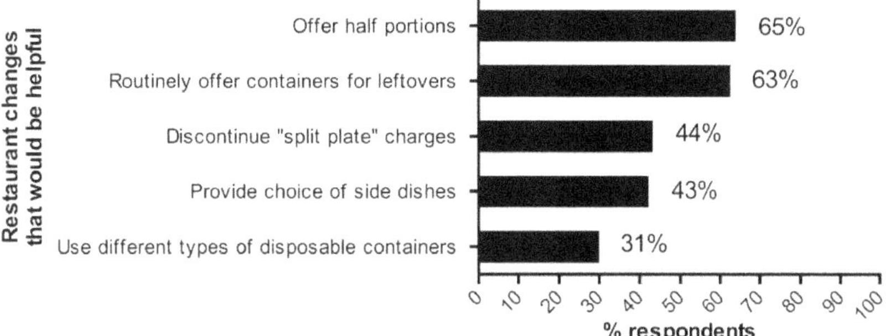


**6.S5** Restaurant changes that would be helpful

# Chapter 7

## A Review of Consumer Food Wasting Behavior

**Brittany Hartmire, Jane Kauer, and Zhengxia Dou**

### INTRODUCTION

Forty percent of all food produced in the United States goes uneaten (Gunders, 2012). The Economic Research Service (ERS) of the USDA estimated that in 2010 alone, 21 percent of the 430 billion pounds of food available at the consumer level was thrown away (Buzby et al., 2014). These 90 billion pounds of food added up to $114 billion in retail value (Buzby et al., 2014), an estimated retail price of $390 per capita per year (Buzby and Hyman, 2012). Consumers threw away 244.5 billion kilocalories, 789 kilocalories per capita per day (Buzby et al., 2014).

The purpose of this chapter is to provide a comprehensive review of current understanding of consumer food waste behavior. Our focus is on the United States, but since the subject is relatively new and research within the U.S. is limited, findings from surveys conducted in the UK or other developed countries are included to shed light. This review compiles the findings of 38 sources, a combination of survey studies from peer-reviewed scholarly journals, newspaper articles, city ordinances, statements by the FDA, think-tanks, NGOs, and government institutions. Most of the sources were published within the last five years. The most relevant studies are listed in Table 7.1.

Findings are organized into the following sections to facilitate discussion: demographic predictors for high food waste production, information about consumer motivations, awareness, and attitudes behind food waste plus a brief legal history of date labels in the U.S., comparison of different research modalities for assessing consumer food behavior, outlines of steps to be taken by consumers, food industry, and government toward food waste reduction, and final conclusions.

### DEMOGRAPHICS AND FOOD WASTAGE

There is limited research on U.S. food consumer behavior, few data exist about the sociodemographic groups wasting the most food within the U.S. However, data on this topic is readily available in other countries, especially due to the work of Waste & Resources Action Programme (WRAP) in the UK. The sociodemographic groups who are wasting the most in the U.S. are likely similar to those of the United Kingdom and other developed economies.

A 2007 study that involved face-to-face interviews with 1,862 UK households found that young professionals, young families, and social housing renters produce the most avoidable food waste (Brook Lyndhurst, 2007). The presence of young professionals (defined as individuals aged 16-34 in full-time work) and school age children are the two most significant predictors for having large amounts of food waste in the UK (Vangarde and Woodburn, 1987; WRAP, 2007). The high amount of food waste produced by young workers is partly due to having poorer cooking skills or time for meal-prep and planning than their older counterparts (Quested and Luzecka, 2014). Children's 'finicky' food habits and parents with greater concern for food safety explain the positive correlation between the presence of children and food waste (Quested and Luzecka, 2014). Furthermore, a WRAP survey found that while large families understandably produce the largest amounts of food waste, they produce less food waste *per capita* than smaller families and single households (Quested and Luzecka, 2014).

This inverse relationship between family size and food waste produced per capita was also found in an online questionnaire comparing the responses of volunteers at two separate academic institutions, 404 from the Joint Research Centre (JRC) in Ispra, Italy and 453 from the Karlsruhe Institute of Technology (KIT) in Karlsruhe, Germany (Jörissen et al., 2015). The two academic groups yielded similar survey responses with few significant differences, suggesting that consumer food wastage is influenced by social class and educational level more than by nationality. While some studies conducted within the U.S. have failed to demonstrate a linear relationship between income and food waste amount at the individual level (Vangarde and Woodburn, 1987), research from the Food and Agriculture Organization of the United Nations (FAO) demonstrates that at the national scale, food waste production is higher in wealthier countries (Buzby et al., 2014). Studies all over the world have supported a correlation between economic concern and responsible food behavior.

Researchers in Finland reported that Finnish food consumers who appreciate deals and low prices generally produce less food waste (Koivupuro et al., 2012). The researchers thought that perhaps the consumers who appreciated low food prices the most might be lured by deals and bargains to compulsively purchase more food than what they would actually eat, but the findings turned out to be the opposite. Research in this area is needed in the United States to

**Table 7.1 Relevant international studies examining retail and consumer food wastage**

| Author | Year | Nation/ Region | Modality |
|---|---|---|---|
| Neff | 2015 | US | Online survey (n=1,010) |
| Buzby | 2012 | US | Calculations based on statistical data of the USDA ERS (n=200 individual foods) |
| Polonsky | 2012 | US | Online survey (n=353) |
| Freedman | 2009 | US | Online survey (n=1,475); Waste composition analyses |
| Tsiros | 2005 | US | Questionnaire (n=270) |
| Bunn | 1990 | US | Face-to-face interviews (n=229) |
| Vangarde | 1987 | US | Face-to-face interviews (n=243); Direct waste analyses (n=243 households) |
| Lee | 2015 | UK | Calculations based on food store data (n=23,299 food samples) |
| Graham-Rowe | 2014 | UK | Face-to-face interviews (n=15) |
| Quested | 2014 | UK | Calculations based on WRAP datasets |
| Giorgi | 2013 | UK | Online survey (n= 5,183); Face-to-face discussion groups (n=113 total participants) |
| Icaro Consulting | 2013 | UK | Face-to-face interviews (n=18); Online survey (n=4,000) |
| Brook Lyndhurst | 2007 | UK | Questionnaire (n=1,862) |
| WRAP | 2007 | UK | Synthesis of WRAP findings |
| Jörissen | 2015 | Italy and Germany | Online survey (n=404 in Italy and 453 in Germany) |
| Chung | 2008 | Hong Kong | Face-to-face interviews (n=4,141); Direct waste analysis (n=1,425 plastic bags) |
| Abeliotis | 2014 | Greece | Face-to-face interviews (n=231) |
| Koivupuro | 2012 | Finland | Online survey (n=380); Food Waste Diary (n=380) |
| Parizeau | 2015 | Canada | Direct waste analysis (n=61 households); Face-to-face interviews (n=61) |

discern whether any differences exist in a more consumerist culture. Yet additional international studies demonstrate that as economic concern rises and household income decreases, responsible behavior for minimizing avoidable food waste production improves. For instance, a 2014 study in Greece, where survey participants were experiencing a severe recession, found that Greek households already practice good habits for reducing avoidable food waste (Abeliotis et al., 2014). In face-to-face interviews, the 231 randomly selected participants reported regularly checking their cupboards, planning meals, and making shopping lists due to economic necessity rather than environmental concerns (Abeliotis et al., 2014). Frugality is likely correlated with responsible food behavior in the U.S. as well, for consumer food loss in the U.S. represents a significant share of U.S. household income levels. The annual value of food loss is nearly equal to 10% of the average amount U.S. consumers spent on food in 2008, over 1% of the average disposable income (Buzby and Hyman, 2012). In 2008 alone, each U.S. household spent approximately $936 on food that was thrown away uneaten (Buzby and Hyman, 2012).

## CONSUMER HABITS, KNOWLEDGE, AND ATTITUDES

Several studies have found that consumers in the U.S. and other countries produce more food waste than they realize (Brook Lyndhurst, 2007; WRAP, 2007; Jörissen et al., 2015; Neff et al., 2015; Parizeau et al., 2015). A WRAP study found that 90% of UK consumers claim that they throw away little food (e.g. 'some', 'a small amount', 'hardly any', or 'none'; Brook Lyndhurst, 2007). If this is true, 10% of UK consumers would need to be discarding nearly all of the food they buy in order to add up to the 6.7 million tonnes of national food waste produced in the UK that year (WRAP, 2007). A study in Hong Kong measured plastic bag waste to evaluate the reliability of self-reported waste data (Chung, 2008). The researchers found that consumers generally estimate their waste production to be 1.3-5 times less than data obtained through direct measurement (Chung, 2008).

The most common reasons consumers report throwing away food are expiration dates, food looking/smelling/tasting bad, mold, or too much time spent in the fridge or cupboard (Jörissen et al., 2015). The most common foods being thrown away are fruits and vegetables, post-prepared leftovers, bakery products, dairy products, and eggs (Brook Lyndhurst, 2007; Jörissen et al., 2015). A few habits make consumers prone to producing greater quantities of food waste, such as shopping at large supermarkets, shopping infrequently, and failing to either plan meals or create shopping lists (WRAP, 2007; Jörissen et al., 2015).

Food consumers generally lack knowledge about simple strategies to reduce food waste. For instance, WRAP found that UK consumers fail to utilize the information found on packaging to maximize in-home shelf-life (Icaro

Consulting, 2013). Consumers care about optimizing shelf-life, but there is widespread ignorance around packaging and the benefits it offers (Icaro Consulting, 2013). In fact, surveys show that UK consumers believe that packaging reduces shelf-life (Icaro Consulting, 2013). And while 90% of UK consumers say that they are 'very' or 'fairly' confident they store food in a way that optimizes shelf-life, two-thirds unpack their food in a way that reduces shelf-life (e.g. removing apples from their bag or piercing holes in their packaging; Icaro Consulting, 2013).

Food labeling is apparently an important factor, as there is widespread confusion around food labels (Icaro Consulting, 2013; Marra, 2013; Abeliotis et al., 2014; Buzby et al., 2014). Throughout the 1970s, the Government Accountability Office (then the General Accounting Office) and the Office of Technology Assessment (OTA) issued reports in Congress about food labeling, and supermarkets voluntarily began adopting open dating systems in response to consumer interest – 'open' labels with calendar dates rather than 'closed' labels with numerical codes only decipherable to manufacturers and retailers (Broad Leib et al., 2013). By 1973, 10 states had adopted regulations regarding open date labels for certain food products, but no federal legislative efforts, mostly proposed amendments to the federal Food, Drug, and Cosmetic Act, gained enough momentum to pass into law (Broad Leib et al., 2013). These efforts were initially opposed by supermarkets for they worried that consumers would only want to purchase the freshest food available on shelves (Broad Leib et al., 2013). But after supermarkets began voluntarily adopting open date labels in response to consumer demand, the National Association of Food Chains argued that such regulations would impose higher costs and discourage food chains from voluntarily adopting progressive programs in the future (Broad Leib et al., 2013). A number of legislative efforts have been made, including Congressman Frank Pallone's (D-NJ) 1999 National Uniform Food Safety Labeling Act and Congresswoman Nita Lowey's (D-NY) 1999 Food Freshness Disclosure Act, with similar bills to the latter reintroduced in 2001, 2003, 2005, 2007, and 2009, but all attempts failed (Broad Leib et al., 2013).

Although the FDA and USDA have the legal authority to ensure that America's food supply is safe, the prevailing practice of food labeling in the U.S. is such that "'expired by', 'use by' or 'best before' dates [are] entirely at the discretion of the manufacturer" with the exception of infant formula (FDA, 2015). While some states and independent food industry actors have responded to consumer demand for greater transparency about food freshness and safety, no uniform date labeling system has been successfully implemented at the national level. As a result, multiple date labeling schemes exist. The inconsistent definitions of food labels lead to premature food disposal, for consumers generally lack knowledge about how to correctly interpret the unreliable wording of date labels (Broad Leib et al., 2013).

Consumer attitudes pose an additional barrier to reducing food waste. A WRAP study found that while about half of food-wasters in the UK are receptive to messages encouraging less food waste, the other half are not (Brook Lyndhurst, 2007). The same study also found that consumers generally do not consider food waste to be an environmental issue (Brook Lyndhurst, 2007). Furthermore, consumers are influenced by social pressures when eating out with others. UK consumers have admitted that asking for a take-home container for leftovers is embarrassing and that the amount of food they eat depends on who they are with (Giorgi, 2013). In addition, consumers are concerned about food safety (Marra, 2013; Quested and Luzecka, 2014), and in places like Japan, where there have been widely publicized food poisoning scandals, there is prevailing distrust of the food industry (Marra, 2013). Furthermore, food consumers in developed economies expect their produce to be cosmetically flawless (Vangarde and Woodburn, 1987; Bunn et al., 1990; Marra, 2013; Buzby et al., 2014). This results in producers and retailers prematurely discarding food with aesthetic imperfections in order to meet consumer demand and expectations (Vangarde and Woodburn, 1987; Marra, 2013).

## ASSESSING CONSUMER FOOD WASTAGE: COMPARISON OF METHODS

Researchers attempting to learn about consumer food behavior have several options when selecting a modality. First, they must decide whether to employ an outside third party to quantify consumer food waste production or rely upon self-reported data provided by the consumers themselves. Each of these methods has different strengths and limitations, and the most valuable studies will combine both.

When food waste is estimated by a third party, studies can form their data via waste composition analyses or calculations based on statistical data of food supply or municipal waste. Data produced through these modalities are more numerically accurate than data produced through self-reporting (Brook Lyndhurst, 2007; WRAP, 2007; Chung, 2008; Jörissen et al., 2015; Neff et al., 2015; Parizeau et al., 2015). However, there are no international standards for either waste composition analyses (Lebersorger and Schneider, 2011) or calculations based on pre-existing statistical data (Monier et al., 2010). Thus comparisons of food waste calculations from different studies, times, or places are difficult and can be skewed. Numerical analyses also neglect the qualitative consumer perspective. These methodologies do not obtain information about consumer habits, motivations, attitudes, or knowledge.

Consumer perspectives can be acquired through questionnaires, interviews, and kitchen journals. Questionnaires make it easy to acquire a large, geographically diverse study sample. They are inexpensive, require little effort from the study participants, and can reach participants from a distance when administered online or over the phone. On the other hand, questionnaires administered online or over the phone exclude rural and low-income households that lack the necessary technological resources. Study participants may also be less engaged and take their responses less seriously when there is no physical interviewer asking questions in a conversation-style setting. Moreover, multiple-choice answers limit the detail of responses and are often subjective (e.g. 'a lot', 'hardly ever', 'sometimes'). What might be 'a lot of waste' to one participant might be 'some waste' to another. This leads to inaccurate information, but it gathers important information about consumer perceptions of individual waste production. Consumers are generally unaware of how much waste they produce (Brook Lyndhurst, 2007; WRAP, 2007; Jörissen et al., 2015; Neff et al., 2015; Parizeau et al., 2015), but they are also apt to underreport their food waste production in an attempt to seem more 'socially acceptable' (Jörissen et al., 2015). The presence, status, and behavior of interviewers can also impose social pressures on study participants, potentially influencing them to alter their answers (Williams et al., 2012; Graham-Rowe et al., 2014). But face-to-face interviews generally produce more detailed qualitative responses, and depending on the environment, may allow for observations of study participants' behavior in a natural setting (e.g. home kitchen, buffet, supermarket).

Kitchen journals are the most accurate modality for collecting self-reported food waste data, but they are expensive and require much time and effort from the study participants (Jörissen et al., 2015). Furthermore, consumers are prone to alter their typical food habits while using a kitchen journal (Koivupuro et al., 2012; Quested et al., 2013). Because these modalities are all useful and limited in different ways, the combination of multiple survey tactics makes for a stronger study.

## WAYS TO LOWER CONSUMER FOOD WASTAGE

U.S. consumers have some capacity to decrease their food waste production. To begin, they can relieve pressure on food retailers and producers by welcoming cosmetically blemished fruits and vegetables into their kitchens. Furthermore, the adoption of several habits may help eliminate excess food: cooking smaller portions; creating shopping lists; checking cupboards regularly;

planning meals; freezing produce; purchasing already frozen rather than fresh produce; and food shopping more frequently. Shopping for food infrequently, which forces consumers to buy larger quantities of food at once, is correlated with higher avoidable food waste outputs (WRAP, 2007). The U.S. Environmental Protection Agency (EPA) also recommends purchasing less food at once, in addition to eating older foods first, including quantities on shopping lists, choosing loose produce over pre-packaged items to better control for quantity, preparing items sooner for convenience, and keeping fruits and vegetable fresh (EPA, 2012). Consumers can also decrease their avoidable food waste production by educating themselves about how to correctly interpret their supermarket's date labeling system, handle food in a way that maximizes shelf-life, and identify when food can still be safely consumed (Broad Leib et al., 2013).

Retailers also have some capacity to impact consumer food behavior. A 1990 California survey discovered that food consumers are more willing to purchase aesthetically unpleasing produce when accompanied with the knowledge of reduced pesticide use (Bunn et al., 1990). Therefore, retailers might encourage consumers to begin accepting cosmetically blemished produce if these food products are accompanied with educational displays. Marketing frozen food products may also encourage responsible consumer food behavior.

Retailers can also apply discounts to incentivize consumers to purchase blemished produce and food products approaching their expiration dates. Research based on personal interviews with U.S. supermarket managers found that while many supermarkets discount beef and chicken approaching their expiration dates, managers fear that discounting dairy or produce would tarnish their store's image (Tsiros and Heilman, 2005). However, interviews with consumers in the same study reveal that such discounts on dairy and produce do not impact store image any differently than discounts on beef and chicken (Tsiros and Heilman, 2005). In fact, consumers in this study were less likely to judge store quality more negatively due to dairy or produce discounts than beef and chicken discounts (Tsiros and Heilman, 2005). In the same study, three versions of a survey asked 300 U.S. consumers about their willingness to pay for a perishable food item with seven (n=85), four (n=93), or one (n=92) day(s) before its expiration date. The majority of consumers believed that item quality and safety decreased as the perishable item's remaining lifespan decreased (Tsiros and Heilman, 2005). Correspondingly, as the food item grew closer to its expiration date, the price that consumers were willing to pay decreased (Tsiros and Heilman, 2005). Economic models for pricing goods with diminishing value already exist (Rajan et al., 1992). Furthermore, evidence shows that price promotions have a greater short- and long-term effectiveness when applied to perishable goods in comparison to other categories (Nijs et al., 2001). Discounting older produce helps retailers both optimize profit and decrease their avoidable food waste production.

There are various ways for retailers and manufacturers to extend product life without having to change packaging or product formulations. Certain food products, such as potatoes, apples, minced and sliced ham, have 'buffers' between the product life specified and the actual time until a product loses its quality or safety (Lee et al., 2015). Retailers and manufacturers are in a position to challenge overly cautious buffers. Furthermore, open life labels (e.g., 'once opened consume within x days') are applied for both quality and safety purposes (Lee et al., 2015). Retailers and manufacturers can help reduce consumer food waste if they stop applying these labels when safety is not of concern. They can also eliminate the use of 'display until' and 'sell by' labels to avoid customer confusion and reduce the premature disposal of safe, healthy food.

Restaurants and convenience food businesses can reduce consumer food waste by being more transparent about portion sizes. A study conducted inside a university all-you-can-eat dining facility found a positive relationship between portion size and plate waste (Freedman and Brochado, 2009). In this study, students were able to take an unlimited amount of pre-bagged servings of French fries which gradually decreased in size over a 5-week period. The study found that as the individual servings of French fries decreased in size, consumers not only consumed less, they left less uneaten food on their plate (Freedman and Brochado, 2009). In a WRAP study involving both consumer questionnaires and qualitative discussion groups, participants reported portion sizes as the most common reason for leaving food on their plate when eating out (Giorgi, 2013). Seventy-four percent of these 5,183 UK consumers reporting their experience eating out at various venues said that their meal was too large (Giorgi, 2013). They expressed a desire for greater transparency and the ability to choose their portion size, especially when smaller portions were priced lower (Giorgi, 2013). The Hospitality and Food Service industry can help prevent consumer plate waste by providing more information and flexibility with portion sizes. Servers can also make a point to offer take-home containers for leftovers, as consumers have admitted it may be embarrassing to ask for one themselves (Giorgi, 2013).

Government has a critical role to play in helping lower consumer food waste. Public education and awareness campaigns can teach consumers about the definitions of date labels, proper food handling, and determinants for whether food is safe to eat. Government can also impose standards and greater regulation on food labels in order to insure their consistency and comprehensibility. A uniform date labeling framework would help consumers not throw away food prematurely (Broad Leib et al., 2013). Educational campaigns about the economic impact of food waste might also encourage consumers to improve their food habits. Learning the retail value of discarded food might impact U.S. consumers to change their food behavior. U.S. consumers are far more likely to be motivated to reduce food waste due to economic concerns rather than environmental ones (Neff et al., 2015).

However, consumers are generally unaware of the environmental impact of food waste (Brook Lyndhurst, 2007). A past study examining the impact of general environmental knowledge and carbon offset knowledge on carbon-specific behavior suggests that environmental awareness campaigns can have a significant impact on consumer food behavior (Polonsky et al., 2012). A random sample of 353 U.S. consumers were given an online survey composed of true/false questions testing for knowledge (e.g., 'All carbon offset programs are government regulated/approved') and seven-point scale questions testing for behavior (e.g., 'I switch brands to ones that are less environmentally harmful') and attitudes (e.g., 'Each of us, as individuals, can make a contribution to environmental protection'). The study found a positive relationship between environmental knowledge and environmentally responsible behavior (Polonsky et al., 2012). Therefore, government efforts to increase consumer environmental knowledge might also result in more environmentally responsible consumer food behavior.

It is important to keep in mind that the strong libertarian sentiment among Americans may make it difficult for government efforts to effectively intervene in the lives of its citizens. Based on telephone interviews with a random sample of 1,017 adults living in all 50 states and the District of Columbia, a 2014 Gallup survey concluded that nearly half (49%) of all U.S. citizens believe there is too much government regulation (Newport, 2014). Nevertheless, some U.S. cities like San Francisco, Portland, and New York have set up composting mandates, and Seattle has recently imposed fines on its residents for failing to comply with its food waste composting mandate (Ferdman, 2015). According to the Seattle City Council, 74% of Seattle residents supported this new ordinance (Seattle Public Utilities and Bagshaw, 2014). Seattle Public Utilities (SPU) predicts that this new composting mandate will divert 38,000 tons of food scraps away from the landfills, located 300 miles away, a hauling distance with significant financial and environmental costs (Seattle Public Utilities and Bagshaw, 2014).

There are positive examples of government collaborating with communities to combat food waste. The local government in the Basque town of Galdakao in Spain recently funded a communal refrigerator, the Solidarity Fridge, where local residents are encouraged to take what they want and leave behind any produce or prepared leftovers that they will not eat in time at home (Frayer, 2015). Furthermore, the Danish government has conducted a series of research studies on national food waste (Halloran et al., 2014). The Danish Ministry of the Environment has also established a voluntary "Initiative Group Against Food Waste" where stakeholders within both the public and private sectors can work together to reduce food waste (Halloran et al., 2014). Local, state, and federal government organizations in the U.S. can engage the general public in fighting food wastage through education, awareness campaigns, incentives and support for research devising innovative interventions.

## CONCLUSION

Much of what we know about consumer food wasting behavior derives from WRAP's research. They have built a greater level of understanding that has served as the foundation for several successful initiatives within the UK. WRAP claims that their initiatives have reduced food waste by 4 million tonnes and diverted 29 million tonnes of waste from landfills between 2010 and 2015 alone (WRAP, 2015). Their six-month Love Food Hate Waste campaign, which cost a total of $270,000, saved households in West London $2.1 million in avoided disposal costs (Parry et al., 2015). Eight dollars were saved for every dollar spent. Similar campaigns implemented in the U.S. may deliver U.S. consumers and businesses comparable savings. We also have much to learn about our own behavior. The application of WRAP research surveys in the U.S. can help us learn more about why and how U.S. consumers are wasting food. More work is needed to examine the impact of geography, age, education, gender, income, race, employment, and family size and type on avoidable food waste in the U.S. International research offers valuable information, but it can only be applied to questions about domestic food waste to a limited extent. Surveys conducted specifically for the United States can offer a deeper understanding for how to reduce consumer food waste while operating within U.S. society and food culture.

## REFERENCES

Abeliotis, K., K. Lasaridi, and C. Chroni. 2014. Attitudes and behavior of Greek households regarding food waste prevention. Waste Manage. Res. 32(3):237-240. http://dx.doi.org/10.1177/0734242x14521681.

Broad Leib, E., J. Ferro, A. Nielson, G. Nosek, and J. Qu. 2013. The Dating Game: How Confusing Food Date Labels Lead to Food Waste in America. Natural Resources Defense Council Report R:13-09-A.

Brook Lyndhurst. 2007. Food Behaviour Consumer Research: Quantitative Phase. Waste & Resources Action Programme (WRAP), Banbury, UK.

Bunn, D., G.W. Feenstra, L. Lynch, and R. Sommer. 1990. Consumer acceptance of cosmetically imperfect produce. J. Consum. Aff. 24(2):268-279. http://www.jstor.org/stable/23859208.

Buzby, J.C., and J. Hyman. 2012. Total and per capita value of food loss in the United States. Food Policy 37(5):561-570. http://dx.doi.org/10.1016/j.foodpol.2012.06.002.

Buzby, J.C., H.F. Wells, and J. Aulakh. 2014. Food Loss – Questions About the Amount and Causes Still Remain. U.S. Department of Agriculture Economic Research Service, *Amber Waves*, June 2. http://ers.usda.gov/amber-waves/2014-june/food-loss—questions-about-the-amount-and-causes-still-remain.aspx.

Chung, S. 2008. Using plastic bag waste to assess the reliability of self-reported waste disposal data. Waste Manage. 28(12):2574-2584. http://dx.doi.org/10.1016/j.wasman.2008.01.002.

EPA. 2012. Food: Too Good to Waste Pilot Toolkit. EPA's Sustainable Materials Management Webinar. http://www.epa.gov/smm/web-academy/2012/pdfs/smm1112_zanolli.pdf.

FDA. 2015. Did you know that a store can sell food past the expiration date? U.S. Food and Drug Administration. http://www.fda.gov/AboutFDA/Transparency/Basics/ucm210073.htm.

Ferdman, R.A. 2015. Seattle is now publicly shaming people for putting food in their trash bins. *The Washington Post*, January 27. http://www.washingtonpost.com/news/wonkblog/wp/2015/01/27/seattle-is-now-publicly-shaming-people-for-putting-food-in-their-trash-bins/.

Frayer, L. 2015. To Cut Food Waste, Spain's Solidarity Fridge Supplies Endless Leftovers. *National Public Radio The Salt*, August 13. http://www.npr.org/sections/thesalt/2015/08/13/431960054/to-cut-food-waste-spains-solidarity-fridge-supplies-endless-leftovers.

Freedman, M.R., and C. Brochado. 2009. Reducing portion size reduces food intake and plate waste. Obesity 18(9):1864-1866. http://dx.doi.org/10.1038/oby.2009.480.

Giorgi, S. 2013. Understanding out of Home Consumer Food Waste, Final Summary Report. Waste & Resources Action Programme (WRAP), Banbury, UK.

Graham-Rowe, E., D.C. Jessop, and P. Sparks. 2014. Identifying motivations and barriers to minimising household food waste. Resour. Conserv. Recy. 84:15-23. http://dx.doi.org/10.1016/j.resconrec.2013.12.005.

Gunders, D. 2012. Wasted: How America is Losing up to 40 Percent of its Food from Farm to Fork to Landfill. Natural Resources Defense Council Issue Paper IP:12-06-B.

Halloran, A., J. Clement, N. Kornum, C. Bucatariu, and J. Magid. 2014. Addressing food waste reduction in Denmark. Food Policy 49:294-301. http://dx.doi.org/10.1016/j.foodpol.2014.09.005.

Icaro Consulting. 2013. Consumer Attitudes to Food Waste and Food Packaging Final Report. Waste & Resources Action Programme (WRAP), Banbury, UK.

Jörissen, J., C. Priefer, and K. Bräutigam. 2015. Food waste generation at household level: Results of a survey among employees of two European Research Centers in Italy and Germany. Sustainability 7(3):2695-2715. http://dx.doi.org/10.3390/su7032695.

Koivupuro, H.-K., H. Hartikainen, K. Silvennoinen, J. Katajajuuri, N. Heikintalo, A. Reinikainen, and L. Jalkanen. 2012. Influence of socio-demographical, behavioral and attitudinal factors on the amount of avoidable food waste generated in Finnish households. Int. J. Consum. Stud. 36(2):183-191. http://dx.doi.org/10.1111/j.1470-6431.2011.01080.x.

Lebersorger, S., and F. Schneider. 2011. Discussion on the methodology for determining food waste in household waste composition studies. Waste Manage. 31(9-10):1924-1933. http://dx.doi.org/10.1016/j.wasman.2011.05.023.

Lee, P., S. Osborn, and P. Whitehead, P. 2015. Reducing Food Waste by Extending Product Life, Final Report. Waste & Resources Action Programme (WRAP), Banbury, UK.

Marra, F. 2013. Fighting Food Loss and Food Waste in Japan (Unpublished Master's Thesis). Leiden University. http://www.fao.org/fileadmin/user_upload/save-food/PDF/FFLFW_in_Japan.pdf.

Monier, V., S. Mudgal, V. Escalon, C. O'Connor, T. Gibon, G. Anderson, H. Montoux, H. Reisinger, P. Dolley, S. Ogilvie, and G. Morton. 2010. Preparatory Study on Food Waste Across EU-27. European Commission Directorate C, Technical Report – 2010 - 54. http://dx.doi.org/10.2779/85947.

Neff, R.A., M.L. Spiker, and P.L. Truant. 2015. Wasted food: U.S. consumers' reported awareness, attitudes, and behaviors. PLoS ONE 10(6):e0127881. http://dx.doi.org/10.1371/journal.pone.0127881.

Newport, F. 2014. Few Americans Want More Gov't Regulation of Business. *Gallup*, September 15. http://www.gallup.com/poll/176015/few-americans-gov-regulation-business.aspx.

Nijs, V.R., M.G. Dekimpe, J.E. Steenkamps, and D.M. Hanssens. 2001. The category-demand effects of price promotions. Market. Sci. 20(1):1-22. http://dx.doi.org/10.1287/mksc.20.1.1.10197.

Parizeau, K., M.V. Massow, and R. Martin. 2015. Household-level dynamics of food waste production and related beliefs, attitudes, and behaviors in Guelph, Ontario. Waste Manage. 35:207-217. http://dx.doi.org/10.1016/j.wasman.2014.09.019.

Parry, A., K. James, and S. LeRoux. 2015. Strategies to Achieve Economic and Environmental Gains by Reducing Food Waste. Waste & Resources Action Programme (WRAP), Banbury, UK.

Polonsky, M.J., A. Vocino, S.L. Grau, R. Garma, and A.S. Ferdous. 2012. The impact of general and carbon-related environmental knowledge on attitudes and behavior of US consumers. J. Market. Manage. 28(3-4):238-263. http://dx.doi.org/10.1080/0267257x.2012.659279.

Quested, T., and P. Luzecka. 2014. Household Food and Drink Waste: A People Focus, Final Report. Waste & Resources Action Programme (WRAP), Banbury, UK.

Quested, T.E., E. Marsh, D. Stunell, and A.D. Parry. 2013. Spaghetti soup: The complex world of food waste behaviours. Resour. Conserv. Recy. 79:43-51. http://dx.doi.org/10.1016/j.resconrec.2013.04.011.

Rajan, A., R. Steinberg, and R. Steinberg. 1992. Dynamic pricing and ordering decisions by a monopolist. Manage. Sci. 38(2):240-262. http://dx.doi.org/10.1287/mnsc.38.2.240.

Seattle Public Utilites, and Bagshaw, S., Councilmember. 2014. Seattle Composting. http://www.seattle.gov/council/bagshaw/attachments/compost%20requirement%20QA.pdf.

Tsiros, M., C.M. Heilman. 2005. The effect of expiration dates and perceived risk on purchasing behavior in grocery store perishable categories. J. Market. 69(2):114-129. http://dx.doi.org/10.1509/jmkg.69.2.114.60762.

Vangarde, S.J., and M.J. Woodburn. 1987. Food discard practices of householders. J. Am. Dietetic Assoc. 87(3):322-329.

Williams, H., F. Wikström, T. Otterbring, M. Löfgren, and A. Gustafsson. 2012. Reasons for household food waste with special attention to packaging. J. Cleaner Product. 24:141-148. http://dx.doi.org/10.1016/j.jclepro.2011.11.044

WRAP. 2007. Understanding Food Waste Research Summary. Waste & Resources Action Programme (WRAP), Banbury, UK.

WRAP. 2015. Our Progress. Waste & Resources Action Programme (WRAP), Banbury, UK.

# *Chapter 8*

## *Variety, Abundance, and Perfection*

### *Exploring the Cyclical Behavior of Waste Creation in the Retail Marketplace*

**Emily M. Moscato and John L. Stanton**

**ABSTRACT**

Food waste is a significant problem at the consumer level. Contributing to the issue are core American values and consumer behaviors surrounding purchasing and consumption. These values and behaviors are reflected in consumer desire for variety, abundance, and perfection in their food. Moreover, these desires are reinforced by retailers and producers in the marketplace, generating a cyclical behavior pattern that negatively impacts food waste creation. Discussions of American values and consumer desires are presented and recommendations for breaking this cyclical behavior around variety, abundance, and perfection are offered. Recommendations focus on consumer education and behavioral nudges primarily at the retailer.

## INTRODUCTION

Food waste, at the consumer level, is a complex issue that has severe repercussions for the marketplace and the planet. In developed countries, the significant proportion of food waste comes at the hands of consumers. The average U.S. consumer wastes more than 20 pounds of food per month (Buzby and Hyman, 2012). Food waste is also on the rise. American per capita waste has increased by 50% since 1974 (Hall et al., 2009). The effect of this food waste on total resources and environmental factors is considerable. Currently, food scraps comprise 21.1% of municipal waste streams; higher than any other single category (EPA, 2014). Organic waste contributes to carbon and methane emissions in landfills and causes 25% of freshwater consumption to be wasted, along with energy and land (Hall et al., 2009; EPA, 2014). Yet the issue is generally unnoticed by most individuals. Consumers do not connect their over-acquisition and under-consumption of foods within their households to the larger problem. Moreover, food waste is considered "part of doing business" by food producers, retailers, and restaurants in their efforts to attract and maintain consumers. The cycle creates a hidden flow of food waste that is sustained by cultural values and interpretations of consumer wants. Shining a spotlight on this *cyclical behavior* is a critical factor in reducing the volume of surplus food, which generates food waste. If consumers, retailers, and producers do not relate their consumption behavior to a larger, more global problem, it will be difficult to bring about any effective change.

The focus of this chapter is to identify the shared responsibility for food waste among consumers and members of the food system value chain. We hypothesize that consumers and food companies are caught in a *cyclical behavior* pattern that has a significant negative impact on food waste. Food companies look to satisfy consumers by offering a variety and an abundance of aesthetically perfect foods to satisfy consumer wants. Consumers react positively and come to expect these offerings. This is then reinforced by food companies through rejecting foods that do not meet aesthetic standards, accepting higher levels of shrink in their business practices, and concealing waste from consumers. Furthermore, the elements of variety, abundance, and perfection carry over to household practices of the consumer, perpetuating the issues of waste. This creates a downward spiral. The more consumers buy or expect variety, abundance, and perfection, the more producers deliver on variety, abundance, and perfection creating an even greater expectation for the same.

The challenge for the industry and consumers is to break the downward spiral of waste. This article examines the elements of waste to better understand how to remedy it. First, American values associated with the want of variety, abundance, and perfection are examined. Second, the roles producers, retailers, and consumers play in this cyclical behavior pattern of food waste creation are investigated. Third, implications and recommendations on reducing food waste among consumers and the value chain are presented.

## AMERICAN VALUES AND AMERICAN WANTS

American cultural values have a significant influence on our consumption behaviors and underscore the want of variety, abundance, and perfection. Ask a foreigner about U.S. grocery stores and a common response is amazement over the vast selection and variety of choices. The depth and breadth of variety on supermarket shelves echo American's strongly rooted connection to freedom of choice. From the very formation of the United States, Americans have fought for this freedom. Recently, freedom of choice is being debated through proposed soda tax bills. Citizens argue that limiting sugary drink sizes or imposing taxes on sodas and snacks infringe on personal choice. In such situations, reactance is a common response. Reactance is the tendency to react negatively to persons, rules, or regulations that are seen as threats to personal freedom. The restricted option – larger soda sizes or less expensive soda – becomes more desirable as an effort to protect freedom and perceived personal control (Brehm and Brehm, 1981). Various strategies to limit purchase behavior to a specific product either by controlled production or tax have high likelihood of failure and can produce untended consequences. For example, one study demonstrated that over a six month period in which a 10% soda tax was implemented, long term consumption did not decrease. Yet, the tax did lead to an increase in beer purchases among beer purchasing households (Hanks et al., 2012).

Along with freedom of choice, conspicuous consumption is a shared value among Americans. Americans have for at least 50 years used it as a way to distinguish themselves from their neighbors. While more associated with automobiles and clothing, conspicuous consumption is also evident in food consumption practices. Specialty food stores, exotic goods, and the emphasis on organic, local, and natural foods lend themselves to conspicuous consumption among a selection of consumers. Broadly, conspicuous consumption is seen in the overabundance and the variety of food on the table, providing evidence that the family has as much or more success than its neighbors. Moreover, conspicuous consumption focuses on appearance, emphasizing aesthetically pleasing products.

Research on shared societal traits consistently demonstrates that the United States rates high on individualism (Oyserman et al., 2002; Matsumoto et al., 2008). Being individualistic implies that consumer desires are placed ahead of the needs of society and the environment. Consumers want the Mackinaw peaches, the British Columbia cherries, the New Jersey blueberries, but in many cases there may not be enough demand to support the sales of full cases of these products. Therefore in order to satisfy these individual tastes, retailers often must accept a high degree of waste in spoilage in order to keep the individual consumer shopping in their store.

Freedom of choice, conspicuous consumption, and individualism are important considerations for implementing changes in behavior. Cultural values themselves do not easily change. Yet, understanding how these values manifest in consumer wants and drive consumption practices aids in creating strategies to help producers, retailers, and consumers make better decisions regarding food waste. Consumption practices that lead to waste manifest in *cyclical behavior* of the marketplace are discussed next.

**Cyclical Behavior**

The *cyclical behavior* in the marketplace around waste creation is evident in the relationships between producers, retailers, and consumers. Producers and retailers believe consumers must have variety, abundance, and perfection to assist their buying process. While consumers are seen as the driving force behind these actions, the roles of the producer and retailers cannot be ignored. To understand what perpetuates this cyclical behavior, three driving factors, variety, abundance, and aesthetic perfection, among three players, producers, retailers, and consumers, are discussed (see Table 8.1). The ideas expressed in this discussion derive from the experiences and observations of the authors. It is meant as a thoughtful exploration of the problem. With the lack of research exploring food channels' behavior around waste, we provide this discussion to entice further research and explore opportunities to implement change.

**Table 8.1 Driving factors of cyclical behavior**

| Factors | Producer | Retailer | Consumer |
|---|---|---|---|
| Variety | Product segmentation | Competitive advantage; one-stop shop | Perceived desire for choice (paradoxical) |
| Abundance | Controlled markets (want of high yields) | Worse to be out of stock | More is better |
| Aesthetic Perfection | Grading | Same price for all items; showcase the best | Aesthetics matter; better value (when price is equal between products) |

## Want of Variety

The desire for variety strikes at a basic premise of marketing for all three players: to create a differential advantage. The search for a differential advantage often leads to the creation of more choices or variety beyond what consumers may be seeking. This is illustrated by the pasta sauce section of the supermarket. Research in the 1980s tested sauce varieties and began segmenting the market based on profiles gathered from these test results. Now consumers can find as many as 40 different varieties of sauces on the shelves, such as extra chunky, zesty, meat, vegetarian in at least two sizes (Moskowitz and Gofman, 2007). While such research is important to uncover consumer wants, these extended, diversified product lines have become extreme. It is the assumption of the authors (based on previous research) that in many cases consumers are not seeking this variety. Brand managers are creating these varieties to have a "differential advantage".

Among retailers, variety is used to demonstrate to the consumers that there is no need to shop anywhere else; the store has what they want. For example, a retailer may carry 25 different stock keeping units (SKUs) of mushrooms, including the same varieties in different sizes and different processing levels (sliced or whole). The modern consumer is often not satisfied with the "same old thing." They are looking for different kinds of items that can fulfill the family desire for food diversity and satisfy the aforementioned desire for conspicuous consumption. A compounding factor when considering variety is the "paradox of choice" (Schwartz, 2004). While consumers enjoy having variety, too many similar choices create buying paralysis. Unfamiliar consumers can easily become overwhelmed in a supermarket that has as many as 15 different varieties of apples. Indeed, the issue of food waste and variety is exacerbated in categories that have the most limited shelf life. Seafood, which may have the most limited shelf life of all products in the store, usually includes great variety. In some cases the leftover seafood can be converted into prepared meals, but in most cases it ends up as food waste. The seafood section deserves careful attention with respect to food waste because of the increased scarcity of available seafood and the continual growth in seafood consumption. If food waste in this category is not dealt with effectively certain species of seafood may no longer be commercially available.

The dilemma is that the consumer's desire for variety is understood to be the driving force behind retailers' desire to maintain high variety in each of its sections, and the producers' desire to satisfy this demand. Yet, one important factor differentiates supermarkets from other retail types when dealing with variety. In most of the retail outlets, most of the profit – if not all the profit – comes from the sale of the products. In the food retailer, a majority of the profit comes from fees paid by manufacturers for space on the shelves. Therefore, the space allocated to the products on the shelves cannot usually be justified in terms of consumer demand. It is not the purpose of this article to discuss the logic (or illogic) of this common strategy but rather to point out that the genesis for variety is not always the consumer. A divergence from this model is the popular chain Trader Joe's which has carved out its own differential advantage by not providing consumers a vast variety. For example, Trader Joe's has only 10 varieties of peanut butter whereas an average supermarket has 40 SKUs (Kowitt, 2010). The limited selection allows for higher bulk purchases and greater savings, which are passed along to the consumer.

**Want of Abundance**

Among developed countries, there is an emphasis on abundance of products and choices. Food retailers attract consumers with abundant displays of produce, bakery items, meats, and fish. These practices are seen as essential to the customer experience and the waste they create a necessary part of doing business (Gunders, 2012). America's culture compounds abundance with an emphasis on "big is better." We do not need to look any further than all-you-can-eat buffets to recognize how the concept of abundance is entrenched in U.S. culture.

Retailers are willing to deal with waste through abundant displays for fear of losing customers. In-store observation has demonstrated when there are only three or four apples left on the shelf, consumers are far less likely to buy those apples as when the shelf is piled high. Yet, while consumers love the supermarket strategy of "pile them high and watch them fly" this often leads to greater food waste as some of the products have to appear at the bottom of the pile and never get sold.

While having only a few items left on a shelf may turn away consumers, the out of stock phenomena is considered a death blow for traditional supermarkets. When consumers are looking for a specific variety of fresh fish, for example, and their favorite supermarket is out of stock, some percentage of those consumers will visit a competitive supermarket. Not only will they likely purchase the fish at this competitor, but all the accoutrements will likely be purchased at the same time. Even worse they might find that the new supermarket is a better presentation than their habitual stop. A bit of waste is a small price to pay to keep customers shopping at your store on a regular basis. Many retailers consider this food waste equivalent to the costs of guaranteeing their products to consumers.

Currently, abundant displays are part of the customer experience; however, that may not be the trend for the future. The methods of marketing that have relied heavily on consumer experience of abundance which similarly yields high levels of waste, may become less important as digital and online shopping for groceries increases. Amazon offers approximately 500,000 food SKUs without any actual displays. For all the consumer knows the 5 pounds of Braeburn apples or 10 pounds of specialty pork sausage may have been the very last on the shelf before the next delivery.

**Want of Perfection**

Research demonstrates that consumers will generally reject food items for their appearance unless they are educated as to their value (Bunn et al., 1990). Even when food is generally seen as lower quality, consumers look for the best aesthetics among lower quality items. In the minds of many consumers, imperfections act as signals for undesired qualities. They raise the question of food safety, including the possibility of mishandling or contamination. Arguably, consumers are socialized to desire this perfection. The want of perfection, especially fresh food, is reinforced from an early age as media (including books, games, and entertainment shows) use aesthetically perfect foods as prototypes. The influence of these prototypes is demonstrated in an interaction documented in a previous research project. In this particular project, a participant had showed her five year old daughter a glossy red apple and an organic red shaded apple. The participant asked her daughter which apple tastes better. Her daughter selected the prototypical glossy red apple, confident that apple was the juiciest. Unquestionably, the bright red apples found in children books to children TV shows have influenced this choice and conclusion. It is the familiar and celebrated choice for produce. Americans want beautiful food, yet not all quality food is beautiful. While this anecdotal evidence indicates that the want for perfection in produce is a learned behavior, it is not universal. In other developed countries, produce that are less-than-perfect in appearance sell with the same propensity as what might be described as aesthetically perfect.

The desire for perfection may be less of a negative factor when dealing with food waste because much of the less-than-perfect foods can be processed. For example, apples to apple sauce, mushrooms to mushroom soup, grains to animal feed, carrots to children's carrots snacks, fruits to cans, and vegetables to soups, frozen foods, or a selection of other products. Yet, as with other such factors related to food waste, food processors will very likely follow the perfection demand to the extreme. Appearances take precedence over other attributes. An illustration of this point happened when the authors had the opportunity to visit a seed farm and discuss with the breeder the characteristics of a specific type of squash. The breeder described how this particular squash variation maintained its color for an extended period of time, was very bruise resistant, and was able to maintain its appearance even in less than ideal shipping and storage conditions. When asked how this particular squash variation tasted relative to the heritage varieties, the breeder was flummoxed and said he had no idea.

Currently in our food system, the want of aesthetics, abundance, and variety has been pushed to excess and has been reinforced throughout the food channel. Strategies to begin the break of the cyclical behavior around waste creation require creativity and can begin with small steps.

## IMPLICATIONS AND RECOMMENDATIONS

Food waste will not have a simple solution, yet positive changes can be made towards reducing this significant global problem. This article recognizes that many forces contribute to the food consumer's propensity to waste food. For this reason, recommendations need to include all three participants in the food channel discussed in this article. The probability of success when focusing on only one player in the food channel will lead to simple frustration on the part of the other channel members. While we assiduously avoid suggesting any one channel participant is the culprit in the problem, a major effort must be directed at consumers.

Critical to changing consumer behavior is education on the significance of the issue and the factors that lead to waste. Consumers are more willing to try imperfect produce when it is accompanied by education efforts regarding its value and benefits (Bunn et al., 1990; Intermarché, 2014). In propagating positive behavior towards reducing waste it is vital to determine the messages that will resonate with consumers and address the areas of variety, abundance, and perfection. While it is beyond the scope of this article to provide direction for this messaging, we do offer recommendations that will help reinforce the messaging. The following recommendations focus primarily on the retailer and are directed towards consumer purchasing behavior. The retailer is chosen as a focus because it acts as a bridge between the producer and the consumer. In addition, retail product offering and presentation has significant influence on consumer decision making.

Many of the suggestions outlined below appear modest, however in most cases they are not easy to execute. On the basis of their business model, supermarkets are reluctant to lower prices outside of scheduled discounts. Yet, price is a clear way to attract consumers to products that are lesser in appearance. Supermarkets can take a lesson from convenience stores by having displays of reduced priced products with close to expiration or diminished appearance near the cash registers. The campaign "Inglorious Fruits and Vegetables" by the French supermarket, Intermarché, provides an example of how a price reduction (by 30%), point-of-purchase displays, and consumer education (including taste tests) can increase the acceptance of less-than-perfect produce. Beyond price, retailers can attract consumers to smaller, damaged, or imperfect foods by providing signage highlighting various uses. Within the produce aisle, examples include: "lunchbox ready" for small apples; "cereal fruit" for ripe, less-than-perfect, mixed fruit; "salad vegetables" or "soup vegetables" for ripe, imperfect, mixed vegetables; and "juicer primed" for ripe, damaged fruits and vegetables. Recipes can be featured with these items to inspire consumers. In the bakery, stale bread can be sold in loafs or bread crumbs as "inspiration starters" with recipe suggestions, such as bread pudding, panzanella, croutons, and meatloaf. Such signage can reduce the consumer's apprehension for purchasing these items. With American tendencies towards conspicuous consumption, choosing "juicer primed" or "lunchbox ready" produce would provide positive social signaling and support health goals.

Retailers can further entice consumers and help control waste by educate consumers (and employees) about the meaning of sell-by dates and other food date labels. Confusion regarding these labels leads to heightened food waste. It is estimated that 91% of consumers have thrown out food based on the sell-by dates because of concerns over food safety, although this label does not indicate these items are unsafe to eat (Broad Leib et al., 2013). Retailers can create endcap displays of complete meals that include products near their sell-by date. A good example is a retailer that puts everything needed to make an Italian dinner in one single location that includes hamburger for meatballs and sausage that may be approaching the sell-by date. Since by 4 p.m. each day, 80% of consumers don't know what they're going to have for dinner, this type of display can not only benefit the potential of less food waste but also be value-added to consumers who can find the convenience of everything they need to make tonight's dinner (Weisenberger, 2014). With the increased emphasis on prepared foods and grab-and-go items, retailers are better integrating resources from specialty departments by using seafood, meats, and dairy items that are close to sell-by dates in dishes for consumers.

It is coordination that helps reduce waste within the store. Retailers can help support reduction in household waste by highlighting ways of preserving foods that are most wasted among consumers. For example, retailers can provide recipes and tips in-store and online for preparing or preserving fresh seafood, dairy, and other items that may be reaching safety expiration. These educational efforts can be supported by visual demonstrations of the amount of food waste created by consumers to drive the issue home. Stores can display infographics, photographs, or have food displays representing the amount of food waste generated by the average consumer. Retailers can also take this opportunity to demonstrate what they do with discarded food, such as donations, composting, or landfill. This is an opportunity to demonstrate that retailers and consumers are both part of the problem and can be part of the solution.

The reduction in variety is an evident way to deal with food waste. It is the obligation of the retailer to demonstrate to the consumer that a wide variety of food is not needed to satisfy demand. A key strategy is to make a major marketing commitment to in-season items. For example, 52% of consumers who bought fish/shellfish in the last three months say they prefer to buy fish that is in season (Frank, 2013). Yet, the concept of seasons is almost foreign to younger consumers today in the global food system. In the past, there was clear delineation of in-season produce (for example, tomato season, the citrus season, the peach season, the asparagus season) and major marketing efforts were made to increase awareness of fresh, peak tasting, seasonal produce. With consumers increasingly concerned with authenticity and transparency, in-season campaigns will have a ready audience and can reduce waste at the retailer and household level.

Another recommendation is to make it clear to consumers that they have choices. Buying fresh foods, rather than frozen or canned, may not be the best choice for consumers given their lifestyles. Frozen or canned foods may in fact lead to much less food waste with virtually no degradation in quality. Most consumers may not realize that frozen food is often frozen and packaged within hours after harvesting, but fresh food may be days or even weeks in the supply chain before it gets to the shelves. Today's technology with frozen foods such as individually flash frozen dishes makes frozen food more appetizing without a loss of nutrients. Even canned food, which has not been in favor for many years, is most criticized for the sodium content, not its nutritional benefits or the impact it has on food waste. Retailers may be reluctant to promote these options since margins are much higher on fresh foods, (Kraushaar, 2014), but this is one way for producers of frozen and canned foods to increase their reputation among consumers.

There are many potential opportunities for retailers, producers, and consumers to affect change around food waste. Here we provide a foundation for understanding the consumer's propensity towards wasteful behavior and how this behavior is reinforced through the actions of producers and retailers. With these reasons in mind, we offer recommendations directed toward changing consumer behavior by leveraging the influence of retailers and producers. These recommendations encompass both education and small nudges that will not solve the entire problem, but will create positive change in the right direction towards reducing the problem.

## REFERENCES

Brehm, S.S., and J.W. Brehm. 1981. Psychological Reactance: A Theory of Freedom and Control. Academic Press, London, UK.

Broad Leib, E., J. Ferro, A. Nielsen, G. Nosek, and J. Qu. 2013. The Dating Game: How Confusing Food Date Labels Lead to Food Waste in America. Natural Resources Defense Council Report R:13-09-A. http://www.nrdc.org/food/files/dating-game-report.pdf.

Bunn, D., G.W. Feenstra, L. Lynch, and R. Sommer. 1990. Consumer acceptance of cosmetically imperfect produce. J. Consum. Aff. 24(2):268-79.

Buzby, J.C., and J. Hyman. 2012. Total and per capita value of food loss in the United States. Food Policy 37(5):561-570. http://dx.doi.org/10.1016/j.foodpol.2012.06.002.

Environmental Protection Agency (EPA). 2014. Inventory of U.S. Greenhouse Gas Emissions and Sinks, 1990-2012. EPA 430-R-14-003.

Frank, J.N. 2013. Fish and Shellfish – US – October 2013, research report. Mintel International Group.

Gunders, D. 2012. Wasted: How America is Losing Up to 40 Percent of its Food From Farm to Fork to Landfill. Natural Resources Defense Council Issue Paper IP:12-06-B.

Hall, K.D., J. Guo, M. Dore, and C.C. Chow. 2009. The progressive increase of food waste in America and its environmental impact. PLoS ONE 4(11):e7940. http://dx.doi.org/10.1371/journal.pone.0007940.

Hanks, A., B. Wansink, D. Just, L. Smith, J. Cawley, H. Kaiser, J. Sobal, E. Wethington, and W. Schulze. 2012. From Coke to Coors: A field study of a fat tax and its unintended consequences. J. Nutr. Educ. Behav. 45(4):S40. http://dx.doi.org/10.1016/j.jneb.2013.04.108.

Intermarché. 2014. Inglorious Fruits and Vegetables. http://itm.marcelww.com/inglorious/.

Kowitt, B. 2010. Inside the Secret World of Trader Joe's. *Fortune*, August 23. http://archive.fortune.com/2010/08/20/news/companies/inside_trader_joes_full_version.fortune/index.htm.

Kraushaar, A. 2014. Perimeter of the Store – US – June 2014. Mintel International Group.

Matsumoto, D., S.H. Yoo, and J. Fontaine. 2008. Mapping expressive differences around the world: The relationship between emotional display rules and individualism versus collectivism. J. Cross Cult. Psychol. 39(1):55-74. http://dx.doi.org/10.1177/0022022107311854.

Moskowitz, H.R., and A. Gofman. 2007. Selling Blue Elephants: How to Make Great Products That People Want Before They Even Know They Want Them. Pearson Education, Upper Saddle River, NJ.

Oyserman, D., H.M. Coon, and M. Kemmelmeier. 2002. Rethinking individualism and collectivism: Evaluation of theoretical assumptions and meta-analyses. Psychol. Bull. 128(1):3-72. http://dx.doi.org/10.1037//0033-2909.128.1.3.

Schwartz, B. 2004. The Paradox of Choice: Why Less is More. Ecco Press, New York, NY.

Weisenberger, C. 2014. More Consumers Choosing 'Grocerants' over Restaurants. *Supermarket News*, May 18.

# *Chapter 9*

## ***Quantifying Food Waste Streams at a College Campus Dining Hall***

**Alexandra Cirone, Elena Crouch, Christine Kim, and Gomian Konneh**

### ABSTRACT

Pilot research was conducted by a team of students in the Politics of Food class at the University of Pennsylvania, with the aim to characterize the different streams of food waste arising from a single dining hall on the university campus. There is little available information about food waste streams at this level of detail. In the present study we collaborated with Penn Dining and Bon Appétit Management to study food waste streams comprising the total waste from an all-you-can-eat style dining facility. This site offers an ideal context, as the dining hall closes between each meal service, thereby allowing for more accurate data collection and reduced interference with kitchen operations. Food waste streams were measured from a single meal period (dinner) over the course of 10 days (weekdays of two weeks), including: unavoidable (kitchen trimmings during food preparation), recoverable (cooked but not served), non-recoverable (service station remains), and plate waste (dumped by individual diners).

Daily plate waste ranged from 110-167 lbs, with a mean of 138 lbs and a standard deviation of 21 lbs; whereas the other three food waste streams (unavoidable, recoverable, and non-recoverable) averaged between 13-17 lbs each. On a percentage basis, plate waste accounted for 76% of the total food waste with the other streams accounting for 7-9% each. Plate waste per person averaged 0.64 lbs (ranging 0.43 to 0.99 lbs), compared to the average 0.45 lbs "perceived" to be wasted by the typical diner surveyed. Our data indicate that over the 10 days a considerable amount of food (17 lbs) could be recovered from this particular dining hall, which would be enough to feed 14 people (one meal is 1.2 lbs, according to Feeding America). Furthermore, the amount of plate waste generated by every two diners would be enough to feed a third person. Unfortunately, plate food can only be delivered to a landfill or composting facility. Our results provide strong evidence that food waste reduction and prevention must focus on the consumers and their attitude towards institutional food. We need to improve our understanding of factors that affect consumer behavior and identify ways to raise awareness, encourage responsible eating habits, and change wasteful behaviors in order to build a sustainable food system.

## INTRODUCTION

### The Problem

With the world population projected to reach 9.6 billion by 2050, we are faced with the daunting task of meeting the global demand for food in an era of dwindling resources (UN News, 2013). While much of this discussion has focused on the need to increase food production, it may very well be the case that the problem is not so much a dearth of available food but rather an excess of food waste. Approximately one-third of all food produced for human consumption goes uneaten; this amounts to a staggering 1.3 billion tons of food waste each year (Gustavsson et al., 2011).

From farm to fork to landfill, food waste is a growing problem with serious economic, environmental, and social consequences. In the United States, food production accounts for 10 percent of the energy budget, 50 percent of land utilization, and 80 percent of freshwater consumption (Gunders, 2012). Forty percent of food remaining uneaten means 40% of the resources used in its production is wasted in vain. The economic value of wasted food in America totals $165 billion each year (Gunders, 2012). Wasted food ends up decomposing in landfills where it constitutes the single largest component of municipal solid waste and accounts for a significant portion of methane emissions (EPA, 2014). Despite these issues, per capita food waste in the U.S. has increased by 50 percent in the last 30 years (Hall et al., 2009). At the same time, more than 49 million Americans are living in food-insecure households (Coleman-Jensen et al., 2013).

Thus, growing concerns about food security, resource conservation, and the economic and environmental costs incurred throughout the food supply chain have highlighted the importance of understanding food waste streams at the global, national, and local levels.

### Current Efforts

The University of Pennsylvania has taken various steps to address the problem of food waste. The most prevalent effort on campus thus far is composting. According to the Green Campus Partnership (2014), all dining halls collect kitchen waste for composting and provide designated compost bins for diners to dispose of food waste. In addition, Mayer Hall also has a compost bin for residents' food waste, the Law School regularly composts leftover food from events and receptions, and the Annenberg Public Policy Center building has a vermicompost bin for waste and leftovers from staff meals. Penn Facilities and Real Estate Services (2014) reports that their compost contractor accepts all food wastes including meat, dairy, eggs, and oils as well as soiled paper, compostable tableware, and yard waste such as leaves. Recycling food waste and turning it into compost has many environmental benefits such as improving soil health and structure; increasing drought resistance; and reducing the need for supplemental water, fertilizers, and pesticides (EPA, 2014). However, composting does not address the fact that the valuable resources that go into our food system are still being lost.

Fortunately, Penn Dining and Bon Appétit Management Company have employed a number of source-reduction strategies to try to stop food waste from occurring in the first place. For example, the dining hall kitchen staff prepares food from scratch in order to make the most of every ingredient. Described by Bon Appétit (2014) as "snout-to-tail and stem-to-root cooking," this practice allows bones and vegetable trimmings that would otherwise be considered inedible to be set aside to make stocks and soups. In addition, meals in the dining halls are prepared in batches at the last possible minute and served in the smallest possible quantities in order to ensure freshness and minimize waste. Furthermore, Bon Appétit reports that their general manager at Saint Joseph's College of Maine pioneered trayless dining in 2005 after finding that consumer waste could be cut dramatically by simply removing the trays from the dining halls. Since 2008, all of Penn's dining halls have adopted the trayless dining program (The College Sustainability Report Card, 2011).

Penn Dining and Bon Appétit have also demonstrated a willingness to work with student groups to reduce food waste. A group of students in the Fall 2013 Politics of Food class delivered a proposal for an institutionalized food recovery program (Dickinson et al., 2014). Since then, Penn Dining and Bon Appétit have been working with Feeding America to conduct a trial for the 2014-2015 school year at the Class of 1920 Commons. The present team builds on this work and continues to collaborate with Penn Dining and Bon Appétit.

**Project Definition**

The present research was conducted with the aim of characterizing the different streams of food waste arising from a single dining hall on the university campus. For the purposes of this study, four food waste streams were identified and measured:

1. *Unavoidable* food waste is generated in the kitchen as trimmings and other parts of food that are inedible, such as bones and rinds.
2. *Recoverable* food waste is food that has been prepared but not yet served, therefore is eligible for redistribution for human consumption.

3. *Non-recoverable* food waste is food that has been served but not taken (i.e., remained at service station at the end of meal period) and cannot be recovered for human consumption due to health regulations.

4. *Plate waste* is food taken by diners but not consumed.

There is little available information about food waste streams at this level of detail. In 2010, Penn Dining and Bon Appétit launched the Scrape Bucket Challenge in order to track the levels of food waste over the course of a week (The College Sustainability Report Card, 2011). While data was collected at every dining facility on campus, only plate waste was measured. Indeed, food waste audits from other university dining halls have focused solely on plate waste (Appendix A). Recognizing the apparent knowledge gap, the present team used a protocol designed to better understand at what point in the dining hall food system the majority of the food is wasted. Our findings can serve as advisory data for Penn Dining, Bon Appétit and other institutions to plan future food waste management efforts.

## METHODS

### Site Description

The study was conducted at the Kings Court English House dining hall. This site offers an ideal context, as the dining hall closes between each meal service, thereby allowing for more accurate data collection and reduced interference with kitchen operations. Before data collection took place, the student researchers met with Patterson Watkins, the Chef Manager at Kings Court, to discuss health and safety regulations. This also provided the opportunity to better understand the day-to-day operations in the dining hall and meet the members of the staff who would be assisting in our research. The food waste audit began on Monday, October 27, 2014 and ended on Monday, November 10, 2014. During this period, food waste streams were measured from 10 weekday dinner services.[1]

### Procedure: Quantitative

At least one member of the research team was present each day through the entire dinner service from 5:00-8:00 pm. *Unavoidable* food waste was set aside by the kitchen staff during meal preparation and was weighed upon arrival. *Recoverable* and *non-recoverable* food waste was each weighed at the end of the meal service. *Plate waste* was weighed as often as necessary based on how quickly

[1] Data collected on Friday, October 31, 2014 was disregarded due to Halloween-related activities in the dining hall (e.g., pumpkin carving, caramel apples, baskets of candy, etc.) that resulted in atypical amounts and types of food waste.

the compost bin filled up throughout the course of the evening – it typically had to be emptied once before the end of service. The majority of the data collection process involved monitoring the compost bin where diners scrape their plates once they have finished eating. Plate waste was closely monitored in order to ensure complete capture of food waste into the compost bins and to record the numbers of inedible compostable items being disposed of along with the food waste.

The majority of the inedible items composted were napkins, apple cores, and melon rinds. Standard weights were obtained for these items and the total weight of inedible items noted during each meal service was subtracted from measurements of plate waste and, when applicable, added to unavoidable waste. A standard weight was also obtained by recording the weight of a full plate's worth of food from the warm food service station. Weights were also obtained for the appropriate containers, trays, and bins associated with each food stream and have been accounted for in the data presented here. In addition, the staff at Kings Court provided data regarding the number of diners that were "swiped in" to each dinner service.

All materials necessary for this research were provided by Penn Dining and Bon Appétit. In accordance with health regulations, members of the research team were required to wear aprons, gloves, and hairnets. A small kitchen scale (Taylor Model TS25 KL, 25 lb scale x 2 oz increments) was used to weigh *unavoidable*, *recoverable*, and *non-recoverable* food waste. A large industrial scale (Fairbanks-Morse, 500 lb scale x 0.5 lb increments) was used to weigh plate waste. A more sensitive scale (OXO, 5 lb) was used to weigh items such as napkins and apple cores.

**Procedure: Qualitative**

A consumer survey was used to collect qualitative data exploring diners' behaviors and general attitudes toward food waste (Appendix B). The survey was created using Google Forms. A link to the survey was sent via email to every resident of King's Court English House and was advertised on slips of paper distributed to every table in the dining hall so as to include diners who were not residents of King Court. A total of 54 responses were received. These responses were automatically converted into a spreadsheet, thus making results more easily quantifiable. In addition to exploring the general characteristics of an average diner, this survey provided information about the average number of plates taken during a typical meal, the amount of uneaten food that is perceived to be wasted by a typical diner, and the reasons for discarding food. In the following section, this data is used to further analyze the *plate waste* measured in the food waste audit.

## RESULTS

### Quantitative

A total of 1,827 lbs of food waste were measured during the data collection period. Values for the daily minimum, maximum, mean, and standard deviation of each food waste stream are presented in Table 9.1. Daily *plate waste* ranged from 110-167 lbs, with a mean of 138 lbs and a standard deviation of 21 lbs. *Recoverable* food waste averaged 17 lbs with a standard deviation of 10 lbs. *Non-recoverable* food waste averaged 15 lbs with a standard deviation of 8 lbs, and *unavoidable* food waste averaged 13 lbs with a standard deviation of 10 lbs. Here, it is worth noting that the minimum weight of *unavoidable* and *recoverable* food waste was zero. This indicates that the kitchen staff has been successful in ordering ingredients and planning meals in a way that meets demands while limiting food waste.

**Table 9.1 Food waste streams measured as lbs/day-meal (as dinner)**

| | Unavoidable | Recoverable | Non-recoverable | Plate Waste |
|---|---|---|---|---|
| Minimum | 0.00 | 0.00 | 1.60 | 110.00 |
| Maximum | 33.80 | 32.00 | 28.50 | 167.28 |
| Mean | 12.99 | 16.90 | 14.69 | 138.10 |
| Standard deviation | 10.37 | 10.11 | 7.74 | 21.18 |

The relative proportions of the four food waste streams are shown in Figure 9.1. On a percentage basis, *plate waste* accounted for 76% of the total food waste, *recoverable* food waste accounted for 9%, *non-recoverable* food waste accounted for 8%, and *unavoidable* food waste accounted for 7%.

### Qualitative

Results from the consumer survey were used to further analyze *plate waste* and explore diners' general awareness and attitudes toward dining hall food waste. First, this data was used to compare the actual amount of *plate waste* per person and the amount of food that a typical diner self-reports.

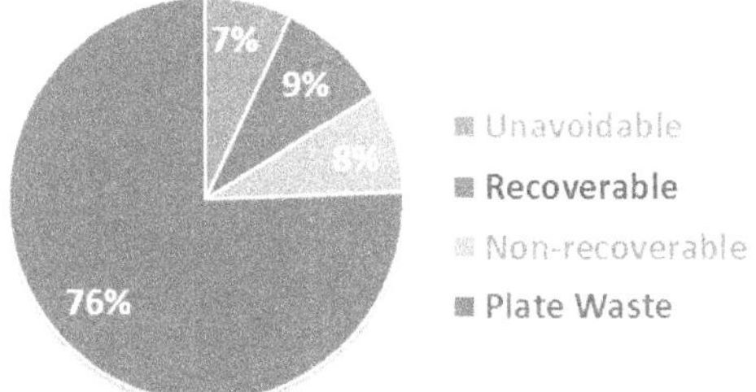


**9.1** Food Waste Streams: Proportion

Over the course of 10 days, a daily average of 226 people ate dinner at Kings Court. The mean daily *plate waste* was 138 lbs, so the average weight of *plate waste* per person was **0.64 lbs**. On average, each diner took 2.3 plates of food and wasted about 1/5 (as self-reported).[2] We measured that a full plate of food weighed 0.97 lbs. Therefore, diner-perceived per capita *plate waste* was calculated to be **0.45 lbs**. The latter measurements were also used to calculate the proportion of food that was taken compared to the amount that was eaten. Our calculation below indicated that about 71% of food taken by a typical dinner was eaten, whereas 29% was discarded as *plate waste.*

$$\left(\frac{0.64 \text{ lbs wasted}}{2.3 \text{ plates}\left(0.97\frac{\text{lbs}}{\text{plate}}\right)}\right) 100\% = 28.7\% \text{ wasted}$$

Figure 9.2 illustrates the survey results for question 11 regarding why diners waste food. Three main themes were evident: personal tastes, too much food served, and too much food taken. The responses were categorized according to the three main themes and the percentage each category made up of the total was calculated.

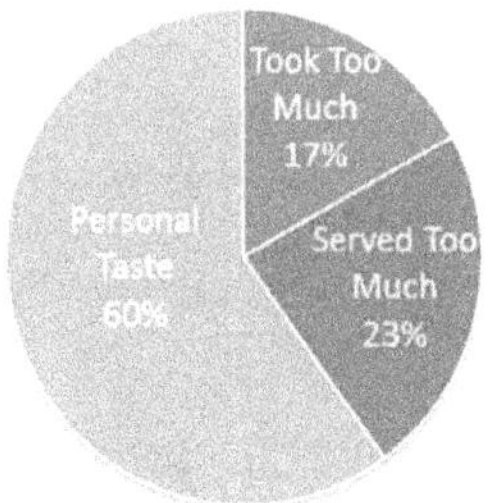


**9.2** Reasons for Discarding Food

[2] This was understood as 2.3 full plates of food, and ⅕ of the total amount of food that is taken.

## DISCUSSION

### Limitations of the Present Study

*Recoverable* and *non-recoverable* food waste were only measured from the comfort station. There are several other stations at King's Court that generate waste, including a salad bar, sandwich station, and dessert line. Our decision to measure *recoverable* and *non-recoverable* food waste from the comfort station was based on considerations for the proposed food recovery program at Penn.

In terms of duration, this was a very brief experiment. Furthermore, this case study only measured food waste generated during the dinner service. In order to gather more complete data about dining hall food waste at Kings Court, data should be collected from all three meal services over an extended period of time, or at least with regularly repeated measurements throughout the year.

Another obstacle was the inability, due to human error, to collect completely accurate data for each stream. Although diners were supposed to dispose of their *plate waste* in the compost bin when they returned their plates and silverware, for whatever reasons they would sometimes scrape their plates into the bins designated for recycling and/or landfill. These losses were not accounted for in the measurement of *plate waste*. Furthermore, *unavoidable food waste* and napkins were disposed of in the compost bin along with *plate waste*. While measurements were obtained for these items, these types of "eyeball" estimates are inherently limited.

The qualitative data could also have been more complete. Because the survey was optional, not all diners participated, especially those who were not residents of Kings Court. As a result, survey responses were based on a very limited sample of diners and subsequent analysis of *plate waste* may not accurately represent consumer behaviors and attitudes. In addition, the wording of our survey may have been misleading, as evident in our calculations of the amount of food that is perceived to be wasted by a typical diner.

In view of these limitations, we believe that our figures underestimate the total food waste at Kings Court.

## CONCLUSIONS AND FUTURE CONSIDERATIONS

The present findings should be considered in light of the current efforts to establish a food recovery program at Penn. According to Feeding America, one meal is roughly equivalent to 1.2 lbs. Our data indicate that a considerable amount of food could be recovered from this particular dining hall. Over 10 days, there was an average of 17 lbs of recoverable food waste daily, which means that this particular food waste stream could have gone toward feeding an additional 14 people. Since Kings Court has a relatively small dining facility, this most likely means that even greater quantities of food can be recovered across Penn's campus. And while plate waste is certainly not recoverable, with 0.64 pounds of plate waste per person, you may also consider that the waste produced by every 2 diners would be enough to feed a third. Instead, 138 lbs of food ends up in the compost bin following each dinner service.

Overall, our results provide strong evidence that food waste reduction and prevention must focus on improving our understanding of factors that affect consumer behavior and identify ways to raise awareness, encourage responsible eating habits, and change wasteful attitudes in order to build a sustainable food system. Education and awareness campaigns may help reduce *plate waste* by encouraging diners to take only as much food as they can finish. Furthermore, this approach will help shed light on the serious social and environmental problems associated with food waste, and it would certainly prove beneficial to engage the entire Penn community in a discussion of food waste.

There is also the need to increase communication between the dining hall staff, the providers, and students, the consumers, across the food system. In general, students take for granted the food providers, time, and resources that are all involved in keeping our dining services running smoothly. Several students expressed that one of the reasons they may take too much food is because they "want to get their money's worth" for each meal swipe suggesting that they put little value on the food offered to them. It is important to find ways to get students to change attitudes and start to identify with their food service, its workers, values and priorities. Increasing communication might also address some of the problems related to serving styles. For example, this could mean making sure students feel comfortable asking for smaller servings, or perhaps ask for a sample of the food before deciding whether or not they would like to take more. Both students and dining hall staff have responded positively to the idea of offering sampling trays in the dining hall.

Penn Dining and Bon Appétit may also benefit from further exploring students' food preferences, since that is the main reason that they are wasting food. For example, some schools conducted surveys of student food preferences in the hope that giving students a chance to contribute to menu planning would reduce the amount of food they discard. According to Kevin Kirwan, director of food services at St. Mary's College, allowing students to submit their own recipes reduced plate waste by 10 to 15 percent (Saphire, 1998). Active engagement can be used to set targets for food waste reduction by creating incentive programs. For example, serving a special meal if students can demonstrate quantifiable reductions in plate waste over a given time period. Doug Brown, then the director of food services at Drew University, came up with a strategy to serve special meals (like a steak dinner) when students could demonstrate quantifiable reductions in plate waste over a given time period. He also tried to communicate the connection between food waste and higher operating costs and, ultimately, higher food plan prices (Saphire, 1998). In the future, it would worthwhile to look into these types of strategies as a way to not only prevent food waste, but also to increase communication between students and dining hall staff.

While our research provides insight into dining hall food waste, future audits are necessary to further understand food waste at institutional facilities such as those at Penn. In addition to identifying and quantifying food waste streams, food waste audits are necessary in order to measure the effectiveness of food waste management efforts. Hopefully, the work we have done this semester is just the beginning.

## ACKNOWLEDGEMENTS

We would like to thank all those who have offered us their guidance and support throughout the course of our work. First and foremost, we owe a special thanks to Professors Jane Kauer and Mary Summers for providing us the opportunity to become involved in this innovative service-research project as part of their Politics of Food class. We are also lucky to have been able to work with Professor Zhengxia Dou, who not only served as a faculty supervisor but also organized the Last Food Mile Conference, bringing together the nation's leading experts on food loss and food waste. Furthermore, we are grateful for the work done by Alyssa Dickinson, Joyce Kim, Yamini Nabar, Conor Nickel, and Cynthia Plotch, who laid groundwork for our project with their proposal for food recovery. In addition, we appreciate Penn Dining and Bon Appétit Management Company for continuously striving to improve their food and food waste management efforts. Mike Frost and Patterson Watkins proved to be invaluable resources as we developed and implemented our food waste audit. The dining hall staff at Kings Court English House also welcomed us into the kitchen and demonstrated a willingness to learn alongside us with patience and understanding during data collection. Finally, we would like to thank all of the other students, faculty, and staff who participated in this pilot study and provided further insight into the problem of food waste at our university. Our experiences here have offered hope that we can continue to work together to build a more sustainable food system at Penn and beyond.

## REFERENCES


Bon Appétit Management Company. 2014. Reducing Food Waste. http://www.bamco.com/timeline/reducing-food-waste/.

Burcham, S., G. Saba, and K. Sing. 2011. Runk Dining Hall Waste Audit. Global Sustainability at the University of Virginia. http://www.globalsustainability.virginia.edu/wp-content/uploads/2014/07/Food_RunkWaste-BurchamSingSaba.pdf.

Clean Plates at State. 2013. Clean Plates at State Food Waste Audit. Michigan State University Residential and Hospitality Services. http://eatatstate.com/content/food-waste/fall2013.

Coleman-Jensen, A., C. Gregory, and A. Singh. 2014. Household Food Security in the United States in 2013. Economic Research Report Number 173. United States Department of Agriculture, Economic Research Service. http://www.ers.usda.gov/media/1565415/err173.pdf.

EPA. 2014. The Food Recovery Hierarchy. http://www.epa.gov/epawaste/conserve/foodwaste/.

Gunders, D. 2012. Wasted: How America is Losing up to 40 Percent of its Food from Farm to Fork to Landfill. Natural Resources Defense Council Issue Paper IP:12-06-B.

Gustavsson, J., C. Cederberg, U. Sonesson, R. van Otterdijk, and A. Meybeck. 2011. Global food losses and food waste: Extent, causes and prevention. Food and Agriculture Organization of the United Nations, Rome.

Hall, K.D., J. Guo, M. Dore, and C.C. Chow. 2009. The progressive increase in food waste in America and its environmental impact. PLoS ONE 4(11):e7940. http://dx/doi/org/10.1371/journal.pone.0007940.

Merrow, K., P. Penzien, and T. Dubats. 2012. Exploring Food Waste Reduction in Campus Dining Halls. Western Michigan University. http://wmich.edu/sites/default/files/attachments/ENVS 4100 Final Project Report - Merrow, Penzien, Dubats.pdf.

Penn Green Campus Partnership. 2014. Frequently Asked Questions. University of Pennsylvania. http://www.upenn.edu/sustainability/resources/faqs.

Penn Facilities and Real Estate Services. 2014. Waste Management and Recycling. University of Pennsylvania. http://www.facilities.upenn.edu/sustainability/waste-management-and-recycling.

Saphire, D. 1998. Getting an 'A' in Lunch: Smart Strategies to Reduce Waste in Campus Dining. INFORM, Inc., New York, NY. http://www.informinc.org/reportpdfs/wp/GettinganA.pdf.

The College Sustainability Report Card. 2011. University of Pennsylvania. http://www.greenreportcard.org/report-card-2011/schools/university-of-pennsylvania/surveys/dining-survey.html.

United Nations News Centre. 2013. World population projected to reach 9.6 billion by 2050. United Nations, New York, NY. https://www.un.org/en/development/desa/news/population/un-report-world-population-projected-to-reach-9-6-billion-by-2050.html.

## APPENDIX A

### Review of Relevant Studies/Reports

#### *"Clean Plates at State Food Waste Audit"*

This report displays the results of various food waste audits at Michigan State University. In Fall 2005, a preliminary food waste audit was conducted in two of the 15 dining halls, one day at each site. However, only plate waste was measured. Between the two venues, a total of 1,411 lbs of food were wasted – an average of 4.48 ounces per person. Since that time, an annual food waste audit has been conducted in the various dining halls on campus utilizing electronic scales that weigh each customer's plate and subtracting a tare allowance for each type of plate. While there isn't much discussion of the research aims or methods in this case, it provides the most in-depth and readily accessible data on dining hall food waste from a university dining hall.

#### *"Exploring Food Waste Reduction in Campus Dining Halls"*

This case study is from a student group at Western Michigan University in Spring 2012. These students set out to perform a waste audit at one of their campus dining halls. However, they only measured post-consumption food waste. Their food audit of took place over the course of single day. Beginning at 7:00 a.m. and culminating at 8:15 p.m., the students weighed the amount of uneaten food that was placed on the conveyor belt for garbage disposal on an hourly basis. During data collection, they also identified and categorized different components of plate waste on a nutritional basis and sorted them into separate buckets. The first bucket contained animal products such as meats, cheeses, and other dairy products. The second bucket contained organic material such as raw vegetables, raw fruits, and salads. The third bucket contained carbohydrates such as breads, pastas, and rice as well as any additional leftovers. This analysis revealed that the majority of food waste was comprised of carbohydrates and leftovers. Furthermore, these students benefitted from being able to draw on data from previous waste audits to compare the effects of specific food service styles (cook-to-order/make-to-order vs. traditional buffet style cafeterias) on reducing food waste in campus dining halls. Contrary to their expectations, this group found that the dining facility that offered made-to-order foods did not produce less food waste than the all-you-can-eat style dining facility.

#### *"Runk Dining Hall Waste Audit"*

This study is part of a series of three food waste audits conducted at the University of Virginia in Fall 2011. Like the students from WMU, the other two student groups were able to compare their measurements of food waste to those obtained in previous food waste audits. However, the student group focusing on Runk conducted a pilot study much like the present research team at Penn. This case study further paralleled our own, as the research site was one of UVa's smaller dining halls and concerned only the dinner service from 5-8 p.m. However, these students only collected data on a single day, and they only measured plate waste. In this case, the research team physically scraped the extra food left on plates into one container, and separated inedible items, napkins, and condiment wrappers into separate containers for garbage or recycling. This type of methodology is especially valuable in food waste audits because it allows a more accurate measurement of plate waste.

## APPENDIX B

### Questionnaire Survey

### Food Waste Streams Survey

**Have you noticed the strange people standing by the trashcans with notebooks this week? That's us, and it would help our research project if you would take a few moments to fill out the survey. Thank you!**

* Required

- 1. If you are an undergraduate student, please indicate your year of study. If you are not an undergraduate student, please specify under "other." *
    - ○ Freshman
    - ○ Sophomore
    - ○ Junior
    - ○ Senior
    - ○ Other:
- 2. Where do you live during the school year? *
    - ○ Kings Court English House
    - ○ other on-campus housing
    - ○ off-campus
- 3. Do you have a dining plan? *
    - ○ yes
    - ○ no
- 4. In a typical week, how many meals do you eat at Kings Court English House dining hall? *
    - ○ 0-1 meals
    - ○ 2-8 meals

- ○ 9-15 meals

- 5. During a typical meal, how many plates/bowls of food do you take? *
  - ○ 1
  - ○ 2
  - ○ 3
  - ○ 4+
- 6. After a typical meal, how much uneaten food is left on your plate(s)? *
  - ○ none
  - ○ 1/4 of the plate
  - ○ 1/2 of the plate
  - ○ 3/4 of the plate
  - ○ all of the plate
- 7. I don't really worry about the amount of food that I throw away. *
  - ○ strongly agree
  - ○ agree
  - ○ neither agree nor disagree
  - ○ disagree
  - ○ strongly disagree
- 8. I don't really worry about the cost of the food that I throw away. *
  - ○ strongly agree
  - ○ agree
  - ○ neither agree nor disagree
  - ○ disagree
  - ○ strongly disagree
- 9. I waste more food eating at a dining hall than I would at home. *
  - ○ strongly agree
  - ○ agree
  - ○ neither agree nor disagree
  - ○ disagree
  - ○ strongly disagree
- 10. If I had more information about the amount of food wasted in dining halls, I would probably make an effort to throw away less. *
  - ○ strongly agree
  - ○ agree
  - ○ neither agree nor disagree
  - ○ disagree
  - ○ strongly disagree
- 11. Why do you typically leave uneaten food on your plate?
- 12. Please give your opinion on the following statement: "food waste is not an issue, as it is natural and biodegradable."
- 13. Do you have any suggestions about how to address the problem of food waste at Penn or in general?
- 14. Please share any additional thoughts about food waste.

# *Chapter 10*

## *Lunch at a Philadelphia Elementary School*

### *What and How Much is Being Thrown Away?*

**Jarrett Stein, Jessica Zha, Ethen Grant, and Jibreel Powers**

**ABSTRACT**

Thousands of meals are delivered each day and served, free of charge, to all students attending Philadelphia public schools. However, there is little data on how much of this food is actually consumed. Borrowing techniques from community-based participatory action research, our intergenerational team of students and practitioners implemented a Science, Technology, Engineering, and Math (STEM) project to quantify and explain school food waste. We directly measured plate waste in a West Philadelphia kindergarten classroom, as well as used surveys, interviews, and photo-voice to measure attitudes and beliefs about food waste among upper grades. We found that approximately 35% of food is wasted in kindergarten classrooms. Furthermore, half of all vegetables served in individual serving containers were thrown out unopened. Among upper grades, vegetables were more likely to be wasted than any other food group, and satisfaction toward school food was low. These findings indicate a potential disconnect between the intent of the Philadelphia school food program and its success in meeting student nutritional needs, and raises concerns about inefficient expenditures in food procurement and waste removal. In the short term we suggest providing evidence-based training to school staff as well as the empowerment of students as change-agents for better promotion of healthy practices. We also advocate for more comprehensive evaluation of food waste and waste reduction strategies in schools.

"We do not inherit the Earth from our ancestors, we borrow it from our children."

Native American Proverb

**BACKGROUND**

Samuel B. Huey School (Huey) is a public school in the School District of Philadelphia. The school is located at the corner of Pine Street and 52$^{nd}$ Street, the main commercial corridor in West Philadelphia. On either side of Huey are corner stores, salons, bars, and barbershops. Directly across the street is the three-square-block Malcolm X Park, a green space with a pavilion and playground.

Huey serves 562 students in kindergarten through 8th grade; 100% of Huey students are economically disadvantaged.[i]

"The lunchroom feels crowded like you're being squished. You can hardly sit anywhere. If you try to sit somewhere you have to try to squeeze in."

Huey Elementary student E.G.

Huey was built in 1908 and underwent major renovations in the 1960's. The building does not have a production kitchen or cafeteria. There is a small "satellite" kitchen in the basement with a freezer and oven. Notably missing is a stainless steel sink. Due to the lack of necessary infrastructure, breakfasts, lunches, and after-school suppers are "pre-plate" – prepared in an off-site commissary, frozen, and delivered via truck to the school each day. As a result, meals served at Huey must be pre-cooked, stored in individual oven-safe packages, and re-heated in 20 minutes. Meals at Huey are consumed either in the gymnasium or in two modified kindergarten classrooms.

Every day the Food Services Division (FSD) of the School District provides 100,000 lunches, 60,000 breakfasts, and 5,000 after-school suppers to 148,000 students in Philadelphia.[ii] In 2001 the FSD had 75 administrative staff. However district budget cuts throughout the decade led to consistent reductions in FSD staff. By 2011 only 18 administrative staff remained. This small staff is responsible for managing 85 million dollars to purchase and serve approximately 28.5 million meals across 302 feeding sites.

In 2011 over 75% of all meals served by FSD were "pre-plate." Although this number has recently declined after FSD restarted "scratch" production in schools with existing full-service kitchens, 180 of the 302 feeding sites are like Huey and do not have the necessary infrastructure.[iii] Given district budget shortages and inadequate facilities, the majority of students in the School District of Philadelphia will continue to receive "pre-plate" meals in the foreseeable future. Currently, the privately-owned Maramount Corporation is contracted by the FSD to provide the food served at Huey and the other "pre-plate" feeding sites in Philadelphia. Figure 10.1 is an example weekly menu for the Maramount-provided food.[iv]

"There's this one lunch, where if they run out of lunches they have this lunch package that has crackers and dragon fruit punch and these kind of vegetable fruit roll-ups and raisins and there's this one can of buffalo chicken. We call it *cat food* because that's what it looks like."

Huey Elementary student E.G.

Home> PA> School District Of Philadelphia > School District Of Philadelphia

| Monday, December 14, 2015 | Tuesday, December 15, 2015 | Wednesday, December 16, 2015 | Thursday, December 17, 2015 | Friday, December 18, 2015 |
|---|---|---|---|---|
| Breakfast<br>RICE CHEX<br>Spiced Grahams<br>Pear Cup<br>Orange Juice<br>1% White Milk<br>Fat Free Chocolate Milk<br>Lunch<br>FUN-DAY MONDAY!"<br>MINI CORN DOGS & CHICKEN FUN MIX<br>Maple Baked Beans<br>Baby Carrots | Breakfast<br>BEEF SAUSAGE ON A BISCUIT<br>Mixed Fruit Cup<br>Apple Juice<br>1% White Milk<br>Fat Free Chocolate Milk<br>Breakfast Cold<br>CHEERIOS<br>Animal Grahams<br>Mixed Fruit Cup<br>Apple Juice<br>1% White Milk<br>Fat Free Chocolate Milk | Breakfast<br>COCOA CHIP LOAF<br>Fresh Banana<br>Blended Fruit Juice<br>1% White Milk<br>Fat Free Chocolate Milk<br>Lunch<br>"NEW ITEM"<br>COUNTRY CHICKEN COMBO<br>Strawberry Applesauce<br>Whole Wheat Bread<br>Fat Free Chocolate Milk<br>1% White Milk | Breakfast<br>WHOLE GRAIN PANCAKES<br>Sliced Apples<br>Grape Juice<br>Syrup<br>1% White Milk<br>Fat Free Chocolate Milk<br>Breakfast-Cold<br>LEMON LOAF<br>Sliced Apples<br>Grape Juice<br>1% White Milk<br>Fat Free Chocolate Milk<br>Lunch | Breakfast<br>VANILLA YOGURT<br>Blueberry Muffin<br>Peach Cup<br>Blended Fruit Juice<br>1% White Milk<br>Fat Free Chocolate Milk<br>Lunch<br>TURKEY & CHEESE ON WHOLE WHEAT BUN<br>Italian Corn Salad<br>Light Mayonnaise<br>CHEF SALAD W/TURKEY HAM |

**10.1** Example school lunch menu for the School District of Philadelphia

Unique to Philadelphia is the Universal School Feeding Pilot Program. Beginning in 1991, the District has used socio-economic data to determine school-wide eligibility for free school meals, rather than requiring parents to apply for their child's individual eligibility. The goals of this program are to:

1. Reduce the administrative cost of managing individual meal applications,

2. Increase access to school meals in low-income communities, and

3. Increase participation in school meals, especially in high schools, by providing universal service and reducing the stigma associated with free school meals.[v]

At Huey, free breakfast, lunch, and after-school supper (for students participating in after-school programs) are available to all students.

This program requires a significant expenditure of energy and money in efforts to fulfill its mission to "prepare students for a quality education by providing the fuel necessary for learning."[vi] While data exists measuring the amount of food served to students in Philadelphia, there is little research quantifying the amount of food actually consumed. In 2013 a survey of 434 School District of Philadelphia employees found evidence that over half the food being served is thrown out.[vii] Given the millions of meals served in Philadelphia each year, this data is troubling and deserves further investigation.

"They call the food *freebies*. It's free food. They don't call the food from the poppy store or corner store freebies because you actually have to pay for it. At school you just grab the food and eat it, you don't have to pay for it at all."

Huey Elementary student E.G.

In the spring of 2014, University of Pennsylvania (Penn) faculty organizers of the Last Food Mile Conference, which focused on the causes and implications of food waste, met with staff at the Netter Center for Community Partnerships to explore the possibility of a youth-led presentation about food waste in a local school. Founded in 1992, the Netter Center is a university-wide center that fosters mutually beneficial collaborations between Penn and the local community; it is funded through university support, government and foundation grants, and private philanthropy. Located 13 blocks away from Penn, Huey is one of five University Assisted Community Schools (UACS) supported by the Netter Center. The UACS partnership provides full and part-time staffing, undergraduate/graduate student interns, and supplies and other resources at Huey to run an after-school program and targeted school day activities, which mainly focus on real-world problem solving and project-based learning in the STEM and Health disciplines. The overall goal of the UACS partnership is to simultaneously improve the quality of life and learning in the school and community while advancing university research, teaching, learning, and service.

The Last Food Mile Conference provided the impetus to integrate the study of food waste at Huey School into an educational project. The approach we used while conducting this investigation borrowed from and was inspired by project-based learning, real-world problem solving, and community-based participatory action research.

Given the various factors that influence the food landscape of Huey and the profound impact that food has on the growth and development of young people, we set out to tell the story of food waste at Huey through the perspective of Huey students. We assembled a research team (three Huey 8$^{th}$ graders, one Penn graduate student, and one Netter Center staff member) to develop and implement the research plan. Our approach followed the UACS model, which recognizes "the school as a community institution that educates young children, both intellectually and morally, by engaging them in real-world, community problem solving."[viii] We also borrowed from principles of Participatory Action Research (PAR)[ix] (described further below), an approach to research that not only involves community members as equal partners, but also aims to inform meaningful action to produce positive change. Our long-term goal is to enable our 8th grade co-investigators to serve as agents of change in their school and city.

## METHODS

"Tell me and I forget, teach me and I may remember, involve me and I learn."

Benjamin Franklin

We employed a mixed methods approach, involving direct measurement, surveys, qualitative interviews, and photo-voice. PAR principles were incorporated throughout the design and implementation to ensure that the data were enriched by both academic and local expertise.

Three 8th graders attending Huey were recruited as co-investigators. They served dual roles as researchers and local experts. Consistent with the PAR goal of balancing academic and community interests when conducting community-based research, student co-investigators were asked to contribute their perspectives on the significance of understanding food waste at schools (see box insert). Study purpose and design were then decided collaboratively, with the adult co-investigators acting as facilitators, allowing the discussion to be grounded in the first-hand experiences of the student co-investigators. Student input was instrumental in determining the feasibility of proposed research methods. We discussed and debated the pros and cons of each possible investigative approach.

**Why is food waste important?**

- There are hungry people, so food should not be thrown out
  - Need to divert wasted food to hungry people
- Wasted $ from taxes when food is thrown out because food costs $ + trash removal costs $

We need food to learn + grow

"It made me feel awesome because I've never been a co-investigator before."

Huey Elementary student E.G.

Student suggestions also determined the content of our survey instrument and interview guide. Once the methods of the study were decided, students were involved in each step of implementation. The individual students divided the analysis so each specialized in one approach.

## Direct Measurement

Student co-investigators chose to focus direct measurement on the kindergarten population because they eat lunch apart from the rest of the school, and there is a larger staff to student ratio, fostering more efficient and accurate data collection. To gather the total weight of food distributed, the research team weighed one package of each food item available for service and multiplied that by the number served to kindergarteners. For foods that had varying weights such as apples, the average weight of three items was used.

"We decided to do the kindergarten class because we decided it wouldn't be as crazy as the older children's classes. If we did it as one whole group, they would've been yelling."

Huey Elementary student E.G.

Weight of empty packages were also recorded and subtracted from the total. Kindergartners were instructed to throw only school food waste into a designated receptacle, since many also brought food from home. The garbage bag was then collected from the designated receptacle and weighed. Packaging weight was subtracted from garbage weight, to determine total weight of food wasted. The garbage was later sorted and unopened lunch items were also counted. Data collection was repeated over three separate lunch periods. Measurements were obtained and recorded by students, using a worksheet that was designed to streamline this process (see Appendix). Students also made direct observations during data collection, which are reported in the Discussion.

## Survey

A three-question survey instrument was developed by student co-investigators (see Appendix) and distributed to $5^{th}$-$8^{th}$ grade classes. The surveys were administered by classroom teachers, and were completed simultaneously by each grade so that no student could take the survey twice. Survey responses were anonymous.

## Photo-voice

Photo-voice is a method of community-based participatory action research engaging participants in discovering the root causes of community problems to collectively address them.[x] Student co-investigators were first engaged in a brief training on digital media techniques (photography, videography, audio recording) and then given the equipment to document the school food experience from their perspective. Their training incorporated additional relevant skills such as interviewing and media ethics. The students also provided commentary to give context and meaning to the documentation.

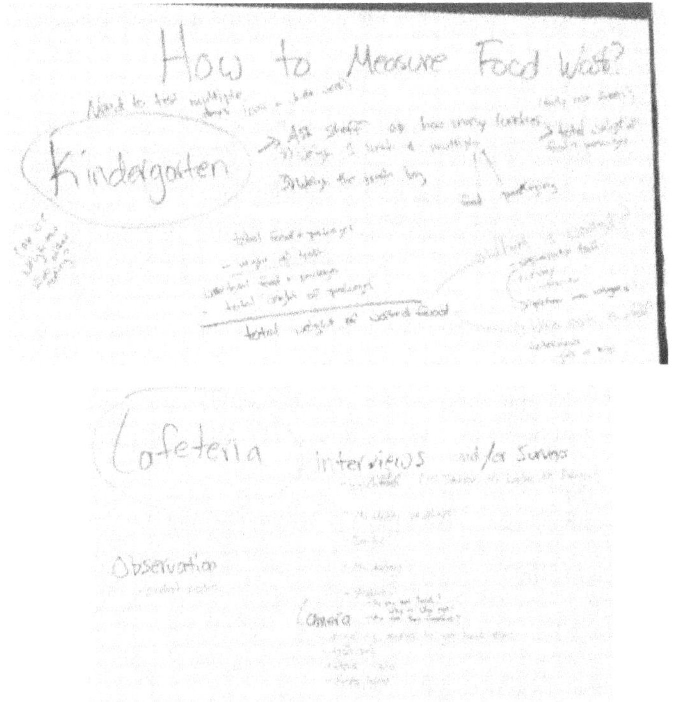


**10.2** Graphic representation of collaboratively planning data collection and analysis methods

## Analysis

While data collection was a true team effort, our student investigators each took individual responsibility for analyzing our findings from the different methods (Figure 10.2). This division of labor allowed us to work more efficiently, and also fostered engagement and ownership in the project itself. Students were then challenged to "learn by teaching" by explaining the analysis to their peers. Adult facilitators worked to match the math and writing involved in the interpretation of data to appropriate grade level standards.

### *Direct Measurement:*

After all direct measurements were recorded, co-investigator SRC[xi] performed the calculations that the team had decided upon in order to find the percent of food wasted. The team then decided together how to best present the data. Co-investigator SRC created a combination of pie charts and bar graphs to show both proportion of food wasted as well as the types of food most commonly thrown away unopened. The mathematics and scientific inquiry involved in this methodology matched the core academic standards in the 8th grade curricula.

***Surveys:***

Once surveys were completed and collected, co-investigator JP entered the responses into a Microsoft Excel spreadsheet. JP and JZ worked together to use Excel to make calculations and graph the data on food waste attitudes and behaviors. Median and mode were calculated where relevant.

***Interviews:***

After comparing the relative benefits and limitations between interviews and surveys, the team decided that interviews would tell an important part of the story that surveys might miss. A targeted list of interviewees was created with the goal of understanding food waste from many different perspectives. This list included:

- Noontime aides in the kindergarten classroom
- School facilities manager
- Principal
- Noontime aides in lunchroom
- Students (intercepted in the lunchroom)

Interview questions were written by student co-investigators and edited by J.S. Co-investigators E.G. and J.S. conducted the interviews. Responses were tape-recorded, and reviewed and transcribed by the team.

## RESULTS

### Direct Measurement

The kindergarten class that was measured in this study had an average of 20 students who received school lunch each measurement day (menus listed in Table 10.1). We found that the kindergarten class wasted 35.4% (by weight) of the school food that they were served for lunch over the three days. Day 2 had the greatest proportion of food wasted (38.7%) and Day 1 had the smallest (33.2%). Food waste measured is presented in Table 10.2.

Across all three days, the items that were most commonly found unopened in the trash were vegetables (see Table 10.3). Over half (22 out of 43) of the vegetables served were thrown away unopened. The least popular item was the peas, 73.3% of which were found unopened in the trash. Fruits were more popular. Only 10.5% of fruits served were found completely intact, including 2 out of 40 fruit cups and 4 out of 17 apples. Three unopened milk cartons were found in the trash, out of 59 milk cartons served. No meat entrée was discarded unopened.

**Table 10.1 School lunch options**

| | | |
|---|---|---|
| **Day 1** | 10/28/14 | Fried chicken fingers<br>or<br>Mac and cheese<br>Bean Salad<br>Carrot Sticks<br>Fruit Cup<br>Milk |
| **Day 2** | 11/12/14 | Chicken nuggets, tater tots<br>or<br>Beef chili and chips<br>Apple<br>Milk |
| **Day 3** | 11/18/14 | Spaghetti, tomato sauce and sliced bread<br>or<br>Hamburger<br>Peas<br>Fruit cup<br>Milk/juice |

**Table 10.2 School food wasted by one kindergarten class during lunch over 3 separate days**

| | # Students | lbs Served | lbs Wasted | % Wasted |
|---|---|---|---|---|
| 10/28/14 | 21 | 33.4 | 11.1 | 33.2 |
| 11/12/14 | 19 | 21.4 | 8.3 | 38.7 |
| 11/18/14 | 20 | 33.6 | 11.9 | 35.5 |
| Total | | 88.4 | 31.3 | 35.4 |

## Survey

A total of 67 students, ranging in age from 9 to 14 (average age 12), responded to our survey. 40.3% of respondents were female. All questions and responses are presented in Table 10.4, and summarized in Figure 10.3.

Over half of respondents reported that they never finish all of their school lunch. Female respondents were more likely than males (66.7% vs. 40%) to report never finishing their lunch. Thirty-nine percent of respondents reported that they are most likely to throw away their vegetables. A larger proportion of males

(43.8%) than females (32.8%) reported that they are most likely to throw away their vegetables.

**Table 10.3 School food thrown away unopened by kindergarteners during 3 lunch periods; same data source as Table 10.1**

| | Served | Unopened | % |
|---|---|---|---|
| Vegetables | 43 | 22 | 51.2 |
| Peas | 15 | 11 | 73.3 |
| Bean Salad | 18 | 8 | 44. |
| Carrots | 10 | 3 | 30.0 |
| Fruit | 57 | 6 | 10.5 |
| Apples | 17 | 4 | 23.5 |
| Fruit Cup | 40 | 2 | 5.0 |
| Milk | 59 | 3 | 5.1 |

On a Likert scale of 1-10, most respondents selected "1" (worst) to rank the school food. Four respondents wrote in "0" even though that was not an option provided. The second most popular response was "5." When separated by gender, the same pattern is also seen in the male responses, with the top two responses being "1" and "5", but among female respondents, the most popular responses were "1" and "10."

**Interviews**

Six adult Huey staff members and four students participated in interviews. The overwhelming perception among all respondents is that a large percentage of school food is wasted, ranging from 40% to "most of it." Similarly, the respondents uniformly reported that the foods wasted most often by students are vegetables.

There was less agreement when participants were asked: *Why do kids waste food?* Some attributed the behavior to an inherent characteristic of the children – that they are simply wasteful. Others connected the behavior of food waste to poor nutritional teaching:

"I think it's because they don't get the required nutrition taught at home."

**Table 10.4 Survey of school food waste among 5th-8th graders**

| | All | | Girls | | Boys | |
|---|---|---|---|---|---|---|
| Respondents | 67 | | 27 | (40.3%) | 40 | (59.7%) |
| Average Age | 12 | | | | | |
| Age Range | 9-14 | | | | | |
| **How many times a week do you eat all of your school lunch?** | | | | | | |
| | | % All | | % Girls | | % Boys |
| **Never** | 34 | 50.7 | 18 | 66.7 | 16 | 40.0 |
| **1 or 2 times** | 21 | 31.3 | 6 | 22.2 | 15 | 37.5 |
| **3 or 4 times** | 9 | 13.4 | 2 | 7.4 | 6 | 15.0 |
| **Every day** | 4 | 6.0 | 1 | 3.7 | 3 | 7.5 |
| Total Responses | 67 | 100 | 27 | 100 | 40 | 100 |
| **What foods are you most likely to throw away?** | | | | | | |
| | | % All | | % Girls | | % Boys |
| **Vegetables** | 48 | 39.0 | 19 | 32.8 | 28 | 43.8 |
| **Milk** | 33 | 26.8 | 15 | 25.9 | 18 | 28.1 |
| **Meats** | 23 | 18.7 | 14 | 24.1 | 9 | 14.1 |
| **Fruits** | 19 | 15.4 | 10 | 17.2 | 9 | 14.1 |

**How much do you like the school food on a scale of 1 (worst) to 10?**

| Score | | | % All | | % Girls | | % Boys |
|---|---|---|---|---|---|---|---|
| | **0*** | 4 | 6.3 | 1 | 3.8 | 3 | 8.1 |
| | **1** | 22 | 34.9 | 10 | 38.5 | 12 | 32.4 |
| | **2** | 4 | 6.3 | 3 | 11.5 | 1 | 2.7 |
| | **3** | 6 | 9.5 | 3 | 11.5 | 3 | 8.1 |
| | **4** | 8 | 12.7 | 2 | 7.7 | 6 | 16.2 |
| | **5** | 9 | 14.3 | 2 | 7.7 | 7 | 18.9 |
| | **6** | 3 | 4.8 | 0 | 0.0 | 3 | 8.1 |
| | **7** | 1 | 1.6 | 0 | 0.0 | 1 | 2.7 |
| | **8** | 1 | 1.6 | 1 | 3.8 | 0 | 0.0 |
| | **9** | 0 | 0 | 0 | 0.0 | 0 | 0.0 |
| | **10** | 5 | 7.9 | 4 | 15.4 | 1 | 2.7 |
| Total Responses | | 63 | 100% | 26 | 100 | 37 | 100 |
| Median | | 3 | | 2 | | 3 | |
| Mode | | 1 | | 1 | | 1 | |

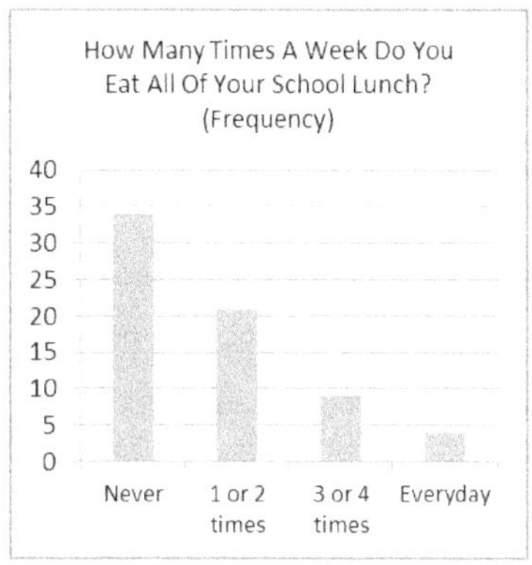


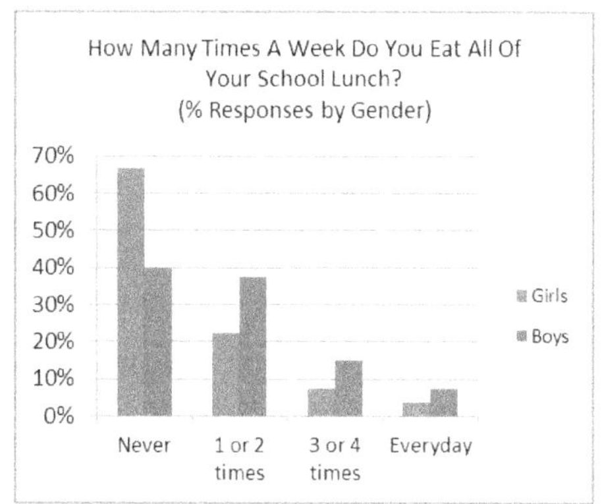


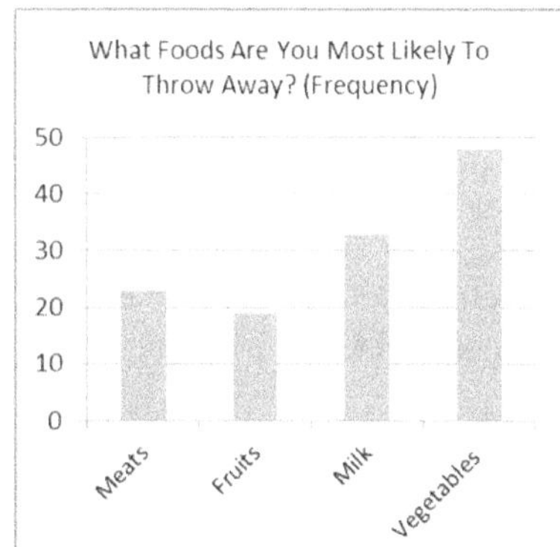


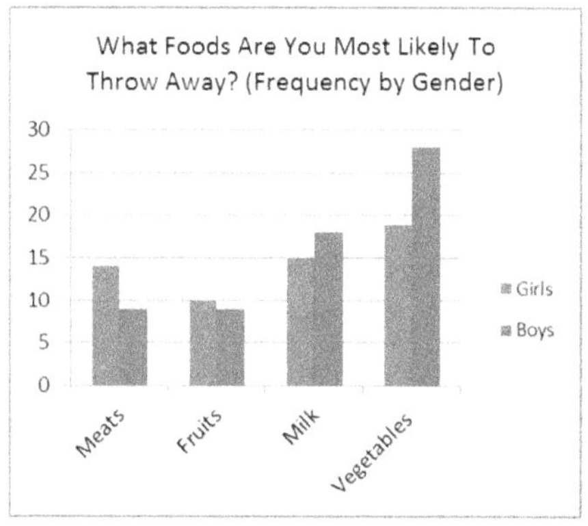


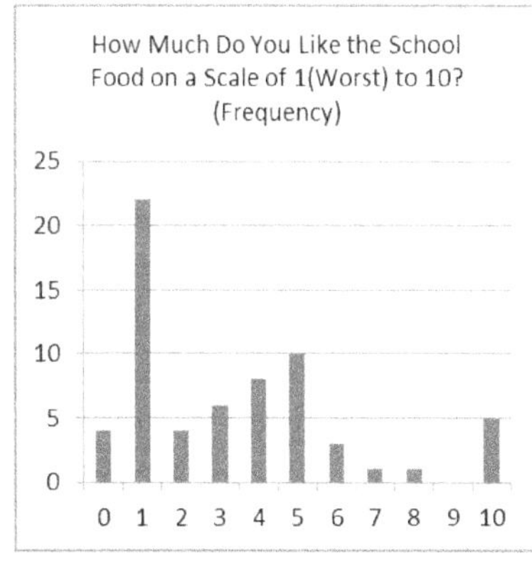


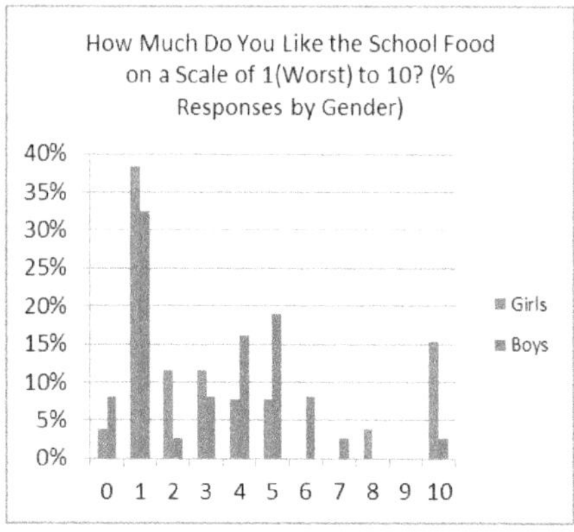


**10.3** Summary of student responses to questionnaire on attitudes and behaviors related to school food

Along with the awareness of food waste among adult staff at the school, there is also a sense of futility and lack of personal responsibility toward the issue. For instance, one staff member commented:

"By law we have to give it to them. If they don't eat it, that's on them."

Overall, the prevalence of food waste was confirmed by interviews. When asked about ways to reduce food waste, only one idea was generated:

"Serve more enticing items. They like the chicken nuggets and the pizza. I mean they can't eat that every day. But that's what they want to eat."

**Photo-voice**

Photos collected by student co-investigators are available at the conference presentation to illustrate the everyday student perspective of food waste: http://repository.upenn.edu/cgi/viewcontent.cgi?article=1033&context=thelastfoodmile .

Photos were taken in a diverse set of environments, including a kindergarten classroom, the school lunchroom, the satellite kitchen, and the dumpster in the schoolyard.

## DISCUSSION

### School Meal Consumption and Waste

The school meal is an extremely powerful eating experience for the child, the school, the community, and the planet. Worldwide, over 368 million children are being fed in schools.[xii] In the U.S., 30 million children receive school lunch in over 100,000 schools every day, costing $11.6 billion in 2012.[xiii] Students qualifying for free meals, including all children in Philadelphia Public Schools, have the opportunity to eat breakfast, lunch, and supper at school, representing over half of the total calories consumed each day.[xiv] The exposure to and the daily consumption of school meals over a child's educational experience (approximately 12 years) present an important opportunity for children to consume the adequate nutrition for development and to learn healthy and sustainable food behaviors.

Our findings provide a snapshot of where this food actually ends up, by examining the consumption and waste at one public school in Philadelphia. Consistently over our three days of measurement in a kindergarten classroom, we found that a substantial (over 30% by weight) amount of food that was served ended up in the trash. Vegetables made up the largest proportion of the wasted food. Our surveys and interviews corroborate that there is at least the perception of a similar phenomenon school-wide.

> Nothing really surprised me. Like with…all of the data like the vegetables being wasted and the food being wasted, I wasn't really surprised by that. Cause I've seen that happen."
>
> Huey Elementary student J.P.

Since we did not attempt to measure food waste directly for the upper grades, we can only make an educated guess about the actual quantity of food that ends up in the garbage at Huey. It is possible that the food waste we measured was unique to the kindergarten setting, and that older students may waste a smaller proportion of food. However, student surveys, qualitative data, and other evidence all seem to suggest that food waste in the higher grades may actually be *worse*. Furthermore, the kindergarten eating environment at Huey is very different from that of the rest of the school. Kindergarteners are served lunch in their assigned classroom seats, and there is a staff person responsible for supervising the class of approximately 25 students and maintaining orderly conduct. There is much less direct supervision for the rest of the school. Therefore, older students have more freedom to behave as they see fit during lunch.

> "Sometimes the more popular boys they wear their hoods because they don't want the girls to see them eat their lunch. The eighth grade girls call them freebies. The 7$^{th}$ grade boys think it will make them look uncool."
>
> Huey Elementary student E.G.

In addition, complex factors such as the social shaming around consumption of "freebies" that is described among older students may serve as additional drivers for food waste among older students at Huey. Findings from our study also suggest a difference in perception of school food between male and female students. Girls were much more likely to report never finishing their school lunch, suggesting that factors influencing eating behaviors may differ by gender.

If there is at least as much food waste school-wide as was found in kindergarten classes, and if vegetables tend to make up a larger portion of the food wasted, then this has great implications for the health and nutrition of the students at Huey. The nutrients being wasted are also nutrients that are not reaching Huey students. Based on student narratives, these lost nutrients are often substituted by snacks and candy from corner stores that surround Huey along the 52$^{nd}$ St. corridor.

"If they weren't always going to the store in the morning like if they ate breakfast but not corner store food for breakfast, they'd have more room to eat vegetables, they're probably going to eat the vegetables or something more."

Huey Elementary student J.P.

It is also well known that students who are well nourished are better able to learn.[xv] In 2014, Huey ranked 142 out of 144 elementary schools in combined Math/Reading Standardized test scores.[xvi] Using standardized test scores to measure academic achievement is, of course, insufficient, and it also takes more than consuming nutrient dense foods for students to succeed scholastically. Nonetheless, ensuring quality nutrition for all students is one important piece of the puzzle to help improve student achievement at Huey, and at all schools in Philadelphia and nationwide.

## Perspective and Role of School Staff

Despite the bare-bones budget that is crippling Huey and the School District of Philadelphia overall, there are many adults on site whose job responsibilities are connected to school feeding – from the three food service personnel, to the noon-time aides (three in the cafeteria, one in each kindergarten lunchroom), to the Principal and Vice Principal who are often found on the loudspeaker directing the lunchtime flow, to the facilities team cleaning and taking out the trash. When speaking with these individuals during interviews and through observations from photo-voice, we recognized that they often placed responsibility for the food wasting behavior on the students ("they are wasteful"), on the parents ("they aren't taught right at home") or on the menu planners at school food services ("the kids don't like the food").

Importantly, from a budget perspective the adult school meal staff at Huey is only responsible for ensuring that the FSD receives adequate reimbursement by the federal government for the meals that are served. This means they focus their attention on the number of meals served, rather than on what happens to the food. There is an opportunity to train staff to also pay attention to food consumption. In order to justify the expenditure of capital on human and material resources to improve student consumption outcomes, more research is needed to quantify the costs of food waste, including hauling excess food waste, as well as the economic consequences of undernourishment on children's mental and physical development.

## The Role of Students as Researchers and Agents of Change

From the beginning of our project-based school food investigation, the members of our research team worked together to plan, organize, coordinate, implement, and analyze every aspect of our study. The role of the adult facilitators was to provide a general structure to assist in goal setting, role delegation, and task completion. They also provided material support as well as mentorship through the research process. To maximize ownership and engagement for the student co-investigator team, all decision-making was made collaboratively, and data collection was completely student driven.

"I didn't know it was going to be this much (work). It was important to do because we got more accurate findings. We got more accurate data for our hypotheses."

Co-investigator E.G.

Over the course of the months working together, our research project transitioned from planning, data collection and analysis to reflection, writing, and planning school waste reduction strategies for Huey. This represents the evolution of our student co-investigators from active learners to agents-of-change in their community.

Our team, particularly the 8$^{th}$ grade co-investigators, generated several ideas to reduce food waste in the cafeteria. Similar to the school staff, the first area students wanted to see improvement is in the quality of the food served (although not necessarily nutrient density).

"We already get pizza but if we got like pizza with crust, or burgers but with lettuce and tomato and cheese and stuff on them, or like actually fried chicken nuggets."

Co-investigator J.P.

Another idea is the installation of a salad bar, noted to be particularly attractive as students can build their own eating experience. Similar to how we found that ownership in the research process improved engagement and learning, our student co-investigators expressed that owning the salad making experience enhances the food itself.

"It's more of your own, and it doesn't feel like everybody has been touching the packaging so you know exactly what you want."

Co-investigator J.P.

A final suggestion on the food service wasn't related to the food, but its paper, plastic, and Styrofoam packaging:

"If the containers looked more presentable and looked better it might persuade more of us into eating them."

Co-investigator J.P.

While all valid suggestions, these improvements will all require some increased spending, and therefore are somewhat unrealistic given the current budget crisis affecting the schools in Philadelphia. Given this fiscal reality, our team also discussed ideas to increase school food consumption that do not require changes to the actual food served or packaging it is served in.

The school meal is a social event and interactions between students and adults are unavoidable. It was noted that adults are too often yelling during eating time and are too rarely encouraging students to eat their meal.

She [kindergarten noontime aide] was yelling at the kids for not eating, and if they wasn't eating then she would like take the food put it back in the crate and they would start crying about the food."

Co-investigator E.G.

To foster a more pleasant eating environment, we determined that adults should engage students in positive marketing and interactive games. When asked what noontime aides can do differently, our co-investigators responded:

"Persuasion. Like trying to convince them, not like trying to trick them, but like trying to tell them that it's actually good. Looks can be deceiving."

Co-investigator J.P.

"They can like play games. Like whoever eats the most vegetables gets a prize or something."

Co-investigator E.G.

These strategies, if thoughtfully planned in collaboration with the noontime aides and with buy-in from the principal, could serve as an important step in the right direction to reduce food waste and enhance both the student and adult experience during school meals. Also, just as students drove the food waste investigation at Huey, they too are poised to leverage their responsibility as role models to encourage waste reduction behaviors.

"They [younger students] basically do everything we do. If they see us, they would do it too."

Co-investigator J.P.

Several of the ideas proposed by our research team have already been gaining momentum in implementation and impact evaluation around the U.S. In 2010 First Lady Michelle Obama's Let's Move Campaign introduced the Let's Move Salad Bars to Schools program, a public-private partnership that resulted in the donation of 2,800 salad bars across the country.[xvii] Little research has been conducted evaluating the impact of this program, but several school districts participating in the program do report increased fruit and vegetable intake among students.[xviii] Children have been shown to be more likely to eat initially disliked vegetables when a peer models the behavior; adult encouragement can be effective in increasing child vegetable consumption, as can older student role models.[xix]

Despite the high prevalence of food waste we found at Huey Elementary School, we are optimistic about the potential for change in both the short- and long-term. Huey is several steps ahead of many schools in its organization and coordination of school-wide health focused activities. A group of committed stakeholders – including teachers, staff, students, parents, and partners meet monthly to plan and review health-related activities. Through its partnership with the Netter Center at the University of Pennsylvania, Huey is pioneering a University-Assisted Coordinated School Health approach to assess needs and leverage assets that exist in the school and community to improve the overall healthiness of the Huey community. This infrastructure of energy, resources, and supports provides a platform for student-driven change to take place.

Our proposed implementation of food waste reduction through student-centered real-world problem solving is pioneering for Huey and for American public schools in general, but this style of change creation is not unprecedented. In Japanese elementary schools, for example, school food looks drastically different than in Philadelphia. Students serve their peers, and are responsible for the setup and the cleanup of the meal. Japan has one of the lowest childhood obesity rates in the world.[xx] While student-led food service is not realistic in Philadelphia in the short-term, the implementation of a food waste reduction plan can be driven by student energy and creativity. Using this approach is a win-win for Huey – it will decrease food waste and increase student engagement in immersive, project-based learning.

## CONCLUSION

Our project has uncovered a complex narrative around school food waste at Huey, which has implications for food waste throughout Philadelphia Public Schools. In our research, we have found that there are many factors that contribute to food waste, and that its consequences are diverse and important – including child health and development, school performance and success, and wasteful use of limited public resources. Our youth-led strategy allowed us to identify many promising ideas for curbing food waste at Huey, including increased peer- and adult-led promotion of healthy eating at school. However, additional research needs to be conducted to quantify the impacts of food waste more broadly so that additional resources can be directed toward implementing these and other food waste reduction strategies.

## NOTES

[i] https://webapps.philasd.org/school_profile/view/1330

[ii] SR Watkins & Associates. 2011. Communities Putting Prevention to Work. Philadelphia Dept. of Health. http://www.phila.gov/health/pdfs/Watkins_final%20report.pdf

[iii] Graham, K.A. 2012. Philadelphia School District Reopens Some Full-service Kitchens. *Philly.com*, November 6. http://articles.philly.com/2012-11-06/news/34931640_1_cafeteria-worker-wayne-grasela-meal-sites

[iv] School District of Philadelphia. 2013. Preferred Meals. http://preferredmealsmenu.com/Site/Menu.aspx?DistrictSchool=PA_A162_APLAN&SchoolName=School+District+Of+Philadelphia

[v] Coalition Against Hunger. 2015. Philadelphia's Universal Feeding Pilot. Issue: Child Nutrition 2015. http://www.hungercoalition.org/summary-universal-feeding-and-CNR-goals

[vi] http://webgui.phila.k12.pa.us/offices/f/foodservices

[vii] Real Food. Real Jobs. n. d. Inside the Lunchroom: Workers' Vision for Real Food and Safe Schools. http://www.realfoodrealjobs.org/wp-content/uploads/PhillySchoolsReport_Final.pdf

[viii] Benson, L., I. Harkavy, and J. Puckett. 2000. An implementation revolution as a strategy for fulfilling the democratic promise of university-community partnerships: Penn-West Philadelphia as an experiment in progress. Nonprof. Volun. Sec. Q. 29(1):24-45.

[ix] Romero González, E., R.P Lejano, G. Vidales, R.F. Conner, Y. Kidokoro, B. Fazeli, and R. Cabrales. Participatory Action Research for environmental health: encountering Freire in the urban barrio. J. Urban Aff. 29(1):77-100.

[x] Strack, R., K. Lovelace, T. Jordan, and A. Holmes. 2010. Framing photo-voice using a social-ecological logic model as a guide. Health Promot. Prac. 11(5):629-636. http://dx.doi.org/10.1177/1524839909355519.

[xi] Co-investigators are referred to throughout this document by their initials. Not all co-investigators co-authored this chapter.

[xii] World Food Programme. 2014. School Meals. https://www.wfp.org/school-meals.

[xiii] U.S. Department of Agriculture. 2013. National School Lunch Program. http://www.fns.usda.gov/sites/default/files/NSLPFactSheet.pdf

[xiv] Gleason, P., and C. Suitor. 2001. Policy Brief: Food for Thought: Children's Diets in the 1990s. Mathematica Policy Research, Inc. Document No. PRO-1-12.

[xv] Troccoli, K. 1993. Eat to learn, learn to eat: The link between nutrition and learning in children." Washington, DC: National Healthy Education Consortium, Occasional paper 7.

[xvi] School Digger. 2013-2014. Pennsylvania Elementary School Rankings. http://www.schooldigger.com/go/PA/schoolrank.aspx?findschool=1899003807

[xvii] Gretchen Swanson Center for Nutrition. 2014. Evaluation of the Let's Move Salad Bars to Schools Initiative: Executive Summary. http://www.saladbars2schools.org/wp-content/uploads/2014/01/Exec_Summary_Eval_Infographic.pdf

[xviii] Harris, D.M., J. Seymour, L. Grummer-Strawn, A. Cooper, B. Collins, L. DiSogra, A. Marshall, and N. Evans. 2012. Let's move salad bars to schools: a public-private partnership to increase student fruit and vegetable consumption. Child Obes. 8(4):294-297. http://dx.doi.org/10.1089/chi.2012.0094.

[xix] Birch, L.L. 1999. Development of food preferences. Ann. Rev. Nutr. 19:41-62.

[xx] Harlan, C. 2013. On Japan's School Lunch Menu: A Healthy Meal, Made from Scratch. *The Washington Post*, January 26. http://www.washingtonpost.com/world/on-japans-school lunch-menu-a-healthy-meal-made-from-scratch/2013/01/26/5f31d208-63a2-11e2-85f5-a8a9228e55e7_story.html

# *Chapter 11*

## *Wasted Food: From Conundrum to Clarity*

**Jonathan Bloom and Steven M. Finn**

### FOOD WASTE: A CONUNDRUM

The statistics on global food waste are staggering. A recent study by the Institute of Mechanical Engineers noted that between 30 to 50 percent of all food produced – between 1.3 and 2 billion tons – is not consumed by humans.[i] Yet at the same time, roughly 800 million global citizens remain hungry[ii] and two billion suffer from micronutrient deficiencies.[iii]

Then again, those statistics are fitting – little about food waste makes sense.

Slow Food founder Carlo Petrini has called the dual flourishing of food waste and hunger a 'paradox.' Others have called the juxtaposition of massive global food waste and extensive global hunger a 'conundrum.' The coexistence of the two is indeed counterintuitive – and quite alarming. One could also describe our squandered food status quo as 'indefensible,' 'unethical', 'unsustainable', and, here's the key one, 'doomed.' Our current levels of food waste are destined to change because we cannot and will not continue wasting food at the current rate while nearly a billion worldwide don't have enough to eat, the global population surges higher, food prices continue to rise, and our waste comes at a significant environmental cost.

The sharpest changes are likely to occur in the developing world, where small technological advances will yield dramatic reductions of losses in the food chain. Yet in the developed world, where 'waste' stems from human decisions somewhere in the food chain, those shifts will require altering both our thinking and behavior. And since such changes don't happen by accident, we must create a plan of action. Doing so will take us from conundrum to clarity.

In the U.S., transitioning from conundrum to clarity will require getting the issue of food waste on the policy agenda. Happily, in September 2015, the U.S. Department of Agriculture and the Environmental Protection Agency announced a goal of reducing food waste by 50 percent by 2030. That dovetails with the UN Sustainable Development Goal that proposes the same result by 2025. And in December 2015, Rep. Chellie Pingree (D-ME) introduced a bill to

Congress with more than 20 provisions aimed at minimizing wasted food at all levels of the supply chain. Pingree's Food Recovery Act, should the bill or parts of it pass, will help the U.S. approach that reduction goal.

Moving from conundrum to clarity will mean expanding wasted food awareness. Addressing food waste must become similar in scope to past initiatives focused on the environment, seat belts, smoking, littering, and drunk driving. At the core, curbing wasted food will mean moving from a culture of abundance regarding food to a culture of responsibility. It will require effective educational efforts on food waste, with special emphasis on the next generation of leaders. We must teach our children that wasting food is not cool! Today, through inaction, we're teaching them the opposite.

Further, we must once again make our culture one that properly *values* food, where actions to avoid waste are not only normal, but expected. We must collaborate broadly and innovate boldly. We will need creative ideas and technologies to continually reduce the excess food we're creating, while we optimize uses for the inevitable excess downstream. Eradicating our ingrained habits and systems that produce more than 30 million tons of food waste annually will not occur without intervention. We need a national framework for change, one that prompts partnerships between government, industry and individuals. Meanwhile, a multi-pronged approach is a must. We will drive this change through the five key steps below.

## RAISE AWARENESS

While wasted food awareness is on the upswing, it remains insufficient. Few people see it as a problem, fewer realize the scale of our wasting, and fewer still understand the consequences. As fellow book contributor Roni Neff's 2014 research found, 42 percent of American consumers had seen or heard something about food waste in the past year. That leaves 58 percent who didn't, and that should provide further targets for awareness raising.

Furthermore, there's a major gap between awareness and action. While 42 percent of respondents had seen or heard something about wasted food, only 16 percent sought information about reducing it. This gap can be closed through further familiarity. While we need to increase awareness, we must do so in a way that prompts meaningful action. Additionally, there's a common misconception – let's call it the methane gap – that throwing away food isn't problematic. The thinking is that organic materials thrown away aren't problematic, when in fact they contribute greatly to climate change through the methane created when they decompose in landfills.

Moving forward, there is reason for optimism. Food waste messaging will soon receive a boost from the Ad Council, a non-profit public service advertising company known for creating the slogans "Only you can prevent forest fires" and "Friends don't let friends drive drunk." The Ad Council together with the Natural Resources Defense Council will launch a wasted food campaign aimed at moms and millennials in Spring 2016. ReFED, a collaborative initiative supported by 13 foundations with an advisory council of 20 businesses and non-profits will release a new report in early 2016. The ReFED Roadmap will provide just that. It will recommend the most effective solutions for fighting food waste, identifying areas and actions that will yield the biggest environmental, social and economic impact.

By setting a goal of 50 percent reduction, the USDA and the EPA have set an aggressive tone in fighting food waste. Yet there remains a need for further federal involvement in raising awareness (in addition to enacting helpful policy). In the UK, for example, the government-funded organization WRAP (Waste & Resources Action Programme) created a successful campaign to curb wasted food called Love Food Hate Waste. With a major assist from extensive media coverage, the campaign helped reduce food waste by nearly 20 percent in five years. In the U.S., the EPA and the West Coast Climate and Materials Management Forum have created a similar campaign, Food Too Good to Waste, and piloted it in more than 25 U.S. cities. Yet the program has languished, as the EPA has largely ignored its potential to raise awareness and prompt change on a national level. There is plenty of room for further USDA and EPA action on the issue of wasted food beyond merely highlighting the best practices of others in the U.S. Food Waste Challenge. Given the recent food waste reduction goal, federal activity must increase.

As a recent cultural phenomenon, wasted food has ascended steadily. In April 2015, the national airing of the documentary *Just Eat It* on MSNBC was a game changer for food waste awareness. The WastED pop-up restaurant helmed by Chef Dan Barber at the well-regarded Manhattan eatery Blue Hill also brought an abundance of attention to the issue. As high profile media and celebrity chefs become increasingly involved in fighting food waste, the awareness tide will continue to rise. Meanwhile, environmental organizations continue to become further involved, with the World Wildlife Federation ramping up its involvement in 2015 and beyond.

While there is plenty of grassroots activism and interest, epitomized by the food recovery organizations and composting companies sprouting in most major cities, it very rarely focuses on the most important part of the EPA hierarchy – reducing excess food. Instead, most activism aims lower in the hierarchy, either in donating or composting food. That's why the Ad Council campaign, which will likely stress the need for behavior change, could prompt a major shift in popular thinking. Even if it does and especially if it doesn't, we'll need to focus more on awareness raising. Knowledge of our current wasted food conundrum is the first step toward getting businesses, institutions, and individuals to shift their practices.

## EDUCATE FOR CHANGE

Convincing adults about the necessity of avoiding wasted food is vital, as they are the ones buying food and planning meals. Still, children must be the primary audience for that message because they will soon become the country's main food purchasers. Furthermore, they are effective educators and ambassadors for environmental issues, often through sheer persistence. Kids are frequently credited with bringing home their school-learned lessons on recycling and littering.

Unfortunately, all too often, we teach kids that throwing away food is perfectly acceptable. Seeing food thrown away daily – and especially taking part – normalizes the practice. And kids are witnessing and tossing a staggering amount of food every day. As Jarrett Stein and Jessica Zha illustrated in Chapter 10, school lunch waste levels in Philadelphia schools vacillate between 35 and 40 percent.[iv]

Such staggering waste stems from several problems, some structural and some tactical. For example, the quality of school food is often lacking, driven largely by poor funding for school lunches. Schools are only reimbursed roughly $3 by the USDA for each free meal served, and that must cover the labor, supplies and contractor fees in addition to the food cost.[v] The massive scale of many centralized school cafeteria prep kitchens does not improve the food's quality, either. Additionally, the scheduling of lunch can prompt waste, as stretched schools often start serving lunch before students may be hungry. Meanwhile, the order of lunch and recess has a major impact on plate waste. And there is seldom enough time for students to eat their entire lunch.

The best way to teach students about not wasting food is by putting the idea into practice. Throwing out less food every day provides tangible lessons, which supersede but also dovetail with awareness campaigns on the issue. One straightforward way to minimize cafeteria food waste is by holding recess before lunch. Students can then work up an appetite, and they aren't as focused on getting outside at the expense of eating. That simple schedule shift results in 30 percent less plate waste, according to a study of Washington elementary schools.[vi]

Not surprisingly, more time to eat means more eating (and less wasting). Longer lunch periods provide a major boost to fighting food waste. A 2015 study compared plate waste among students with fewer than 20 minutes for lunch with those who had at least 25 minutes. Students in the former consumed 12 percent less of their entrée, 11 percent less of their vegetable, and 10 percent less of their milk.[vii]

Decreasing the amount of food students throw away at school will certainly help them form healthier habits around wasting food. Yet not throwing away *any* food teaches an even better lesson. That's why redistribution of unopened, unwanted food items – milk, whole fruit, yogurt, packaged produce, or snacks – is essential. The practice can be accomplished by a simple "sharing table" for intra-cafeteria redistribution. Other times, though, those foods can be donated to hunger relief agencies through a not-for-profit organization. Virginia-based Food Bus provides a simple model for sharing excess cafeteria foods with those in need. Their scheme, applicable nationwide, usually supplies an extra refrigerator to participating schools for storing donated food from Monday through Friday, when it is redistributed. Indiana-based Food Rescue has helped schools learn waste avoidance through donation since 2007.

The 1996 Bill Emerson Good Samaritan Act stipulates federal protection from liability when food is donated in good faith. Yet, the urban legend that such donation is forbidden in schools still pervades, prohibiting much participation in school food redistribution. In 2013, the USDA reiterated that donating National School Lunch Program food is not only condoned but encouraged.[viii] That clarification is starting to prompt more and more schools to start donating served but unwanted milk, whole fruits, and unopened packaged foods like yogurt, baby carrots, and apple slices. For example, Food Bus has seen its membership grow rapidly since launching in August 2012 to now include more than 28 schools in five states.

There will always be a decent amount of school food unfit for donation, though. And that is where composting or anaerobic digestion become vital. Providing a third bin for organic waste creates yet another learning opportunity for students and adults alike. For example, all 20 schools in the Chapel Hill-Carrboro City Schools (CHCCS) district of North Carolina are composting in the cafeteria. CHCCS Sustainability Coordinator Dan Schnitzer reported a smooth adoption process, even amongst kindergarteners, despite initial fears that students wouldn't be able to sort their organic waste from recycling and garbage. In fact, the younger students tend to do the best job, Schnitzer said. All it takes is implementing a system, teaching students both how and why to compost, and letting them learn by doing.

Education through composting, whether in a school cafeteria or elsewhere, is powerful. When individuals are confronted with the massiveness of their school or home's uneaten food, it resonates. Such hands-on learning will impact young people and adults. Meanwhile, a K-12 curriculum and education initiative would provide a useful companion to in-school waste minimization efforts. There is a real opportunity to create such a program similar to the one currently piloting in Italy, France, and Britain by the European-based International Food Waste Coalition.

## VALUE OUR FOOD

The most productive way to 'educate for change' involves not just targeting young people, but also reestablishing Americans' gratitude for food. In short, we need to learn to value our food again. Throughout much of America's early history, the mostly rural, farming population certainly appreciated what they produced. And hard lessons learned during the Great Depression and Second World War further cemented that food was not to be taken for granted. Yet those values gradually eroded, as the difficult circumstances of the 20th Century faded in years and memory. And being disconnected from our food's creation has further diminished our appreciation for food. By 1970, 70 percent of Americans lived in urban settings.[ix] It is no coincidence that American food waste has increased 50 percent from 1974 to 2006.[x]

Further exacerbating the wasted food problem, food has become increasingly cheap. Americans spend just 10 percent of personal income on food, less than any other nation and an all-time low for the U.S.[xi] It's worth noting that this is an artificial cheapness, fueled by commodity crop subsidies and not reflecting future healthcare costs stemming from our poor diets facilitated by those federal subsidies. Meanwhile, America continues to produce twice the amount of food necessary on a calories per person basis. We see food everywhere we go, and it's not terribly dear. Those factors combine to create a mindset that we don't have to be careful with our food. And so we aren't, prompting the society-wide wasting of 40 percent of the available food.[xii]

Getting Americans to once again value their food will face the aforementioned barriers. Yet, doing so will be a vital part of the change. If our culture values its food, it will naturally follow that our wasted food will decrease. How might that happen, though? Macroeconomic factors will surely play a role. American (and global) urbanization will likely continue apace, but food won't always be so cheap. Food prices will continue to grow as arable land shrinks and global population increases. Income growth is unlikely to keep pace, so we're soon approaching a time when food will be increasingly dear for all economic classes. That will alter our collective psychological approach to food and prompt less waste.

Rising food prices will prompt more hunger concerns, yet that grave problem can only be solved by tackling poverty. Because nobody wants to increase the number of food insecure Americans – now 49 million! – hopefully cultural changes, not economics, will be the dominant factor in getting Americans to revalue food. That will require a collective return to the kitchen and an increased *connection* to our food. Wasting food becomes much harder when you've had a role in growing or preparing it. And buying food from the person directly involved – yes, following that familiar 'shop local' adage – never hurts. Similarly, improving the quality of the food we eat will help. Sustainably-raised foods are more easily treasured. Whether local, humane, organic, or conventional, the future of meat consumption likely means spending the same amount (or more) for a smaller piece of meat. And as animal protein occupies a smaller percentage of our plates, we'll waste less of it.

Here is where fighting food waste dovetails nicely with the sustainable food movement. Those two concerns have different goals, but shared paths. And if the wasted food movement can learn anything from the sustainable food one, it is to frame its message positively. While sustainable food thinkers often rail against "Big Ag," they also suggest tangible, positive actions. Whether they are buzzwords or ways of life, we can aspire to eat local(ly), sustainably, humanely, or even slowly. Instead of telling Americans not to waste, the messaging must provide a proactive alternative. Convincing a person to do something is difficult; convincing them *not* to do something is near impossible. Instead of 'Don't Waste' it must be more like 'Treasure Your Food.' More 'Love Food,' and less 'Hate Waste.' The environmental non-profit organization Sustainable America is onto something with its website called I Value Food. Similarly, the Food Too Good to Waste initiative hits a perfect tone that is well worth echoing.

## COLLABORATE AND PARTNER

Collaboration is a natural way to leverage increased awareness on the importance of wasted food reduction and translate that awareness into meaningful action. Food recovery partnerships are powerful and result in triple bottom line wins – hungry people get needed meals, environmental impact is minimized, and donor organizations benefit through tax deductions, improved community relations, and improved employee morale.

These partnerships can vary in mission from high-level education to actual food recovery and in size from small to large and local to global. They can also include informal grassroots groups and large public and private organizations. The tangible nature of food recovery work is incredibly rewarding and tends to bring out the best in people, leading to an increased sense of community. Whether food is recovered for people, animals, industrial use, or composting/energy recovery, we all win.

A notable example at the local level involves the Rook Farm, a Pennsylvania sweet corn grower. When approached for a possible donation arrangement with a regional food bank, the farm's owner listened patiently and discussed the operational and logistical issues before commenting that he knew that many people were hurting, and that if his family could help, they would. The operational hurdles turned out to be easily overcome, and the farm recovered tens of thousands of pounds of corn. An adjacent farmer, Bob Solly at Solly Farms, agreed to a similar partnership to donate a portion of his apple crop, initially agreeing to allow a gleaning crew on site to pick them. Epitomizing this can-do collaboration, Bob eventually used his own staff to pick the apples – allowing the food bank to reallocate its resources elsewhere. Later, upon learning that the neighboring farm was donating corn, he used his own truck to move the apples there to facilitate pick-up by the food bank. Bob noted that he had long wanted to grow broccoli but was in need of a special piece of equipment to do so. He decided that the partnership with the food bank was the justification that he needed to make the purchase – and he has been growing and donating excess broccoli ever since.

Rolling Harvest, a Pennsylvania-based food rescue organization, has had tremendous success in partnering with local farmers to capture and redistribute extremely high-quality produce (much of it organic) to pantries, shelters, and senior centers. Rolling Harvest is committed to improving the availability of *nutritious food* at partner sites – so its work has the twin benefit of reducing food waste and improving the health of community members. The organization has collected more than 500,000 pounds of locally-produced fruits, vegetables, and meats since inception in 2009 (more than 2.5 million servings). Perhaps most impressive is the organization's ability to galvanize multiple stakeholders in the community to reduce food waste and put it to the ultimate beneficial use. There is palpable and spreading excitement about its work: donor farms have grown from five in 2009 to 31 in 2015, while hunger relief sites served have increased from 12 to 63.

Partnerships on college campuses are energizing young people (i.e. future leaders) about the importance of food waste reduction as well. Food Recovery Network (FRN) has a simple, powerful focus: utilize student volunteers to capture edible food from college dining halls and deliver that food to local relief organizations. Founded in 2011, FRN has quickly established 160 campus chapters and has recovered more than one million pounds of food in that short time. Its "Fighting Waste, Feeding People" mantra is impactful and is catching on.

The Campus Kitchens Project (CKP), a group under the D.C. Central Kitchen umbrella, links service and leadership to food waste reduction. The organization partners with 49 high schools, colleges, and universities across the country. Student volunteers develop partnerships to recover food and transport it to kitchen facilities, where they organize the personnel to prepare meals and distribute them to those in need. CKP seeks to develop leaders to develop *systemic* change; its philosophy is to *teach* students about poverty, *reach* the least fortunate in the community, *feed* the hungry, and empower young people to *lead* the change. In other words, the organization promotes "student-powered hunger relief." CKP has rescued more than 4.6 million pounds of food since 2001, saving food that would have been discarded and converting it to more than 2.6 million meals for the needy.

Also on the college level, Drexel University's Food Lab, a student-run group in the Center for Hospitality and Sport Management, has partnered with the EPA and a local chain grocery store to create easy recipes from blemished produce – a significant aid for small relief organizations with limited knowledge of preparing healthy meals that are so critical for its constituents.

Several larger partnerships are effectively capturing food that would otherwise go to waste. The Society of St. Andrew (SoSA), for example, runs America's largest gleaning network. Throughout 2014, more than 31,000 volunteers gleaned fruits and vegetables from orchards and farm fields across the U.S. In addition, SoSA organizes truckload drops of excess or cosmetically imperfect potatoes and other produce items in community parking lots, where volunteers bag the material and efficiently distribute it to local pantries in need. The organization provides a valuable service for its partners. For example, SoSA helps truckers, who often have no outlet for rejected payloads, by receiving their cosmetically imperfect goods. This not only helps feed people, but also enables trucking companies to avoid landfill disposal costs. (A startup called Food Cowboy provides logistics support for that same need.) SoSA leverages its logistical strength, partnering with farms, manufacturers, and trucking groups to capture large amounts of excess food in tight time windows. As a result, the organization has captured and distributed more than 23 million pounds of food (about 70 million servings) through the first ten months of 2015. Yet as great as that is, SoSA also displays a running counter of the amount of food wasted annually in the U.S. (currently over 120 billion pounds) – showing just how far we have to go in this area.

Corporations are increasingly recognizing the value in partnerships to reduce food waste. Sodexo, for example, provided initial funding for the Food Recovery Network, and is also a partner with Campus Kitchens Project. Ahold USA forms partnerships to address food waste at all levels of the EPA hierarchy through its Responsible Retailing strategy. In 2013 alone, Ahold donated the equivalent of 12 million meals and diverted more than 19,000 tons of food waste. That same year, the company donated more than three million pounds of protein to local food banks. Ahold USA partners with farmers to divert excess food to animal feed, while other food remnants are diverted to composting and digestion operations.

Panera Bread reduces food waste and helps reduce hunger by donating unsold bread and baked goods to local hunger-relief partners daily through its Day-End Doughnation Program. The company provides clear instructions to make it easy for local non-profits to participate. Panera goes further, building on this work through its partnership with Feeding America and by feeding the hungry through its non-profit Community Cafes.

The Food Waste Reduction Alliance (FWRA) – a collaboration between food retailers, restaurateurs and manufacturers that was launched in 2011 – is working to reduce food waste at is source while diverting excess food to food banks. FWRA member Darden Restaurants, for example, has partnered with Food Donation Connection to donate more than 62 million pounds of surplus food since 2003. Food Donation Connection is an innovative organization that manages food donation programs for food service companies; the group is funded by a small portion of the tax savings that it generates for its donor partners. YUM Brands donates more than ten million pounds of food annually in a similar partnership.

Globally, UK-based WRAP partners with governments, businesses, and charitable organizations to utilize resources more sustainably. WRAP is promoting a goal of cutting food waste in the UK in half by 2025 (a reduction of 15 million tons) and has quantified both the financial savings (to individuals and tax authorities) and the environmental savings in terms of reduced water consumption and reduced greenhouse gas emissions. WRAP has partnered with UNEP and FAO in developing a global food waste guidance tool to assist governments and organizations in developing strategies to reduce food waste.

FUSIONS (Food Use for Social Innovation by Optimizing Waste Prevention Strategies) is a four-year project funded by the European Commission. FUSIONS is a collaborative effort of 21 partners from multiple sectors (business, universities, and knowledge institutes) spanning 13 countries and includes WRAP, FAO, Deloitte, Ahold, the University of Bologna, and Feedback. The group is tasked with creating "a shared vision and strategy to prevent food loss and waste across the whole supply chain through social innovation." More specifically, FUSIONS seeks a more "resource efficient" Europe, with a 50 percent reduction in food waste and a 20 percent decrease in resource inputs to the food chain by the year 2020. The group has received support from numerous organizations across Europe and is leading meaningful change.

Clearly, partnerships of all sizes can and are having an impact on food waste – both locally and globally.

## INNOVATE

Thankfully, there is burgeoning momentum behind food waste reduction at this time. This stems from an increased recognition of both the urgent need to optimize resources to sustainably feed the planet and the many benefits (social, environmental, and financial) that can result from such efforts. The EPA's food waste reduction hierarchy is becoming a well-recognized symbol, and news stories about the level of food waste in developed countries are a daily occurrence. This momentum is fueling a wave of entrepreneurial solutions that warrant attention and support.

Notably, companies such as LeanPath and Winnow Solutions are addressing food waste at the source – the most important part of the EPA hierarchy, helping food service and restaurant kitchens reduce waste through technology platforms that highlight waste. CropMobster helps reduce food waste by providing an on-line portal that efficiently connects farmers and producers with excess food to individuals and non-profits that can put it to good use. AmpleHarvest.org uses technology to efficiently connect American gardeners with excess produce to local food pantries; the gardeners deliver the food themselves, thus reducing food waste without costly infrastructure. Hidden Harvest addresses the millions of pounds of produce that are left unharvested in California fields, employing low income workers at above average wages to glean that food and redistribute it to thousands of needy individuals in the community each month.

These efforts are advancing the well-established efforts of organizations such as City Harvest, D.C. Central Kitchen, Food Donation Connection, and Rock and Wrap It Up to recover and redistribute excess food.

Corporations are finding innovative ways to reduce food waste as well. After implementing a food waste reduction campaign, institutional caterer Bon Appétit was able to reduce food waste by 30 percent in its cafes by educating chefs on portion sizes and prepping techniques, implementing a daily waste management program, and educating consumers on food waste. The company looks for additional opportunities to effectively utilize food scraps, such as animal feed and composting, and looks outward to source "cosmetically challenged" produce from suppliers that would otherwise be discarded through its Imperfectly Delicious program. Within stores, Ahold USA uses blemished fruit and unsold cooked chickens in salad bar items, and reduces the price of day-old bakery goods (and blemished produce) to increase sales instead of discards. Campbell Soup helps its community through an innovative partnership with peach growers and the South Jersey Food Bank. Blemished and irregular peaches that are unsuitable for market are purchased by the Food Bank at low cost for the production of salsa, giving farmers some marginal revenue in lieu of disposal costs. Campbell donates the manufacturing and packaging costs, and its employees donate their time to pack and box thousands of jars of the salsa – with profits from the sales providing a revenue stream for the Food Bank's operations. It's another example in which social, environmental, and financial "wins" are achieved. Finally, beyond all of its food donation efforts, Ahold USA has constructed its own anaerobic digester to produce energy from excess food that is unsuitable for other uses.

Odd-looking or non-homogeneous produce is a facet of food waste in critical need of innovative solutions. Each year we waste billions of pounds of cosmetically imperfect produce – fruits and vegetables with minor blemishes (though still perfectly good to eat) that either never leave the farm or are quickly discarded by food retailers each day due to picky consumers jaded by abundance. Spend a little time observing operations at your local food retailer, or take a look at some dumpsters behind the store, and you'll likely get a sense of the extensive waste of fresh produce that occurs daily. Fortunately, French retailer Intermarché's recent "Inglorious Fruits and Vegetables" campaign elevated global awareness of the vast waste of less-than-perfect produce, and well-intentioned individuals are starting to chip away at this opportunity. Hungry Harvest, launched in 2014 near Washington, D.C., sources surplus produce and provides its customers with a weekly delivery of fruits and vegetables at a price point significantly below that of the average CSA package, along with free delivery. In Northern California, two founders of the Food Recovery Network have continued that ugly-produce-CSA model in a new start-up – Imperfect – which seeks to create a market for perfectly edible fruits and vegetables that are "cosmetically-challenged." The organization offers weekly deliveries of all-fruit, all-veggie, and mixed boxes of imperfect produce at prices that are 30 to 50 percent of that in retail stores. Back in Washington, D.C., Fruitcycle makes healthy snacks from produce that would otherwise go to waste and Misfit Juicery makes, you guessed it, juice from similar produce.

Further innovation is found in California, where Robert Egger's LA Kitchen partners with farmers and produce companies to capture unsaleable produce that they then transform into healthy snacks and meals for the needy in the Los Angeles area. At the same time, LA Kitchen provides culinary training and jobs for community members as part of its mantra that "neither food nor people should ever go to waste." On the East Coast, Daily Table provides healthy food options to residents of a Boston-area food desert through an innovative store concept. Founded by Doug Rauch, Daily Table minimizes wasted food by obtaining donations – often food nearing its sell-by date – from a large number of food businesses and converting that food into healthy (and competitively-priced) options for community members. Finally, British retailer Sainsbury's has collaborated with Google to create a Food Rescue tool that allows consumers to type in the contents of their refrigerator and receive recipe suggestions in return. In addition to instilling a "use versus discard" mindset, the site reinforces the idea that consumers can save money by using up their leftovers. And once again, everybody wins.

Events are another innovative, tangible way to raise awareness and change mindsets. In New York City, Josh Treuhaft's Salvage Supperclub drives home the message of just how much food is wasted by serving gourmet meals in a unique setting. Partnering with a trained chef, the Salvage Supperclub sources food that would otherwise be discarded and turns it into a gourmet meal with outdoor seating in a (clean) garbage dumpster. Patrons get a great meal while getting exposed to the scope of the food waste problem, and proceeds go to food rescue operations in NYC. In the UK, Tristram Stuart's Feedback organization hosts "Feeding the 5K" events throughout Europe and beyond. Upon selection of a site, excess food is captured from numerous local partners, cooking infrastructure is arranged, and numerous volunteers creatively turn that food into a feast for thousands of local residents in a festive, music-filled atmosphere – showing the power of harnessing excess food.

These efforts are having an impact, yet there remains plenty of room for innovation and, hopefully, investment in several areas. Date labeling is in need of a major overhaul. Packaging can be enhanced to extend a product's shelf life. And transportation systems and related cold chain infrastructure are ripe for improvement to minimize food losses prior to market in less developed countries.

## CONCLUSION

When it comes to wasted food, there is near universal agreement that we must change our ways, and do so quickly. The only real question we face is what that change looks like. How much of it will be voluntary versus mandated? How much will be top-down versus bottom-up? And how quickly will those changes have real impact? Regardless of the answers, the changes must be multifaceted, broad and deep, local and global. They must include all stakeholders throughout the supply chain (consumers, producers, retailers, transporters, regulators, governments, NGOs, and non-profits) as there are no mutually exclusive actions in combating wasted food.

Food businesses are in a key position to lead this change, as efforts to reduce food waste can improve their bottom line while also increasing employee morale and consumer opinion. They can also push sustainability initiatives through the supply chain, demanding waste-reduction initiatives from suppliers that can achieve a multiplier effect throughout the food system. Yet, there is also a key role for supportive government policy. Entrepreneurial initiatives to reduce food waste should be supported by responsible, far-reaching, holistic policies that reflect circular thinking. There is evidence that this is occurring in certain parts of the world.

The European Union, for example, recently set a goal to reduce the waste of edible food by 50 percent by 2020. The EU supports FUSIONS, the above-mentioned coalition of 21 organizations across 13 European countries that focuses on food waste across the supply chain. And Expo Milano provided a six-month global stage in which a majority of the world's countries addressed food security issues, including food waste, under the "Feeding the Planet, Energy for Life" theme. Against that backdrop, the U.S. just joined other countries in adopting its own 50 percent food waste reduction goal. For maximum effectiveness, this work can't happen in silos, it must cross borders through a unified global coalition focused on minimizing wasted food.

And, last but far from least, the slumbering consumer in the U.S. (and other developed countries) must be a key part of the solution. We all must heed the scope and scale of food waste and change our wasteful ways accordingly. Overcoming the culture of abundance and our own apathy that leads to so much needless waste will be essential.

These are all critical steps on a sustained path to reducing wasted food: Raising awareness. Educating for change. Properly valuing our food. Collaborating and partnering. Innovating for solutions. If enough actors in the food supply chain heed these factors, we will move from conundrum to clarity on wasted food in America. But clearly, for myriad ethical, environmental, and economic reasons, we need to act sooner rather than later.

## NOTES

[i] Fox, T. 2013. Global Food: Waste Not, Want Not. Institution of Mechanical Engineers. London, UK.

[ii] FAO, IFAD and WFP. 2014. The State of Food Insecurity in the World 2014. Strengthening the enabling environment for food security and nutrition. FAO, Rome, Italy.

[iii] FAO. 2014. Micronutrients. http://www.fao.org/food/nutrition-sensitive-agriculture-and-food-based-approaches/micronutrients/en/.

[iv] Bergman, E.A., N.S. Buergel, A. Femrite, and T.F. Englund. 2003. Relationships of Meal and Recess Schedules to Plate Waste in Elementary Schools. National Food Service Management Institute, Univ. of Mississippi, NFSMI Item Number R-71-03. http://www.nfsmi.org/documentlibraryfiles/pdf/20080225034510.pdf.

[v] USDA Food and Nutrition Service. 2015. School Meals – Rates of Reimbursement, School Year 2015-16. http://www.fns.usda.gov/school-meals/rates-reimbursement.

[vi] Ibid.

[vii] Cohen, J.F.W., J.L. Lahn, S. Richardson, S.A. Cluggish, E. Parker, and E.B. Rimm. 2015. Amount of time to eat lunch is associated with children's selection and consumption of school meal entrée, fruits, vegetables, and milk. J. Acad. Nutr. Dietetics 116(1):123-128. http://dx.doi.org/10.1016/j.jand.2015.07.019.

[viii] USDA Food and Nutrition Service. 2015. Join the US Food Waste Challenge FNS-485, February. http://www.usda.gov/oce/foodwaste/webinars/K-12/Tips_Resources_for_Schools.pdf.

[ix] US Census Bureau. 2012. 2010 Census of Population and Housing, Population and Housing Unit Counts, CPH-2-5. U.S. Government Printing Office, Washington, DC: U.S. Census Bureau. pp. 20–26.

[x] Hall, K.D., J. Guo, M. Dore, and C.C. Chow. 2009. The progressive increase of food waste in America and its environmental impact. PLoS ONE 4(11):e7940. http://dx.doi.org/10.1371/journal.pone.0007940.

[xi] USDA Economic Research Service (ERS). 2015. Percent of Consumer Expenditures Spent on Food, Alcoholic Beverages, and Tobacco Consumed at Home, 2014. http://www.ers.usda.gov/data-products/food-expenditures.aspx.

[xii] Hall et al.

# *Part Three*

## *Food waste reduction, recovery, and recycling*

# Chapter 12

## *Food Handling and Waste Reduction in the Food Industry*

**David Masser, Dave Stangis, Yasmin Siddiqi, and Andrew Shakman**

## INTRODUCTION

In the modern, farm to table food system, food typically undergoes handling and processing before being made ready for human consumption. Food loss and waste occur at every step of the way throughout these procedures. Some of the loss is unavoidable due to natural causes such as dehydration or biological shrinkage. Other losses, attributed to our contemporary food culture, can be redirected to other purposes. For example, remnants at flour mills are generally considered unfit for human consumption but these remnants provide valuable resources for feeding livestock (see Chapter 15). Then, there is the avoidable food loss and waste that can be minimized or even prevented by implementing better management coupled with improved technologies. For any entity in the food industry, be it food manufacturing or foodservice, reducing avoidable food loss and wastage not only fulfills its social and environmental obligations but also improves the business bottom line. Less waste means lower cost, and oftentimes greater profit. Although not widely publicized, many business organizations recognize the importance of this route and are creating ingenious ways of reducing food waste while lowering costs and addressing environmental sustainability. The four case studies presented here serve as models that illustrate innovative ways of reducing pre-consumer food waste.

## "NO POTATO LEFT BEHIND"

### David Masser, President of Sterman Masser Incorporated

Sterman Masser Inc. is an 8th generation family-owned agricultural business headquartered in Sacramento, Pennsylvania. What started as a 100 acre farm, shipping 7 million pounds of potatoes annually, has grown into a company shipping over 300 million pounds of potatoes, onions and sweet potatoes each year, with over 300 employees and 4,600 acres farmed. The company's centralized location (located within a day's drive to over 30% of the U.S. population) has opened doors to growth, innovation, and the ability to provide over 3.2 million four-ounce servings of convenient, pre-packaged product to consumers per day. The company's growth and ability to touch so many consumers has created challenges related to food loss and waste, issues related to resources and food sustainability, and ultimately to hunger and food security in the U.S. At Sterman Masser we have tackled these challenges head-on with a vision and structure that can make a difference in the world and will last for generations. The Sterman Masser Inc. motto for our initiatives is "No Potato Left Behind." Our efforts to reduce food waste can be grouped into four categories discussed below.

**Precision Agriculture and Best Management Practices**

Reducing food waste begins with the implementation of precision farming practices. At Sterman Masser Inc., we are fortunate to have one of the most state of the art, technologically advanced farming operations on the East Coast. Precision farming is the application of technologies and agronomic principles to manage spatial and temporal variability associated with all aspects of agricultural production for the purpose of improving crop performance and environmental quality and reducing waste. Sterman Masser Potato Farms utilizes several conservation tools to ensure full implementation of precision farming cornerstones. To begin, the farm manager schedules *crop rotations* in all 4,600 acres, along with *contour stripping*, to ensure minimal erosion issues. This is key to keeping nutrients within the potato plant and out of the streams, ensuring larger yields of higher grade potatoes. In addition, the use of *low-pressure irrigation* during the hot summer months ensures that potatoes are kept well hydrated, with minimal water use. We are one of the few growers in Pennsylvania employing such water conservation related irrigation techniques.

**Integrated Pest Management**

*Integrated Pest Management* or IPM has enabled us to move away from pesticides as the primary way to produce a pest free potato crop. It has saved the company money, reduced potato loss due to insect and disease infestations, and allowed us to produce a better yielding, damage free crop. Our 100% Good Agricultural Practices certified potato crop is nurtured along during the growing season with the use of *GPS precision nutrient management* practices. This technology enables us to track each and every part of an individual potato field to determine sections of the field where plant health is poor, and then target specific plant health additives to only that portion of the field. Sterman Masser Potato Farms' equipment and machinery line is fitted with this technology, from the potato planters to the sprayers to the actual harvester. With the world's population pegged to reach 9 billion by 2050, our precision farming practices are allowing us to provide more food, with fewer acres and input costs, less pesticides, nutrients, and water – in addition to reducing food waste in the field.

The second major criterion for reducing food loss and waste is directly tied to making the most out of the land you farm. At Sterman Masser Inc., our use of GPS based systems affords us the opportunity to maximize efficiencies in the field and reduce the need for pesticides, fertilizers, labor and equipment. It also allows us to focus in on every single potato in that field to ensure that every potato is nurtured with the highest level of care – thus reducing waste. We use *disease-resistant varieties* of potatoes that are not susceptible to the wide range of diseases that plague Pennsylvania growing conditions. We select well-drained strips for potato production, and even trade ground with other farmers, to ensure that our spuds are grown only in the finest ground. This helps to maximize yields and quality, but we have a long way to go. For 2015, Sterman Masser Inc. has invested in a new self-propelled four-row air separator harvester, rock crushing equipment, and the most advanced tillage equipment on the planet. This technology upgrade (and large capital investment) is all aimed at producing a bruise- and cut-free crop that is highly marketable and produces little or no waste.

**State-of-the-art Facilities to Handle Potatoes**

Sterman Masser Inc. believes that the first key step in reducing food loss and waste starts in the field. Once the potatoes are harvested and loaded on the truck for transport, they head to our environmentally-controlled facilities for storage and processing. While there, significant additional steps are taken to reduce and eliminate food waste. It begins in our extended-life storage facilities, where the spuds are treated like eggs. Temperature, humidity, light, air flow and sprouting are all controlled through state of the art technology and refrigeration equipment. When we say "treat the potato like an egg," we mean it. We try to foster an environment where every Sterman Masser employee values every potato, and handles it delicately like an egg. Taking this to the extreme, we utilize water to transport potatoes out of the cold storage facilities and into our state-of-the-art packing facilities (in both Sacramento, PA and White Pigeon, MI). Once in the packing shed, potatoes move through a series of washers, conveyors, hoppers, grading tables, baggers and palletized stacking, all designed to reduce bruising and eliminate waste. In addition, a recent capital investment into an electronic and visual quality inspection machine has enabled us to sort out poor potatoes at the front end of the plant, immediately redirecting them to the supply chain for which we can maximize what remaining value exists within them.

All of this care and nurturing of our potatoes, from the field to the bag, is critical because of expectations of quality from today's fresh potato consumer. Today's fast-paced families demand their potatoes to be visually appealing, fresh, taste good, and they want us to add value to them. The grocery stores and food service companies that we sell table grade potatoes to have very high quality standards. *If we are unsuccessful in properly growing, harvesting, handling, and packing our potatoes the opportunity for rejection increases dramatically.* This can have a substantial impact on the amount of food waste generated. One thing is clear; every year presents a different challenge and can create more or less off-grade product and waste.

**Processing Facilities for Off-grade Potatoes**

In 2003, Sterman Masser Inc. made the bold move to tackle a tough issue we faced: potato waste volume and the negative impact it had on table grade potato prices. Potato chip plants, as well as packing sheds such as ours, had nowhere to go with off-grade potatoes. Such off-grade potatoes would usually end up flooding the market and depressing prices, or filling up field dumping sites. We knew there had to be a better way to deal with this waste stream. Along came Keystone Potato Products, a state-of-the-art dehydration and fresh cut plant that we have had in operation since 2005. Keystone utilizes the entire potato crop, extends shelf life for potatoes that are not table grade, and allows for the nutrients in "waste" potatoes to ultimately make it to the consumer for their nourishment and health. *Off-grade raw product is shipped to Keystone and turned into potato flake, potato flour, and diced/sliced/peeled potatoes.* Our customers then use these products to create breading, instant mashed potatoes, soups, potato salads, and even potato vodka! In 2008 a third dehydration drum was added to take the off-grade residual potato mash to make an ingredient for pet food! *Over 60 million pounds of waste potatoes are used annually to produce over 10 million pounds of finished dehydrated flake and flour! But it doesn't stop there. After the entire process of dehydrating or fresh cutting potatoes at our plant, we are left with four types of peel and mash waste. Instead of landfilling this food waste stream, Keystone provides it to Pennsylvania's beef and swine industry as a feed ingredient for various feed rations!*

**Maximizing Environmental Stewardship**

In addition to all that Sterman Masser does to reduce and eliminate food waste and food loss, we have also made significant investments into our companies to become more "green." We recently installed an eight-acre solar farm, which produces over 1,100 megawatts of green power annually. This "farm" supplements 40% of the facility energy requirements. Our freight company, Masser Logistics Services, has made significant steps towards becoming one of the most fuel efficient fleets in Pennsylvania. Wind drag reduction technology, auxiliary power units, and increased fuel efficiency have helped us lower costs and reduce spoilage during freight movement. Our previously mentioned dehydration plant utilizes methane gas from a neighboring landfill to fulfill all its steam power needs. And a recently installed water treatment plant is taking 25 million gallons of acid mine discharge-impacted water from an underground mine and treating it, removing all iron and manganese from the water. Anaerobic digester technology is then used to treat the water after it circulates through the plant, returning drinkable water back into a local stream. Finally, over 100 tons of dirt a year is collected from our washing plants and returned to the fields from which it came. As you can see, whether it is recycling metal wickets, poly bags and cardboard in our packing shed, or converting broken pallets into landscape mulch or home heating fuel, we are committed to doing our part to reduce all types of waste within the Sterman Masser family of companies.

"No Potato Left Behind." That is really what we are about. For eight generations the Masser family has been striving to feed people. We take great passion and pride in the fact that at the end of the day, that truly is our mission. We are farmers, and *as farmers, whether growing onions, milking cows, producing pork, or picking apples, our focus has been and always will be to reduce waste and feed as many people as we can.* At Sterman Masser Inc., we utilize every potato, and every part of every potato. Of the over 300 million pounds of potatoes that move through our facilities each and every year, over 84% of them are fresh pack, 10% is processed products, 3% is pet grade and the rest (3%) goes to animal feed. That is quite a success story and something we are very proud of. We are not done, It is our belief that there is much more to do to eliminate food waste, and ensure that food loss from our industry is driven out. We are ready for the challenge.

## "JUST PEACHY"

**Dave Stangis, Vice President, Public Affairs and Corporate Responsibility, Campbell Soup Company**

At Campbell, our businesses and our people are guided and inspired by our Purpose – *Real food that matters for life's moments*. Our Purpose affirms our connection to the core values that have inspired trust in our company for 146 years as well as the evolving values and priorities of new generations of consumers. It's based on three unwavering beliefs. First, we believe food should be delicious, accessible and affordable – all three – without compromise. Second, we believe in the power of food to connect people. Finally, we believe that what we do every day matters; that we have a fundamental responsibility to safeguard the planet's natural resources and help our communities thrive.

Reflecting our firm commitment to sustainability and our heritage as a leading food producer, we recognize that food waste is a significant issue that our industry and our company must address to help ensure a sustainable food supply for the world's growing population.

"Just Peachy" is one of our innovative projects that achieve the dual purposes of reducing food waste and fighting hunger in our communities.

In 2011, the Food Bank of South Jersey met with Campbell's Community Affairs team to discuss its search for new, sustainable revenue streams in the face of growing community needs and cuts in federal funding. A series of meetings led to the genesis of "Just Peachy," a joint program in which Campbell converts undersized or blemished peaches from local farms into peach salsa that is then sold in local stores, with the proceeds going to the Food Bank.

Under the program, the Food Bank buys the peaches for pennies on the dollar from farmers who would have otherwise paid waste disposal companies to haul the peaches to landfills. Campbell donates the manufacturing and packaging costs and our employees volunteer to box and pack the jars for retail distribution. A number of Campbell vendors contribute ingredients and packaging materials.

Sales of "Just Peachy" provide a sustainable source of funding for the Food Bank's hunger relief programs, with proceeds of $250,000 over the last three years, while reducing food waste by thousands of pounds each year. We produced more than 60,000 jars of "Just Peachy" salsa in 2014 and will have produced close to 200,000 by the end of production in 2015. The effort has been recognized by many external groups for its impact, engagement and sustainability.

"It's a win for local farmers, a win for the community, and a win for the Food Bank of South Jersey. The revenue generated from this program will help fund our programs in South Jersey, where thousands of adults and children face hunger every day."

Val Traore, the Food Bank's CEO

In fact, a large amount of food waste generated in the food processing industry is unfit for human consumption, such as vegetable peelings and various remnants. Converting such waste material into energy is highly preferable to the conventional way of dumping it at the landfill. In 2013, a $10 million biogas power plant was constructed adjacent to Campbell's manufacturing complex in Napoleon, Ohio, the home of our largest manufacturing plant. Using cutting-edge technology, the biogas power plant turns potato peelings, tomato remnants, and other organic waste from the manufacturing plant into methane, which is then used to fuel two 1.4-megawatt turbines that generate electricity for Campbell. The biogas facility also uses organic waste from nearby farms and food processors to produce an estimated 22.7 million kilowatt-hours per year of renewable electricity and low-pressure steam. Approximately 450 tons a day of vegetable rinds, eggshells, dairy by-products, and other natural materials are anaerobically digested in enclosed containers to produce the methane. The biogas plant accomplishes two goals: it diverts 35 to 50 percent of Campbell's waste away from Henry County landfills and provides Campbell's Napoleon operation with 25 percent of its electric power needs. Under a 15-year agreement, Campbell purchases 100 percent of the power that the biogas plant generates. We expect the new biogas technology to improve the Napoleon operation's recycling rate to approximately 95 percent.

The "Just Peachy" and biogas projects are two examples of how Campbell is fulfilling our Purpose and expanding our efforts to reducing food waste. Campbell also donates approximately 7 to 8 million pounds of food each year that would otherwise be wasted. The donated food goes to Feeding America, a nationwide network of member food banks that combats hunger in communities across the country. Our other initiatives include separating food waste from packaging after processing, increasing diversion to animal feed, and composting. Our future plans include more comprehensive food waste audits at our manufacturing operations. We are also exploring our regional disposal options and ways to optimize diversion for donations while ensuring that we have enough to support additional biogas capacity. We have made encouraging progress, but we realize that we have more work to do as our company seeks to play an important role in addressing this challenge across the value chain.

## PLASTICS FOR EXTENDED SHELF LIFE

**Yasmin Siddiqi, Global Marketing Manager, DuPont Packaging**

Food loss and waste occur all along the way from farm to table. During industrial processing, distribution, sales and at homes, food wastage can occur due to packaging while the food is being processed or shipped. The $16 billion U.S. food packaging industry has an important role to play to sustainably and cost-effectively reach the goal of ensuring that every piece of food packaged can be consumed and not wasted.

*Packaging can contribute critically in reducing food waste, through protecting food, extending perishable shelf life, and enabling portion control and efficient dosage.* Within this framework. I will share examples of how packaging can help ensure that healthy nutritious food is readily available and reduce food loss and waste.

Let's look at an example from the meat industry where both food waste and sustainability can be improved.

While meat is not one of the most wasted foods, it does have a significant environmental and economic impact. Estimates vary, but on average, nearly 50 pounds of greenhouse gas is emitted for every pound of beef produced that reaches a typical consumer's kitchen.

This means that every time you throw away one pound of meat due to it turning brown or smelling bad, you're throwing away 50 pounds of greenhouse gas – which is the equivalent of burning three gallons of gasoline without going anywhere.

The packaging industry has embraced this issue and today offers a number of solutions. One solution is to use *high performance films* that seal through grease and other contaminants. This helps ensure packaged meat stands up to the rigors of transport, reducing loss during distribution.

Another solution is to use *shrink films and barrier shrink bags* to replace tray-and-lid alternatives. These can keep the meat fresh longer, lower overall cost and reduce packaging waste.

While shrink bags have been historically used to transport meat on the bone from the slaughterhouse to retailers or restaurants, the trend now is to use shrink bags for retail packaging because they improve color, reduce odor and help meat maintain its texture – all while reducing packaging waste.

In the packaging industry we need to strengthen our story and help consumers understand why their meats and other foods are packaged. Who wouldn't opt for the packaging if they knew that it helps reduce waste and protect the environment?

Another key challenge facing the packaging industry is ensuring that we find affordable solutions that protect food and keep it fresher longer. This is especially important when developing *packaged foods for the developing regions.*

In India, 75 percent of milk is collected in an 'unorganized' fashion, leaving consumers at a high risk of buying contaminated or adulterated milk. Parakh Agro Industries in India developed a low-cost method to package ultra-high temperature (UHT) processed milk in aseptic flexible film pouches. *This affordable five-layer barrier film solution helps ensure milk stays fresh without refrigeration for 90 days*. This innovation earned a gold award in the 26th DuPont Awards for Packaging Innovation. Mr. Vinay Nalawade, the director of the Packaging Business said their aim is to ensure nutritious milk can be delivered to both cities and remote areas across India.

The wide geographical area and unavailability of good, refrigerated transportation in India led the innovators to think differently about the challenge and to come up with the solution where the pouch does not need refrigeration and the milk does not require boiling after the pouch is open. This technological innovation will change the lives of many people in the outlying areas and significantly reduce the amount of milk that is spoiled during transport.

For the global food packaging industry, we must challenge ourselves with a goal of ensuring every piece of packaged food can be consumed. We need to institutionalize collaboration as a way of moving forward. By that I mean bringing all the parts of the value chain together to understand and transform market-facing insights into ideas and then into cost effective and sustainable packaging solutions.

## YOU MANAGE WHAT YOU MEASURE: REDUCING FOOD WASTE AND FOOD COSTS

**Andrew Shakman, Co-Founder and President/CEO, LeanPath**

Within the U.S. out-of-home restaurant and foodservice industry, operations typically throw out 4-10% of food purchased before they even reach a consumer's plate. This pre-consumer food waste adds up to at least $9-$23 billion dollars in the foodservice industry every year. At the same time, the environmental impact of the food that's being thrown out is tremendous – consuming precious resources during production and ending up in a landfill, emitting methane, a potent greenhouse gas.

Reducing pre-consumer food waste is one of the best ways to boost an operation's bottom line while running a more sustainable and socially responsible business. Perhaps the easiest and quickest place to cut costs in any foodservice operation is at the garbage bin. The food waste in that bin represents inefficiency, plain and simple, as the operator is paying for each waste item four times:

- on the invoice for the food purchases;
- in labor costs to prepare items;
- for energy and water consumption during prep and clean-up;
- and in hauling and disposal costs.

Pre-consumer food loss and wastage in restaurants and other foodservice places is oftentimes driven by overproduction, spoilage, expired items, excessive trimmings and other culprits. At LeanPath, we have developed innovative systems to help business operators of all sizes cut pre-consumer food wastage and reduce food-related operational cost. Our approach integrates the human behavior and management dimension with technologically-enhanced automated recording systems (Figure 12.1), which can be summarized into six key elements below.

1) **Focus on source reduction:** The Food Recovery Hierarchy diagram created by the U.S. Environmental Protection Agency (Chapter 18) clarifies the order of priority in food waste reduction. Source reduction is the highest priority and involves preventing and minimizing food waste before it is generated. By engaging a foodservice operation to avoid waste altogether, operators can maximize the financial savings from food cost reduction and get the greatest environmental benefit by addressing both the upstream and downstream environmental impact of food produced.

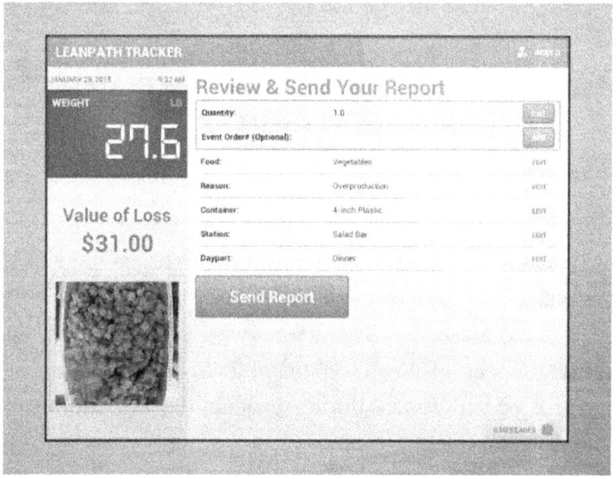


**12.1** Example of LeanPath's automated tracking systems allowing users to see the estimated monetary value of what they are throwing away, which helps raise awareness and change behaviors

2) **Create a waste prevention culture:** Every facility has different menus, customers, and expectations and each requires a custom approach. However, there is one universal principle: *treat the problem as a behavioral challenge rather than a simple checklist* to be implemented or a forecasting issue. Even if an operator had the most complete checklist or accurate forecast, food wastage will still occur. That's because no team executes with robotic precision and customer demand is never totally predictable.

Instead, we must recognize that each person interjects some measure of personal experience, work ethic, prioritization and judgment to execute in a manner that makes personal sense. To be effective, a food waste reduction program requires that every member of a team thinks about food waste every day, in the same way we need them to think about sanitation, safety, customer service and quality. In this way, we need to *put food waste on the "scoreboard" as a key measure of a foodservice operation's success.*

It's important for foodservice operators to speak about food waste in almost every employee communication, starting with new employee orientation and continuing in staff meetings, on communication boards, and with employee recognition. If an operator focuses on the topic, so will his or her team.

3) **Provide positive feedback, not blame:** Want to see a shockwave of anxiety advance through the kitchen? Try walking over to the waste bin, pulling out a healthy head of lettuce, and asking loudly "who threw this away?" Actions like that, whether by a manager or a chef, only create a negative, fearful environment and encourage staff to hide the waste that inevitably occurs. Instead, operators need to train staff to recognize that all large and complex operations have some measure of food waste. There's no reason to assign individual blame when the system itself is often at fault. It's important to get everyone – managers, supervisors, and chefs – to agree not to penalize others for identifying waste, even if a specific person created the waste, Instead, we must *create a blame-free, continuous-improvement mindset in which every team member receives positive feedback for spotting waste and suggesting ways to avoid it next time.* Food waste may be "bad," but staff engagement with reducing food waste is unequivocally "good."

4) **Track food waste every day:** We all know "we manage what we measure" and that the things we measure improve. Operations collect information about customer satisfaction, sales, purchases, food temperatures, cash balances and numerous other metrics, yet most don't have a regular metric for food waste. They may have information from an occasional food waste audit, but lack detailed information about specific waste items, reasons, sources, dates, trends, and values.

When we consider that food represents one of the top two cost items in any foodservice operation, there's a clear need for daily metrics. Without them, reducing food waste remains an untended goal. To address this gap and get the information an operator needs to make changes, facilities should start measuring all pre-consumer food waste on a daily basis. This discipline has roots in process improvement models such as Lean and Six Sigma.

Beyond simply generating useful information, the data collection process reinforces efforts to create a "food waste aware" culture. Operators should make a hard and fast rule: no pre-consumer food waste enters a waste bin without being recorded. They can use a paper-based system or bring in an automated food waste tracking system (which improves speed, accuracy, reporting ease and consistency); but they would be wise to measure, one way or another.

5) **Set specific goals for improvement:** If an operator does not know where he or she is heading, it's unlikely they will reach their intended destination. Once managers have access to detailed, daily food waste data, they should set specific goals for improvement. Operators should communicate these goals to all staff members and share news regularly about progress.

6) **Realize food waste reduction needs ongoing attention.** Food waste cannot be solved permanently in one stroke, nor can an operator expect to maintain gains without ongoing effort. Food waste solutions aren't difficult to develop. The hard task is making sure those solutions get implemented every day despite emergencies, changing priorities, and staff turnover.

Regardless of the type of foodservice operation, it is important to start focusing on food waste prevention without delay. The financial and environmental consequences of wasted food are too great to ignore and too important to defer.

**An Example**

In the past 10 years, LeanPath has worked with hundreds of foodservice operators helping them cut down food wastage while achieving financial gains with simultaneous environmental and social benefits. The outcome obtained at Michigan Technological University is briefly described below as an example; interested readers can find more client case studies at http://www.leanpath.com/case-studies/.

Michigan Technological University (Michigan Tech), located in Houghton, Michigan, has approximately 2,100 students on their unlimited suite meal plan, offered at three residence halls. Many of the dining hall staff members noticed that there was a lot of food waste. To address it, in the fall of 2009, they launched a pilot program at one of the residence halls implementing the LeanPath automated food waste tracking system to track and record all pre-consumer food waste. At the end of the six-week study, they had experienced over a 50% reduction in pre-consumer waste, and made the decision to implement the LeanPath program across all three dining halls as well as at the retail food operation.

Michigan Tech continues to track pre-consumer food waste daily using LeanPath and they review their dashboard reports every Monday. They use the information to discuss the biggest losses, why it's happening and what they can do to fix it. Across the four locations where food waste is tracked, Michigan Tech has cut its waste in half and is saving about $1,000 per week in reduced food costs.

"This has been the single biggest thing that involves all of the people in our department – staff, management and students all have a part in this. It's been a great tool to help us implement culture change."

Kathy Wardynski, Manager of Purchasing and Process Improvement

# Chapter 13

## *Food Rescue to Feed the Hungry*

### *National, Regional, and Local Success Stories*

**Karen Hanner, Cathy Snyder, Steven M. Waldmann, and Leah Oppenheimer**

## INTRODUCTION

The food system in the United States is complex, diverse, and multi-dimensional. Institutional giants e.g. General Mills co-exist with any number of small growers who sell their produce directly to consumers through local markets or roadside stands. The variety of food available to consumers is immense, from fresh and raw to countless processed commodities and brands as valued added products. The food supply chain can be short, straight from farm to table; or long, with multiple layers and players involved in processing, packaging, storage, and distribution before the product can reach the end user. Food loss and waste occur every step of the way in the food system – at the farms, the local markets, the processing sites and manufacturing plants, retail stores, food service places and homes. Some losses are systemic owing to industry standards e.g. grading or customary quality specifications; some losses are irregular and unpredictable due to market conditions and variability; yet other losses may result from legitimate reasons because of rules and regulations that are in place to protect consumer interests, public health, and food safety.

Food that is excluded from the supply chain does not necessarily mean it is unfit for human consumption. Fresh produce out-graded due to appearance (shape, size, or color not fitting set specifications) has the same nutritional attributes as the perfect-looking ones. Food that is pulled out of distribution because of incorrect labeling or approaching the expiration date still retains its healthy, wholesome, and nourishing nature. Leftover items in the field or at the end of farmers' markets can help fill the nutritional gap for low-income families. The most desirable and best use of such food is to feed people, all other alternative uses (e.g. feeding animals, biofuel generation, or composting) should be of lower priority. Fortunately, billions of pounds of food is recovered each year in the U.S., thanks to hundreds of organizations big and small and millions of people young and old who are engaged in various food rescue and hunger relief efforts and activities. Examples of such efforts include the most common and happening-everywhere activities of field gleaning (dubbed "America's 2nd harvest"), the less visible but critically important effort of food recovery from grocery stores (oftentimes organized by food banks equipped with necessary infrastructure and facilities), and the rescuing of mislabeled food commodities for the best use (owing to the tremendous drive and organizational power of Feeding America, the nation's largest food and hunger relief non-profit agency). In fact, the impact of individual as well as organized efforts in food recovery goes beyond mere waste reduction and hunger relief, with additional and profound benefits of connecting people, strengthening communities, and moving information. What follows below are stories of four organizations, each operating at a different scale and platform and offering its unique approach and perspectives.

## FEEDING AMERICA – A NATIONWIDE NETWORK OF FOOD BANKS LEADING DOMESTIC HUNGER-RELIEF

### Karen Hanner, Managing Director, Manufacturing Product Sourcing

Feeding America distributes 4 billion pounds of product annually, of which 2.5 billion pounds is recovered food. This perfectly safe and edible food would have been sent to a less desirable waste alternative, including being landfilled. Feeding America's network of 200 food banks has worked with partners across the food industry since the 1970's to ensure food that might otherwise have gone to waste, is used to feed those in need. Over the years the food banks have developed facilities, partnerships and infrastructure to pick up billions of pounds of donated food, mostly from manufacturers and retailers, and distribute it to agencies serving their communities. Most of this product has been packaged food that is approaching the end of its shelf life. In recent years, a new focus has been placed on rescuing perishable food from all points in the supply chain.

In America, 49 million people, or 1 in 6, are food insecure meaning they do not have consistent access to enough nutritious food to feed their families. In looking to the future, there will be even more challenges to provide adequate amounts of food to feed the growing population. These are urgent issues that the food industry wrestles with. From sourcing food more locally to enhancing crop yields through genetic modification, the viewpoints are diverse. *But the one commonality across all perspectives is that the first step should be making sure we eat what we are already producing.*

It has been the philosophy behind food banking in the U.S. for over 35 years.

In this country, it is estimated that more than 30% of all food produced ultimately does not end up being consumed by people. Some goes to feed animals, some is used to make energy, and some ends up in landfill. All of these paths for uneaten food result in the ultimate waste of the nutrient value of the food and the environmental resources invested in its production.

Donation is a valuable way that many food industry partners use to ensure that safe, wholesome food that is not sold, is still able to be consumed by men women and children so that they benefit from the nutrient value and resources invested in producing it. Donation has the added benefit of helping fight hunger in our communities. Feeding America's combined mission of ending hunger while making sure that any safe, edible yet unsold food across the supply chain is used to help feed families in need fuels innovative and collaborative programs with our industry partners as well as government agencies. Through these programs (briefly summarized with examples below), our network of food banks works with a range of partners throughout the food system in the endeavor of saving food and fighting hunger in this country.

**From the Field:**

- Rescuing passed fields of Minnesota sweet corn

800,000 lbs of fresh, ripe ears of Seneca corn that would have been passed fields due to changes in yield quantities and timing vs. processing capacity were instead captured by Second Harvest Heartland in partnership with Hunger Free Minnesota. The corn was harvested, loaded into totes, cooled and shipped to food banks across the country with the help of donated labor and equipment by Cargill, General Mills and SuperValu. IRS code allowed the farmer a nominal fee to cover the cost of harvesting while still protecting the opportunity for the donor to receive enhanced tax benefits as a donation. Read more at http://goldenvalley.patch.com/articles/general-mills-works-to-repurpose-excess-corn.

- Investing in equipment to sort green beans from field waste

Second Harvest Food Bank of Middle Tennessee has partnered with Hughes Farms, one of the largest green bean producers in the country, to pioneer a sorting, chilling and packing procedure to rescue millions of pounds of beans from going to waste along with field residue resulting in an innovative line reconfiguration that can be expanded to other fresh fruit and vegetable processors. http://youtu.be/2V53LMm2ISM.

- Turning excess Jersey peaches into salsa

When 80,000 lbs of ripe hand-held fruit couldn't be distributed and eaten fast enough by food banks before spoiling, Campbell's stepped in to provide a recipe, additional ingredients, line processing time and glass jars/lids to create Peachy Salsa – a delicious added-value product in a shelf-stable form that extended the shelf life of the product significantly. Also see "Just Peachy" in Chapter 12.

- Packing excess carrots and green beans grown by farmers on unused land

Another innovative program created by Del Monte and the food bank in Madison, WI enables farmers to support their communities by growing vegetables on open land then working with a can supplier and the Del Monte processing plant to pack for the food bank. Additional crops were added as the Field to Foodbank program expanded. Read more at http://youtu.be/ki7OYNdhGB4.

- Using technology to connect surplus produce with food banks

On July 8, 2015, Feeding America launched the Produce Matchmaker (PMM), a new produce supply and demand matching platform, designed to better connect member food banks and donors of fresh produce. Created based on food bank input, the PMM addresses the most common pain points in the produce sourcing system and introduces new functionalities requested to increase speed, transparency, and access to a high volume of produce at the lowest possible landed cost. The Produce Matchmaker has already seen an increase in total pounds of produce sourced relative to the same time last year. For more information please see http://www.feedingamerica.org/hunger-in-america/news-and-updates/hunger-blog/new-produce-matchmaker.html.

**From Manufacturers:**

- Capturing the rejected pieces of peaches due to size and/or color from a canning plant

Del Monte has been capturing the pieces of peaches that are out of spec for their branded product line which had been previously gone into the plant waste stream, and packing them for food banks. The IRS tax code allows a nominal fee to be provided to the manufacturer to compensate for the additional resources needed to package and label the product for distribution to those in need while still allowing the product to be donated and qualify for enhanced tax benefits.

- Providing code date extensions

Many times food is safe and wholesome after the code-date. Additionally, confusion around the wording used to communicate dating leads to food loss. Providing food banks with code dates extensions allows Feeding America to distribute product nearing the end of its shelf life, but still safe and valued.

- Rescuing line waste

By scheduling regular pickups at food production plants, food banks have gained access to ongoing supplies of nutritious food that is out of specification or over production and had been previously directed to animal feed or landfill. Still in food-grade totes, food not subject to USDA inspection is then packaged and labeled by volunteers at the food banks for distribution to clients or to be served at congregate feeding centers.

Innovative programs that have become best practices for companies to follow include:

  - Slim Jim ends and pieces from Con Agra
  - Underweight pizzas from Nestle
  - Underweight granola bars from Kellogg's
  - Cereal overrun from multiple donors

- Episodic opportunities for food diversion from animal feed or waste include line start-up or change-over:

  - Conversion from crunchy to creamy peanut butter
  - Out of spec sliced meats as newly installed slicing equipment was calibrated
  - Mis-shapen tortillas as line starts up

**From the Grocery Store:**

Rescuing perishable food from grocery stores has become a Feeding America food bank core capability as donation programs are now in place with over 16,000 stores representing over 1.2 billion pounds of food that had previously gone into dumpsters. Recent innovative practices by leading retail partners have expanded the categories able to be donated to include rotisserie chicken by creating an in-store protocol that cools and freezes them to fulfill safe handling requirements.

**From the School Cafeteria:**

- Reducing wasted fruit from school cafeterias

A food bank in Rochester NY was approached by the school system because they saw how much food was wasted by kids at lunch time. The food bank worked with a local orchard which was supplying the apples to the schools, to bring apples into the food bank for slicing and packing into individual kid-size bags because children eat sliced apples better! Apples going into the garbage after lunch decreased significantly – because the students were eating the nutritious fruit! Read more at https://image-base.wistia.com/medias/0l2zy8t6pg.

**From Food Service:**

- Each year, billions of pounds of waste are generated from local food service outlets like convenience stores, restaurants, and hotels. Today, very little food waste generated at the local level is being recovered for human consumption, as the foods are generally highly perishable and come in smaller volumes, creating operational and economic viability barriers.

- Feeding America is responding to this, with its newest food rescue initiative, specifically targeting these local food service outlets. Through shared-value partnerships with new donors, Feeding America is investing in the creation of an enhanced food rescue infrastructure within its network of 200 food banks and 60,000 agencies. This new initiative will empower local donor-agency relationships, with the oversight of Feeding America food banks, and enable the sourcing of an estimated 500M incremental meals annually. Investments include Food Safety Manuals and SOPs for local donation programs, and the leveraging of technology to enable food banks to more effectively manage donations and ensure efficient recording and receipting of locally donated pounds. This newest channel of food rescue will expand Feeding America's value as a waste reduction partner for food service and convenience store donors, helping ensure that more food produced within these important sectors is not wasted, but instead, consumed by those in need.

**As a Result of USDA and FDA Recalls:**

USDA and FDA protocols for recalled products have historically always required affected product be dumped or destroyed. In the case of product with allergens missing from the label, Feeding America has worked with both regulatory agencies to establish approved protocol for food banks to accept this mis/unlabeled product and to correctly add the necessary ingredient/allergen information. This protocol has successfully helped divert hundreds of thousands of pounds of product annually from landfill as food bank volunteers follow carefully documented processes established by donating manufacturers. Examples include: granola bars, frozen vegetables, frozen entrees, cereal and others.

As the leading hunger relief organization, Feeding America's work with the USDA/FSIS office has revised previously restrictive labeling guidelines to open new opportunities for donation of product that had been going into the waste stream:

- Economically altered product

New guidelines for donating product that has a missing ingredient or differing proportions of ingredients than declared on the label now require only that correct information and "not for sale" be communicated on bill of lading instead of prior guidelines which required each individual item be labeled accordingly.

- Retailer exemption for bulk product

Feeding America food banks with clean rooms have been approved to accept excess or out of spec product in totes or containers too large to distribute to clients or use in congregate feeding sites and to pack into smaller containers and label for client distribution.

- Out of spec imported produce

Blanket approval has been given by USDA for imported produce that is safe but falls below grade required for the U.S. retail market to be donated to food banks providing an economically preferable alternative to previous alternatives of returning to port of origin or destroying.

Feeding America food banks have numerous examples of working under the new USDA guidelines to rescue large quantities of valuable product from going to waste including pizza, frozen side dishes, beef patties, and canned soup.

**In Summary**

The need to avoid nutritious food going to waste coupled with the need to feed today's hungry as well as the growing population inspires Feeding

America and our partners. Collaboration across the industry has generated innovative processes, many of which are the basis for the best practices shared here. It is through the implementation of what's working now combined with a continued focus on additional innovative collaborative practices that we will achieve these two critical goals.

The opportunity is now for the food industry to fully embrace a commitment to achieve zero food waste goals while simultaneously ending hunger. Insuring unsold food is consumed by those in need through donation is the first step.

For more information on Feeding America, visit FeedingAmerica.org; for food rescue and donation opportunities, contact Feeding America at khanner@feedingamerica.org.

## ROLLING HARVEST – A WORKING MODEL FOR LOCAL FOOD RESCUE, DISTRIBUTION, AND HUNGER RELIEF

### Cathy Snyder, Founder and Executive Director, Rolling Harvest Food Rescue

Rolling Harvest "rescues" a variety of locally-grown foods from farms and farmers' markets and delivers them free of charge to area food pantries and various hunger-relief sites in Bucks County, Pennsylvania and the surrounding region. Rolling Harvest has established relationships with 27 partner farms and markets and more than 48 distribution sites, including food pantries, soup kitchens, domestic violence and homeless shelters, group homes for the HIV-affected, disabled veterans associations, low-income senior housing and community senior centers. More than 80 volunteers donate their time and energy providing a variety of services. Over 2,150,000 servings of food have been provided to food-insecure families within our service area.

**How We Started**

In 2009, Cathy Snyder began to volunteer at a food pantry in Hunterdon County, New Jersey. There, the food items she helped hand out were typically mac and cheese, canned soups, bruised bananas, etc., whereas fresh locally-grown fruits and vegetables were simply out of reach financially for both the food pantry and the receiving families. Soon after, a bountiful farmers' market opened up less than a mile away. Snyder realized that she was able to enjoy the taste and benefits of the market products simply because she had a car and disposable income, a luxury many of the food pantry clients were lacking. Determined to find a way to enable these families to gain access to the fresh and bountiful food produced locally, Snyder founded Rolling Harvest Food Rescue.

Many local farmers and food producers wanted to share their excess vegetables and fruits with area families in need, but lacked the time and staff it would take to ensure effective distribution. On the other hand, food pantries and other hunger-relief agencies often lacked volunteer help to collect the bounty from local farms. It became apparent from the start that approaching farmers at the Farm Markets was not the way; one particular vendor explained that she wanted to help and donate some of her produce, but that she had to hold on to everything she was selling for the next day's market, after which she would be doing her weekly quality control review. If volunteers could come to her farm Wednesdays at 11 a.m., she would be in a position to share so much that was still perfectly good and appealing, but unsalable. We did. And we filled half a van there that day.

That has been our model ever since.

Rolling Harvest Food Rescue now fills this nutritional and logistical gap by collecting fresh (mostly organic) produce and meats from these generous farms when it is most convenient for the farmer and delivering them to hunger-relief sites. Our work is spreading into Hunterdon and Mercer Counties in New Jersey as well.

**What We Do and How**

*On a typical day*, we have two or three scheduled pickups, dispatching volunteer drivers to farms in the most logistically efficient way. The produce is transferred into crates that have been donated by local plant nurseries and then delivered to nearby hunger-relief sites in time for their operating hours. We also get many last-minute calls, emails and texts from farmers who have leftover produce from the previous day's markets, or have picked an excess that they know they will not be able to sell. We then dispatch drivers, often within just a few hours, to do the pickups from these partners and then store in a large, centrally-located donated walk-in cooler for the next day. Having the cooler has enabled us to store fresh produce overnight, and has changed our model to be even more efficient, with greater impact. We are now able to schedule two weekly mass distributions at partner food pantry church parking lots. This is open to all our sites to come collect from us at a convenient time and location, with thousands of pounds of rescued food heading out to an average of 30 sites in less than an hour.

During the harvest season, these *scheduled weekly pickups* and distributions (Figure 13.1) account for about 40% of our annual food volume; another 40% from last-minute opportunities and the remaining 20% from our Gleaning Program. The relationships are based on trust and mutual respect, and our focus is always to be the least intrusive and demanding of farmers' limited time. We have a designated pickup space in the farmers' walk-in coolers or packing areas (we have also been entrusted with the entry codes to many of our food pantry partners).

**13.1** Coordinating distribution of fresh produce

Gleaning in farmers' fields is becoming a growing part of our weekly activities, and is responsible for ever-increasing yields for donation. We have a quick-response harvesting team ready to help within just a day's notice. This is an essential service that ensures no viable food is left in the fields, or is threatened by unexpected weather conditions. Rolling Harvest volunteer gleaners are ready to pick even during a heat wave, frost or right before a storm. We provide the extra labor that the farms typically lack. From our gleaning events has come an extra opportunity to support the dedicated, hard-working farmers – we now offer volunteer "Friends of the Farmers" days when extra hands may be needed for weeding and tending to the crops.

"Every farmer knows that one of the heartbreaks is that not all of the great food that you have sells at markets. Rolling Harvest works really had to make our produce be as beautiful when they deliver it as when they pick it up for donation."

Maria Nocolo, Gravity Hill Organic Farm (Titusville, NJ)

Volunteers are at the core of the operation. More than 80 volunteers now contribute their time and talents during the harvest months from May through December. They provide a wide variety of services – drivers pickup and deliver, help facilitate our nutrition education, set up Free Farm Markets, work in all types of weather for the essential gleaning, handle IT issues, marketing and so much more.

To make our operation and services effective and efficient, it is also important to have a working knowledge of the receiving sites so as to tailor to their specific needs and conditions. Some food pantries are large, with hundreds of volunteers, walls of freezers, refrigerators, adjacent storage areas, while others are bare bone operations with no more than a backyard storage shed, no refrigeration or space for excess food items. Therefore, it is essential to make sure that no site receives more than they can easily distribute. For this reason, we insist that each hunger-relief site we share food with has a back-up plan to partner with a nearby church, senior center, or other similar program that can take what is left over from that pantry's daily operation. This has resulted in impacting even more local families who would not have any fresh food to benefit from, and helps to further our reach and build connections within the community.

Also, by knowing the clients better we can target specific foods (e.g. jalapenos, kohlrabi) that might not have a broad appeal to areas/groups we know they will be most enjoyed. All of these measures are in place to help us not become part of the problem of food waste, especially considering the quality, nutrition and deliciousness we are now able to provide from our farm partners.

**Developing Relationships, Building Communities**

Our partner farms are local small family operations ranging from 5 to 300 acres. Many are financially constrained themselves but passionate about sharing their products with the hungry in their towns. When we began, our focus was to provide the link between where good food is grown and where it is most needed, but we soon realized that our food rescue work means so much to the farmers as well. These farmers, especially the organic farmers whose practice is much more labor-intensive and costly, hate to see the fruits of their labor plowed under, or left to rot, or left behind because of some minor cosmetic imperfection that makes them unsalable. We are humbled by the farmers' gratitude for the dependable services we provide; we make every effort to keep the farmers informed and connected to where their produce has been delivered and for whom it is really making a difference. A few of our 27 partner farmers are now even growing extra for us to distribute to our 48 sites in great need. Farmers are asking us what vegetables are needed, and are growing such in-demand, nutrient-dense foods like broccoli, sweet potatoes, cauliflower, melons, onions, kale and carrots – the very foods that many families really want but can't afford. Word is spreading; awareness is growing. Recently, we were approached by several home gardeners who wanted to turn their lawns and fields into gardens to grow food for donation.

We have also reached out to local schools, engaging high school students in our gleaning opportunities and involving them in transporting gleaned and donated food to food pantries. We have implemented a "fresh produce only" model of school food pantry food drives. Most well-meaning people donate canned or dried goods that are part of a basic diet – foods like canned green beans, pasta, cereals and such. We ask people to dig a little deeper into their pockets and really help food pantries by providing essential sources of nutrient-dense foods like tuna, oatmeal and peanut butter. It is easy and novel to hold a food drive project that consists of providing fresh produce, especially in the harsh winter months when nothing fresh is grown locally. We offer this new model of food drives to schools, religious groups, Boy and Girl Scouts and businesses, who are now donating bags of sweet potatoes, potatoes, apples, oranges, bananas and other vegetables and fruits that can be distributed to families in a short turn-around time and therefore do not need refrigeration.

To reduce our own food waste (excess and leftovers), Rolling Harvest offers weekly Free Farm Markets at several food pantries and community meal programs. Under the guidance of our Nutrition Educator, food pantry families enjoy a real farm market experience, with cooking demonstrations, tasting delicious dishes incorporating that week's seasonal offering, and can take home what they need to easily recreate the dishes with recipes we provide. *Our recipes are designed and developed for financially struggling families, and require no expensive or unusual ingredients.* They are healthy, low fat, low salt, flavorful ideas that have made a big difference in encouraging food pantry clients to incorporate fresh veggies into their diets, with Rolling Harvest supplying the most expensive ingredients – the produce for free. We also give our farmers the opportunity to be guests at our Free Farm Markets. The interaction between "customers" and farmers is rewarding for both the growers and recipients, connecting people to where their food is coming from.

We learned as we grew. In the early days of our food rescue efforts, we would just drop off what was donated. On some occasions, we returned to the same pantry days later and found some of the donated produce still sitting there, wilting and wasting. But now, as we build relationships and expand our operation to include supplying many different community-feeding models, we focus more on finding the right fit. We take the time and effort to label many of the lesser-known vegetables such as mustard greens, kohlrabi and garlic scapes, and have devoted increasing resources towards nutrition education, for example, offering cooking demonstrations for the variety of greens we receive.

**Beyond What We Do**

Nationwide, up to 40 percent of food grown for humans is never eaten but lost in the process, according to a National Resource Defense Council report (http://www.nrdc.org/food/files/wasted-food-ip.pdf). Production losses on farms are greatest for fresh produce, with approximately 7 percent of planted fields in the United States not being harvested each year. The main reasons perfectly edible but unsalable crops are left behind are: (i) Growers may plant more crops than there is demand for in the market in order to hedge against weather, pest pressure and uncertain demand from customers; (ii) Workers are trained to selectively harvest, leaving behind any produce that will not pass minimum quality standards in terms of shape, size, color, and time to ripeness; (iii) Produce may not be harvested because of damage caused by pests, disease, and weather; (iv) Food safety scares, labor shortages, and economics.

At the other end of the food chain, 49 million Americans are food insecure and many families are struggling to put fresh and healthy food on the table due to lack of access and/or financial constraints. As shown in Feeding America's *2014 Hunger in America* survey of participating recipient families among its 58,000 food programs nationwide, client households frequently face difficult decisions and are confronted with choices between paying for food and paying for other essentials such as gas, housing, medical needs, etc. (http://frac.org/pdf/food_hardship_2014.pdf).

At Rolling Harvest Food Rescue, we are committed to helping our partner hunger-relief sites provide better and healthier food choices for their client families throughout Bucks County, PA and into Hunterdon and Mercer Counties in NJ. And we want to be part of the larger conversation about food waste and food-inequality. Let's join forces to no longer have it just be about feeding our neighbors who are struggling financially. In this country of such abundance, let's put more of our efforts and engage more people towards feeding them *well*. Well-nourished children perform better academically; workers with healthier diets can be more productive. Our experience at Rolling Harvest can serve as a working model for others to follow. For more information, please visit www.RollingHarvest.org.

## GLEANING AMERICA'S FIELDS

### Steven M. Waldmann, Executive Director, Society of St. Andrew

The USDA estimate of 133 billion pounds of food being lost each year in America is a gross underestimation because the figure does not include food wasted at the farm level where many more billions of pounds of healthy, fresh produce are left in farmers' fields after harvest for a variety of reasons, or the food that is excluded from sales for human consumption at packing facilities where harvested food goes through another "grade-out" process.

The reasons why this massive wastage of fresh, healthy food is happening are multiple, and the topic has been discussed elsewhere (e.g. Chapters 5, 6, 8 and 9 in this book). A common notion is that American consumers have a *perceived image* of what an apple, a head of broccoli or tomato is supposed to look like. Consider your last or the next trip down the produce isle of the grocery store. What you would not see is variation in size, shape, or other cosmetic factors for the same produce. As a result, fruits or vegetables that do not fit the image of perfection are left in the field or graded out. In addition, occasional market conditions may dictate that it is not worth the farmer's cost to harvest it. Also, unexpected bumper crops can create an oversupply of particular produce types, leading to wastage.

The enormous food wastage presents an opportunity to feed the hungry families in America. The Society of St. Andrew's End Hunger Programs, www.endhunger.org, bridge the very gap between millions of pounds of fresh produce that is otherwise wasted and the millions of Americans who need it.

SoSA organizes a gleaning network that comprises volunteers, growers, and distribution agencies in all 48 contiguous states. Each year, 35,000 to 40,000 people participate in our field gleaning activities; about 20 million pounds of fresh and nutritious food is salvaged this way and given to food pantries or families in need. Our food rescue efforts also include a Potato & Produce Project. We salvage tractor-trailer loads of potatoes and other produce that are rejected by commercial markets or factories due to slight imperfections in size, shape, sugar content, or surface blemishes. Instead of sending these rejected loads to landfills, we intercept and redirect them to food banks, soup kitchens, local churches, etc. Overall, SoSA's End Hunger Programs recover and redistribute 30-40 million pounds of fresh produce annually that would otherwise go to waste.

## BRIDGING GAPS AND CONNECTING PEOPLE USING TECHNOLOGY

**Leah Oppenheimer, AmpleHarvest.org**

### The Situation

More than 42 million Americans are engaged in gardening, growing fruits, herbs, and vegetables. Bountiful harvest creates excess. Excess fresh produce oftentimes rots if not consumed in a timely fashion. Meanwhile, millions of families across America, both those chronically and recently economically challenged, have come to rely on food pantries (also called food shelves, food closets, food cupboards or food banks in some areas) to help put food on the table. These food pantries, over 33,000 of them across the country, operate based on charitable donations (food and money). The food they handle is typically canned or processed products. Every food pantry would love to have fresh and nutritious local produce to help their clients.

### The Innovation

AmpleHarvest.org was born in 2009 to bridge the gap and connect gardeners with surplus fresh and healthy produce with local food pantries. Built on a specialized Google Maps platform, an online registry was created for food pantry listings around the country. Putting in a zip code, city, or address, the user would see all of the food pantries in the surrounding radius of however many miles chosen, much like how people typically search for a movie theater or a restaurant. The directory also shows the pantry's hours for receiving donations, driving instructions, and any store-bought items the pantry might be in need of.

### The Impact

Currently, 7,455 food pantries across all 50 states are registered in the AmpleHarvest.org directory to receive sustainable and recurring supplies of freshly harvested, locally grown food (many for the first time) from area growers. By the end of 2011, food donated via AmpleHarvest.org's technology platform exceeded 20 million pounds.

Now through AmpleHarvest.org, numerous food pantry clients are introduced to new and fresh varieties of food they might have not had access to previously. Children are learning in real life that peas come in pods and not cans and carrots are normally sweet and crunchy, and that apples do not normally come pre-sliced in cellophane. Gardeners across America enjoy the satisfaction of knowing that they are helping their neighbors in need by reaching into their backyard instead of their back pocket. Alongside, the community waste stream is reduced, so is the carbon footprint, as excess food is donated instead of being thrown away.

"I can tell you how much those who utilize the food pantry enjoy the opportunity to take home some fresh produce. The turnips were flying out the door yesterday. Yes, that's right. These are the same turnips (except these are organically grown) that sit unloved in the grocery store produce section. Keep that fresh produce coming. It's good for the whole country because healthier people means less money going to the hospitals."

Katherine Meyers, Newport Food Pantry, CA

"I wanted to let you know that a representative from one of our local farmers/community garden contacted me today desiring to donate excess vegetables to our Fresh Market. He was made aware of us through the AmpleHarvest.org website. I'm very happy that your organization has a presence on the internet, and that you are connecting local farmers and food pantries."

Crystal Robinson, Breakthrough Urban Ministries, IL

"Because of AmpleHarvest.org, our food pantry got lots of lovely, delicious grapefruit the first part of 2012. This cool man even washed & bagged the fruit. And before that I thought no one knew we were in existence. Thank you AmpleHarvest.org."

Patricia A. Wilson Sircy, Meal Mania Food Pantry, FL

**Behind the Scenes**

Backyard agriculture has its grassroots spread deep and wide. "Victory Gardens," a movement led by President Wilson and the First Lady in an effort to help the home front during WWI, was significant enough that by the end of the war, American backyards were supplying 40% of the produce consumed in the country (Barnes, 2013). More recently, backyard gardening again became hip across the nation, a trend propagated in part by the First Lady Michelle Obama as a part of the "Let's Move!" campaign. According to the National Gardening Association (2014), about 42 million Americans, 35% of all households, engage in growing food – either at home or community gardens. This is an increase of 17% since 2008, although community gardening has grown by 200%. Successful first-time gardeners find out how explosive a tomato plant can be, and quickly come to realize that friends cannot be bought with extra cucumbers that need a new home. Unlike an overstock of canned beans, fresh produce has a very short life to be passed back and forth.

Food pantries in America are expansive, totaling 33,500 across the 50 states. For the first time, these hunger-relief agencies are linked with gardeners in the area who have surplus of fresh and healthy food to share with those in need through free service provided by AmpleHarvest.org.

**13.2** WWI National War Garden Commission Poster SHSND# 10935-P107. Source: State Historical Society of North

"Within one hour of registering Community Resource Center on the AmpleHarvest.org website I received a call from a local family of four with 10 orange trees. I spoke with the mother of the family and she said that until she heard of AmpleHarvest.org her family was spending time cleaning up rotten fruit off the ground. Now her family can spend time harvesting fruit to give to low income families in their community. Since speaking with her, she has dropped off 8 large bags full of locally grown oranges."

Sarah, Community Resource Center Food Pantry, CA

AmpleHarvest.org is about connecting people (gardeners, food pantry volunteers and the families they serve) and maximizing the use of resources (food and the natural resources embedded in it) through moving information. On average, food pantries number 670 per state. But only 140 per state have registered with AmpleHarvest.org so far. We envision an America where millions of gardeners share their excess harvest with those in their community, while helping fight hunger and malnutrition in their own community. And we endeavor to connect each and every food pantry with these gardeners. Our goal is ***No Food Left Behind***.

# Chapter 14

## *Food Too Good to Waste*

### *An EPA Project on Household Food Waste Reduction*

**Thomas O'Donnell**

**ABSTRACT**

The *Food Too Good to Waste* (*FTGTW*) program directly messages consumers and food waste prevention advocates while providing specialized tools that people can use to reduce the amount of food that they throw away. Now, after four years in the pilot phase, the results are clear. Small communities of people gathered together to implement, share, and support one another can consistently reduce their edible food waste by up to 60% and their overall food waste by 20% or more. Once mastered, new behaviors for household food management continue to deliver for as long as they are practiced. Behavior change tools help people to shop efficiently, store food properly, manage left-overs, and minimize waste. The environmental, social, and economic benefits of reducing this amount of food waste on a national scale are enormous. For example, benefits accrue to the environment in reduced greenhouse gas emissions, to society by contributing to healthier communities and reducing landfill waste, and economically by stretching household budgets. A family of four that masters the food waste prevention practices taught using *FTGTW* can save $1600 each year. The time is ripe for community groups from around the country to pick up these simple tools, enjoy the group activities required for learning them, and then to reap the benefits of knowing that households and communities throughout the country are collectively making a measurable difference in sustainable food management in the United States.

## BACKGROUND

In the United Sates, 31% of edible food, equivalent to 133 billion pounds, that reached retailers and consumers in 2010 was never eaten (Buzby, 2014). The retail value of this food was estimated to be $161 billion. Consumers were the single largest source of this waste, totaling 90 billion pounds. We spent $110 billion on food that we never ate! Figure 14.1 shows the prevalence and partitioning of food waste in the manufacturing and processing (concentrated), grocery store (scattered), and in the restaurant and household sector (dispersed).

The fact that the greatest amount of waste is dispersed among so many households creates a particularly challenging situation.

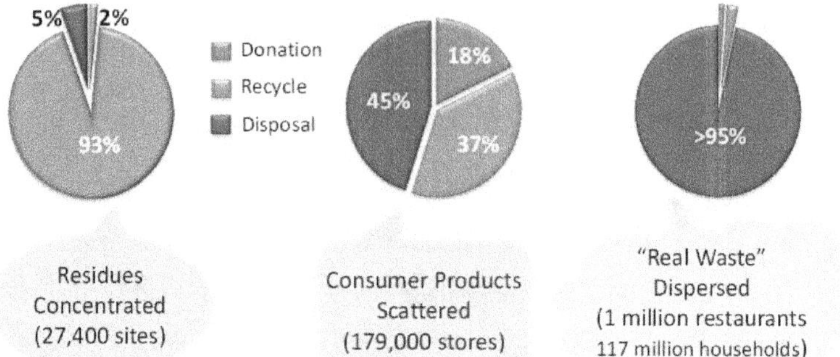


**14.1** Food loss and waste from the U.S. supply chain. Source: Dou, 2014; http://repository.upenn.edu/thelastfoodmile/sessions/session/2/

Clearly, consumer-based source reductions at homes and restaurants are crucial to stemming the problem of food waste in the United States. Recognizing this need, the U.S. Environmental Protection Agency (EPA) created a pilot program designed to learn how to help consumers reduce their food waste. It was evident from the Agency's solid waste characterization studies that a focus on reducing the amount of food reaching landfills must target waste that comes from households, otherwise we lose an important opportunity to encourage sustainability and decrease health and environmental impacts of food waste. The USEPA *Food Too Good to Waste* (*FTGTW)* was the name given to this pilot project.

## PROJECT DESCRIPTION

### Targeting Household Food Waste

The environmental, social, and economic impacts of food waste are most effectively addressed by source reduction or avoiding waste in the first place. However, proven methods for prevention are not necessarily obvious, easy, or quick, which is the reason the *Food Too Good to Waste* Program offers fresh opportunities in this most important area of food waste management.

Drawing on experience and insight from household waste prevention programs in the United Kingdom[1], the *Food Too Good to Waste* initiative for household waste prevention was born.

---

[1] The United Kingdom's *Love Food, Hate Waste* http://www.lovefoodhatewaste.com/

The initial purpose of the *FTGTW* program was to help consumers make small changes in the way they manage their food purchases so they would not waste as much. A community-based social marketing (CBSM) approach was selected as the foundation for the program.[2] Strategic objectives are achieved by introducing new activities and tools that remove barriers and encourage desirable new behaviors. Changes were sought in the ways that people shop, prepare and store food, and manage left-overs.

**Pilot Projects**

Beginning in 2012 and continuing today, pilot projects from around the country are collecting data and information about the performance of the initiative. Work by Buzby (2014) demonstrated where the problems existed and that most of the loss was food staples (Table 14.1). These losses should be avoidable. The early objectives for *FTGTW* were to determine if the targeted behavior changes are occurring and to what extent they resulted in less waste. Secondary objectives consisted of estimating costs to implement individual community projects, environmental benefits, and how projects can be integrated into existing programs.

**Table 14.1 Consumer Food Losses by Type of Food. Source: Buzby et al., 2014**

Estimated Percent Food Loss in the United States, 2010

| Commodity | Losses from Food Supply* | | |
|---|---|---|---|
| | Retail | Consumer | Total |
| | | Percent | |
| Grain products | 12 | 19 | 31 |
| Fruit | 9 | 19 | 29 |
| Vegetables | 8 | 22 | 30 |
| Dairy products | 11 | 20 | 31 |
| Meat, poultry, and fish | 5 | 22 | 26 |
| Eggs | 7 | 21 | 28 |
| Tree nuts and peanuts | 6 | 9 | 15 |
| Added sugar and sweeteners | 11 | 30 | 41 |
| Added fats and oils | 21 | 17 | 38 |
| Total | 10 | 21 | 31 |

Another approach tested by the EPA with *FTGTW* used newer methods to activate more robust consumer behavior change. Community-based social marketing methodologies (USEPA, 2012) were integrated into the program in ways that would help remove barriers to changing food handling behaviors while helping people appreciate and embrace new ways of managing the food that they

[2] The West Coast Climate and Materials Management Forum

purchase. The tools that are described in this paper are an outcome of this strategic approach. CBSM tools are not just used for awareness and information but also to evaluate effectiveness. Questions such as the following can be answered: was there public commitment, were prompts followed, was social media leveraged, was the delivery effective, did the program operate at the community-level, was there direct contact with people involved, and did the toolkit work to produce results and change behavior?

**Program Development**

*FTGTW*'s main message is that "Food is too essential to be thrown away." Specific outreach tools are available for education and for organizing workshops to explain a series of tools to give people options for working on new behaviors (Table 14.2). People are asked to measure their food waste using tools and materials provided to them. This step is important because nearly everyone tends to underestimate the amount of food they throw away. After measuring their food waste, participants are shown simple tools to shop smarter, store food properly, prepare food to last longer, and how to enjoy leftovers. Households commonly experience lack of planning in the different foods they purchase and then learn of their other automatic habits that lead to waste, for example, eating preferences of some of the family members or dislike of leftovers. Cosmetic changes to fruits and vegetables are an everyday occurrence; a problem that can be reduced with new knowledge about proper storage of perishable foods. Armed with new awareness of how much and why food is wasted in their own homes, consumers can use any of the tools they like to begin wasting less.

**Table 14.2 *FTGTW* Strategies, Benefits, and Barriers**

| Behavior or Strategy | Individual Benefit | Barrier |
|---|---|---|
| Get Smart: measuring your food waste | Waste aversion | Time, habits, 18-30 age group with variable eating habits |
| Smart Shopping: buy what you need | Waste aversion and saving money | Time, habits, 18-30 age group with variable eating habits |
| Smart Storage: keep fruits and vegetables fresh | Waste aversion, healthy eating, saving money | Time, knowledge, space |
| Smart Saving: eat what you buy | Waste aversion | Gratification, knowledge, convenience |
| Smart Preparation: prep now, eat later | Convenience, healthy eating, saving money | Knowledge, skills, space |

## Behavior Change Toolkit

Suggestions for completing a *FTGTW* project in any community are provided in detail at the West Coast Climate and Materials Management Forum's website.[4] The tools can show a person why food is wasted in their homes, how to reduce the waste, how to recover foods for composting or reuse, and how much money is saved in the process.

**14.2** Templates used in the Behavior Change Toolkit

Templates (Figure 14.2) are used to introduce each of the tools needed to understand food management behaviors and how to change them for the better. Developing a shopping list involves a couple of simple tasks that help identify what is actually needed for the meal purchases and what is already on hand. Buying only what you need saves money and using what is already in the house is gratifying. This is a good way to keep home pantries and refrigerators tidy and organized.

Not all fruits and vegetables should be stored in the same way. Some are best kept in the refrigerator, others in just a cool place. Some vegetables stay fresh the longest if they are kept in water just like other plants. The Smart Storage strategy explains while showing results right away because they see their produce staying fresh longer.

The Eat Me First tool is based on organizing and labeling. Too often uneaten foods get lost in a refrigerator and spoil. This fun container label adds a touch and feel to the idea of putting leftovers in a single place so they are noticed and eaten before they go bad. The week-end refrigerator clean out that leads to so much wasted food is thereby no longer as prominent.

There are two basic units of measurement that reveal the story. The first is the amount of food, edible and non-edible, that is thrown out each week, and the grocery bill. Throwaways are weighed or the volume is estimated using the bin-scale shown in Figure 14.2. That label can be sized to fit on any container. The weekly edible food lost teaches what is actually going uneaten and what can be used for composting or other reuse. The cost of the wasted food can be estimated based on these measures to help people understand how much money they can potentially save by using some or all of the project tools. People are asked to measure their food waste in this way each week while the project is going on. Four to five weeks of data have been shown to be adequate to make the point and encourage people to continue using their new knowledge, skills, and these tools after the project is finished.

## CASE STUDIES

*FTGTW* was piloted by seventeen different organizations in 15 communities in the United States between 2010 and 2014.[3] Table 14.3 summarizes some of the characteristics of the test groups. This project diversity

**Table 14.3 Test Groups that Piloted the *FTGTW* Program[4]**

| **Number of pilot project campaigns** | 17 |
|---|---|
| **Time period** | 2012 through 2014 in all seasons |
| **Partners** | Community solid waste departments and non-profits including housing groups |
| **Location** | Urban and rural across the United States |
| **Project Scale** | Small pilots with some significant multi-media |
| **Target populations** | Families with children, young adults, general population, various income categories |
| **Outreach and engagement** | Varied with partner resources, all involved direct engagement |

[3] The West Coast Climate and Materials Management Forum

helped clarify how people in different environments and locations would respond, react, and report the benefits of their efforts. Importantly, the time spent on pilots helped to further the objectives of learning how to achieve an overall, scalable, and long-lasting reduction in food waste at the household level. Two of the projects are discussed in this report. King County, Washington was one of four initial adopters of the pilot in 2012 and has completed three years of work. Refinements and experiments tried by this group offered valuable experience to more recent participants in the program. The Rhode Island Food Policy Council, for example, initiated their first pilot in 2014; hence, was able to apply lessons that were learned by earlier practitioners, while providing new insights from working in a multi-lingual community with a higher and lower income residents. Evaluation of each pilot will be available at the West Coast Climate and Materials Management Forum as the evaluations are completed in 2015, although some of the results of early pilots are currently available.[4]

**Case 1: King County, Washington**

King County, the home of Seattle, contained nearly 790, 000 households according to 2010 census data. These were made up of about 68% white, 6% black, 15% Asian and 9% Latino families. The median family income was $87,010; whereas per capita income was $29,521. About 10% of the population were below the poverty line, significantly better than the national average. Over 90% of residents have a high school degree or greater.

The first *FTGTW* project was started in 2012 with 47 families with children in 4th grade that were recruited by email. EPA partnered with an elementary school and a consulting firm that specializes in social marketing programs. The pilot took place over five continuous weeks. The first week was used to measure the amount of all food that was wasted, setting a baseline for subsequent comparison. Food waste reduction strategies were then introduced in the second week. One tool was introduced each week thereafter with the help of daily tips given by the teachers. By the end of the five weeks families were *wasting an average of 30% less food*. However, many of the households did not make measurements for the entire period. Only 15 out of the 47 initial households completed the 5-week project. Several lessons were learned, the most important is that people will reduce their food waste if prompted, incentivized, and encouraged. It was demonstrated that children will bring home the message to parents. The group also learned that the length of the pilot is important and that

---

[4] FTGTW pilot project summaries: http://Pilotprojectsummaries.com/sites/westcoastclimateforum/files/related_documents/pilotdescriptions.pdf

for many families five weeks may be too long and actually not necessary to get the message across and the tools introduced and used. Participation was lower than expected, which may have resulted from a sense that the measurements and reporting were complicated. Overall, this early pilot led to modifications in many different elements of the program including recruitment, ease of using tools, making measurements, and reporting.

In 2013 the group tried another approach designed to see if participation could be scaled up to many more people. An extensive website specifically about the project was created that included all of the strategies, tools, and templates. This was supported by an ongoing social media campaign. Residents were encouraged to register on the site, work through the tutorials including video showcases, and to register their baseline and results. In spite of all of the effort expended in this way there was little documented participation. The results suggested that *personal intervention is essential*; people working with people at least at the early stages of a project like this seemed to be a requirement for any meaningful success.

In 2014, the campaign involved more direct contact with people who might be interested in a pilot and regular support to those who chose to participate. A *cash value gift certificate* was offered to people who completed the entire program. Outreach recruiting at grocery stores and farmers markets tapped into interested and engaged residents. Anywhere from 200 to 250 people could be reached at venues like these on a single day. Cohorts of 40 persons or less were involved and the program was completed in four weeks. The retention rate was high (75%) because of a structured intervention using weekly emails, short weekly surveys with prizes, new video links that were interesting to watch, and support from morning talk shows on the local NPR affiliate, and other media support from NBC, Fox, and newspapers. Fifty-three households participated and reported *27-39% reduction of edible food waste*. The 2014 outcomes reinforced conclusions from the previous two years and added new insights to developing a more effective program. For example, *incentives were necessary to enlist participants*. Support from the project team was intensive and costs can become significant. A shorter 4-week project timeline helped to keep people involved without reducing the desired outcome of less food waste.

**Case 2: Rhode Island Food Policy Council**

The Rhode Island Food Policy Council was launched in late 2011 for the express purpose of improving the "capacity, viability, and sustainability" of the local food system. Members of the Council's Healthy Environmental Work Group partnered with the EPA in 2014 to pilot *FTGTW* among 40 households for the purpose of reducing food waste. Based on earlier pilots, the project leaders decided to do a practice run within their working group so they could identify any parts of the program that should be customized for Rhode Islanders. Following that phase, which got feedback from 10 households, a one-year project was implemented with people from their networks listserve – a group of people expressing some interest in food issues, and households from both an upscale apartment complex and the lower income Providence Housing Authority.

Implementation consisted of two weeks of baseline development followed by four weeks of implementation. Recruitment was accomplished by just networking because the State is so small. *FTGTW* was also translated into Spanish to help reach bilingual families. Interactions were very personal with much of the program content presented face to face early in the pilot at gatherings that included cooking demonstrations. Overall each group met three times. The first meeting at week 1 started the pilot and also incentivized families by giving them their own scale and bin for the weekly food waste measurements. They were also given their own code number for entering data, which gave a sense of uniqueness and relevancy to the project. Two weeks later everyone came together to share experiences with their baseline measurements of food waste, ask questions, and for introduction of the program tools. The group met once again at the end of the project to talk, tell stories, and discuss how to continue with the process on their own.

Some anecdotal results included an awareness of the sensitivity of food insecure families that don't always have enough to eat. People in these test groups preferred the Eat Me First, Prep Now Eat Later, and Smart Shopping tools. People were very interested to learn about the environmental benefits of the program particularly as they relate to landfill issues. They also shared how they liked networking with their friends on their own because the project was useful, important, and fun. That enthusiasm was so high that the Working Group is designing a train-the-trainer model in the hopes of steadily increasing the program – Rhode Island does have a zero-waste goal.

As an outcome, participants in both the upscale and lower income demographics *reduced their edible food waste by 48-55%* (Figure 14.3). Combined with the early "friendly and listserve participants", average reductions among this diverse set of households was a noteworthy 60%. Retention levels were high and the progression of reductions each week was generally consistent. Figure 4 shows a declining waste trend towards 0.5 pounds per person per week after only 6 weeks.

This group also spent some time and effort to understand what portion of the wasted food was avoidable or preventable. The distinction is that preventable food waste is edible, perhaps representing uneaten potions or extra food that was not even served. Unavoidable food waste represents parts of food that would not be eaten anyway, for example, some bones, prep scraps, or damaged produce. Although a small sampling of participants, these measurements reported that 64% of food waste was likely preventable.

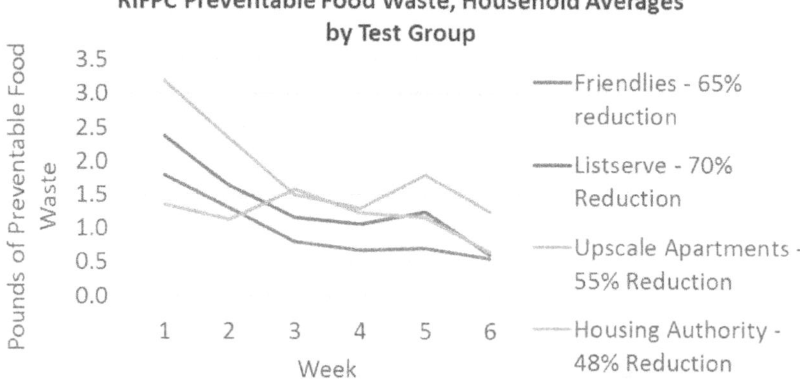


**14.3** Edible food waste reduction among Rhode Island *FTGTW* pilot project participants[4]

## DISCUSSION AND CONCLUSIONS

It is evident after three years of *FTGTW* pilot projects that meaningful amounts of food waste reductions can be accomplished at the household level. This is very encouraging from an environmental, social, and economic perspective because households are the leading source of food waste in the United States (Buzby, 2014). Although the 17 *FTGTW* projects (Table 14.4) have only been with small groups of people, they have been completed across the country with community groups that represent considerable geographic, income and cultural diversity.

All of the projects were successful in providing new insights into the way behavior change programs that reduce food waste can be adopted. *Household reductions of up to 60% of edible food and 15-25 % of overall food waste are consistently demonstrated.* Even with just 0.5 pound of food waste reduced per person per week the impacts nationwide would be very impressive. For example, if 25% of the population wasted 0.5 pound less edible food each week the impact would reach 2,047,000,000 pounds per year. In theory, that represents enough to feed the 49.1 million food insecure people in the United States with one meal per day for over a month. The USDA has estimated that reducing food waste in the range discussed could save a 4-person household $1,600 per year, money that could be injected into the economy in any number of ways. From the perspective of greenhouse gases, taking this amount of food out of landfills could reduce methane emissions by thousands of tons. Helping people value their food by wasting less does not appear to have any significant down-side; on the contrary, new behaviors to sustainably manage household food enhances social and environmental goals and helps bring people together around an issue that is fun and important.

**Table 14.4 Communities and Organizations Participating in *FTGTW* Pilot Projects**

| | |
|---|---|
| **King County, WA** | **San Benito County, CA** |
| **Boulder County, CO** | **Seattle, WA** |
| **Honolulu, HI** | **Santa Monica, CA** |
| **Oakland, CA** | **Chula Vista, CA** |
| **Gresham And Or Metro, OR** | **Thurston County, WA** |
| **Oak Park, IL** | **Iowa City, IA** |
| **Minnesota Pollution Control Authority, MN** | **Rhode Island Food Policy Council, RI** |
| **University Of Denver, CO** | **State Of Vermont** |
| **Sustainable Jersey City, NJ** | |

Much has been learned about how to be successful with a *FTGTW* community-based project.

Importantly, the program can work in all types of communities because it is the personal interactions that are important. As shown by the King County, Washington and Rhode Island Food Policy Council pilots, the outcome – reducing food waste – is always positive. Preventable food waste generation trends, normalizing for more waste when fresh produce and food is eaten, are consistently downward. How the longer-term behaviors and outcomes fare remains to be tested but a similar program in the United Kingdom, which has been operating since 2007, demonstrates continued success. In fact, the UK *Love Food Hate Waste* program is responsible for reducing household food waste to such an extent that it has been measured on a national scale (Quested, 2014). Once people adopt new behaviors that improve their way of life they continue using them. The pilots were reported to be manageable and there has been no negative feedback from participants. Responses also show that people are interested in saving money but also in learning about the environment and how they can contribute to local issues like landfill waste management and lessening global concerns like climate change.

A lot of lessons were learned about starting a *FTGTW* community program and recruiting the optimal participant group size. Organizers have found that outreach and workshop events create lively conversations and participation because the food topic is so generally interesting. The key is to raise awareness by becoming involved with the community directly and often. Presenting the strategy and tools requires attention and enthusiasm but once individual barriers to change are recognized people will select a tool(s) and try it out. The practices are quickly learned and so are the results. Individuals tend to continue when they earn postive feedback. Project leaders are very effective when they know about things that are important to the participants and introduce information and tools that match. Sometimes it is a matter of helping with recipes for culturally preferred ingredients and meals. *FTGTW* has all of the elements for success that become even more effective when individualized as shown by the Rhode Island Working Group projects. Based on the pilots, new projects can and should be successful but possibly at different levels.

Different ogranizations estimate different amouts of food waste in households, an admittedly difficult metric to resolve. The EPA estimates that a household generates about 2.5 pounds of food waste per week. *FTGTW* household pilots have measured similar amounts. Measuring edible versus non-edible waste varies based on the definition and technique for measurement, for which there is no standard. The Rhode Island group combined edible and probably edible foods and reported that this represented 64% of the waste stream during their pilot. It is entirely reasonable to predict and expect success and even set a waste reduction goal of 50% of edible foods, or even reach for 0.5 pound per person per week, if that makes sense to the participants. Having people actually measure their waste by weight or volume is important; otherwise the goals and measures can't be quantified. This step also gives participants a lot to talk about throughout the project.

Every group was interested in trying to estimate a budget for their project. The costs of course vary with the type of program with higher values associated with indirect activities like website development and management or extensive presonal outreach. Figure 14.4 shares a budget provided by the Rhode

Island Working Group that helps to illustrate several factors to consider in budget development for cohorts of people in the 40 person or less range.

| **Labor Support:** | | | |
|---|---|---|---|
| **Labor Category** | **Labor Hours** | **Hourly Rate** | **Price** |
| Network Coordinator | 14 | $42.50 | $595.00 |
| Outreach Coordinator | 17 | $40.00 | $680.00 |
| FTGTW Project Coordinator | 69 | $25.00 | $1,725.00 |
| **Total** | | | **$3,000.00** |
| | | | |
| **Materials and Supplies Support:** | | | |
| **Printing of CBSM Tools for 40 particpants:** | **Each** | **Qty.** | **Total** |
| Food storage guide (2-sided) | $0.50 | 80 | $40.00 |
| Shopping list (2-sided) | $0.50 | 80 | $40.00 |
| Stickers for hard containers | $2.00 | 40 | $80.00 |
| Weigh to reduce instructions (1-sided) | $0.50 | 40 | $20.00 |
| Tracking sheet bag instructions (1-sided) | $0.50 | 40 | $20.00 |
| *Printing sub-total* | | | *$200.00* |
| | | | |
| **Incentives for 4 cohorts of 10** | | | |
| Measurement containers | $5.00 | 40 | $200.00 |
| Scales | $30.00 | 40 | $1,200.00 |
| Gift cards (only for 3 cohorts, not pre-pilot) | $25.00 | 40 | $1,000.00 |
| Food for kick-off & wrap up workshops | $50.00 | 8 | $400.00 |
| *Incentives sub-total* | | | *$2,800.00* |
| **Total:** | | | **$3,000.00** |

**14.4** Example of a *Food Too Good to Waste* pilot budget. Source: Rhode Island Healthy Environment Working Group, 2014

Project team labor costs are shown for the six week effort in an amount that coincendentally matches costs for incentives. Printing costs were relatively low. The greatest labor costs were for the coordinator who spent nearly full time on the project. Materials and equipment costs were the largest for scales and gift cards, which were incentives. This project relied on networking to gather recruits.

Projects that required aggressive outreach for recruiting have experienced large labor costs although still completing the project for under $15,000.[5] The USEPA *Food Too Good to Waste* Program, currently available through the West Coast Climate Action and Materials Management Forum, promises to be one tool that can have a significant impact on sustainable food management. The foundation for the program is at the consumer, household level, which has the additional impact or raising awareness of important issues to the public in great numbers. In closing, the co-leadership of the Last Food Mile Conference concluded that three initiatives targeted directly at consumer food waste should be stressed: improving understanding of consumer food behavior, better measurement of food waste, and setting reduction targets (Dou, 2014). Although some of these objectives apply to all food waste sources, reducing household food waste has lowered national scale food waste disposal in the United Kingdom. The first indication of the same outcome has recently been reported for the United States where food waste reaching municipal landfills in 2012 decreased for the first time in history (EPA, 2014). Perhaps this can become a trend led by households.

## REFERENCES

Buzby, J. 2014. Overview of Food Loss in the United States. *The Last Food Mile Conference.* http://repository.upenn.edu/thelastfoodmile/sessions/session/34.

Buzby, J.C., H.F. Wells, and J. Bentley. 2013. ERS's Food Loss Data Help Inform the Food Waste Discussion. Economic Research Service, U.S. Department of Agriculture. *Amber Waves*, June 3.

Dou, Z. 2014. Conclusions and the Road Ahead. *The Last Food Mile Conference.* http://repository.upenn.edu/thelastfoodmile/sessions/session/2.

King County. n.d. http://your.kingcounty.gov/solidwaste/wasteprevention/too-good-to-waste.asp.

Quested, T. 2014. Household Food Waste: Quantification, Understanding and Reduction Campaigns, abstract. http://repository.upenn.edu/thelastfoodmile/sessions/session/22/.

U.S. Environmental Protection Agency. 2012. Food: Too Good to Waste Pilot, EPA 910-R-12-006.

U.S. Environmental Protection Agency. 2014. Municipal Solid Waste Generation, Recycling, and Disposal in the United States Detailed Tables and Figures for 2012. February 2014.

---

[5] See Table 11 in http://www.epa.gov/osw/nonhaz/municipal/pubs/2012_msw_dat_tbls.pdf

http://epa.gov/epawaste/nonhaz/municipal/pubs/2012_msw_dat_tbls.pdf.

WRAP. n.d. http://www.wrap.org.uk/content/solutions-prevent-household-food-waste.

**Disclaimer** This article was written in the author's personal capacity, and not on behalf of the U.S. Environmental Protection Agency (EPA). Any views expressed in this report do not necessarily represent those of the United States government nor the EPA. Mention of trade names or commercial products does not constitute endorsement or recommendation for use.

## *Chapter 15*

# *Food Waste as Animal Feed*

**James D. Ferguson**

## ABSTRACT

In 2012 the U.S. disposed of 227,603,579 tonnes (250,890,000 U.S. tons) of waste to municipal solid waste sites. Food waste was 14.5% of the total waste disposed. This waste was largely from households, retailers, and restaurants and represented edible food refusals (plate waste), peels, seeds, pits, bones and trimmings, outdated and spoiled food, and edible food rejected for aesthetic reasons. However, food waste occurs all along the food chain from unharvested crops and fruits, transportation losses, and losses during food processing in addition to losses from end users and retailers. A significant amount of processed waste from manufacturers is utilized as animal feed. About 13.6% of edible food produced is used in animal feeds, the largest component represented by cereal grains. By-products of the food processing, distilling and brewing industries comprise about 25% of animal feeds, providing important sources of energy, protein, and fiber. Reducing waste across the food chain will entail a concerted effort to improve harvest, processing, retail sales and restaurant and home use.

## INTRODUCTION

In 2012 the U.S. disposed of 227,603,579 tonnes (250,890,000 U.S. tons) of waste to domestic municipal solid waste (MSW) sites. Food waste was 33,048,740 tonnes (36,430,000 U.S. tons), 14.5% of the total. A significant amount of non-food solid waste is recovered in recycled products, 34.5%, whereas only 4.8% of food waste was recovered from waste that was delivered to landfills, increasing the proportion of food waste deposited in landfills to 21.1% of MSW. Food waste represents a significant loss of edible food and a significant burden on MSW sites. Environmental consequences include the physical space required for land disposal, greenhouse gases produced from decomposition of food waste, and potential seepage of organic effluent if landfills are not properly lined. Methane from decomposition of waste may be recaptured at landfills and utilized for useful purposes. The USDA Food Recovery Hierarchy prioritizes recovery of food from farm to fork (www.epa.gov/foodscraps). In the United Kingdom a concerted effort has been made to reduce avoidable food waste from consumer households to reduce the burden on MSW disposal sites. From 2007 to 2012 household food waste was reduced by 21%, 1.3 million tonnes (1.43 million U.S. tons), through a concerted consumer education program conducted by the Waste & Resources Action Programme (WRAP).

Food waste needs to be categorized into sources and type. Food waste can be generated by the food processing industry, which produces many human inedible food residues, and by the retail, restaurant and household sectors, which produce both edible and inedible food waste. The food processing industry produces co-product residues only partly consumable by humans, such as pomace and pulp from juice and sauce production, high fiber residues from flour, starch, and nut production, high protein meals from oil production, residues from the brewing and distilling industries, and animal renderings from meat production. Other products associated with food processing that are unacceptable for human consumption include stems, leaves, and pits which may be collected for animal feed or disposed of through land application or landfill disposal. A major amount of co-products produced by food processing are extremely useful as animal feed and are only partially acceptable for human consumption. In addition in the food processing industries there are rejected vegetables, fruits, nuts, milk, and animal products that are unacceptable for human food and can be a source of animal feed. Edible food waste more typically arises from retail, restaurant, and individual households. Some waste is inevitable from food preparation, such as peels, skins, trim fat, and bone. However much of the waste from this sector is edible and arises either due to unconsumed prepared food (plate waste), or food that has spoiled or is deemed unacceptable due to appearance or loss of quality. It is this food waste that has a high likelihood of entering municipal landfills.

In the USDA hierarchy of food recovery, the number one priority for reducing food waste is "Source Reduction" to reduce surplus food produced. Number two is to use surplus food to feed hungry people, which suggests that a misallocation of food contributes to waste. Third on the list is to use surplus food to feed animals, fourth is to use food waste for industrial purposes, and fifth is to compost food waste and then return the compost to soils. Last is to send food waste to landfills. The ability to capture food waste streams from industry, retail, restaurant, and households into animal feed has different degrees of difficulty. This review will focus on the use of food waste and residues as animal feed.

Animal feeds may be produced from edible food waste, expired or uneaten food products, or from residues of the processing industry. Utilization of these food waste streams as animal feed can reduce MSW disposal and be used to produce high quality animal protein as a human food source. However, there are constraints on capturing food waste as animal feed. These include efficient collection points for food waste, efficient transfer to feed mills or directly to farms for incorporation into animal feed, nutrient quality of food waste, acceptability of food waste as an animal feed, and safety of animal products and food waste as animal feeds.

## THE ANIMAL FEED INDUSTRY

A survey of the world feed industry by Alltech (2014) reported that there were 26,240 feed mills in the world producing 963 million tonnes (1,062 million U.S. tons) of animal feed. The number one producer of animal feed was China at 189 million tonnes (208 million U.S. tons) from 9,500 feed mills, followed by the U.S. with 169 million tonnes (186 million U.S. tons) of animal feed from 5,236 feed mills. World-wide, poultry feed is the number one product at 444 million tonnes (489 million U.S. tons) (Table 15.1). Swine are the second largest feed consumer at 243 million tonnes (268 million U.S. tons) followed by all ruminants at 196 million tonnes (216 million U.S. tons).

**Table 15.1 World Feed Production in 2014 for the Situation in 2013. Source: Alltech, 2014**

| Rank | Animal Group/Item | Million tonnes | Number |
|---|---|---|---|
| 1 | Poultry | 444.0 | |
| 2 | Swine | 243.0 | |
| 3 | Ruminants | 195.0 | |
| 4 | Aquaculture | 34.4 | |
| 5 | Pets | 20.7 | |
| 6 | Equine | 12.4 | |
| | Total feed mills | | 26,240 |
| 1 | China | 189 | 9,500 |
| 2 | US | 169 | 5,236 |
| 3 | Brazil | 67 | 1,237 |

The major products used in animal feed are summarized in Table 15.2 based on a paper by Capper et al. (2013). Forage crops comprise a major component of herbivore (primarily ruminants) feedstuffs and are not consumed by humans. Other feed items are co-products or waste products of the food processing industries and have variable value as human food sources. Cereal grains are the major food group used in animal feeds which could be consumed by humans.

**Table 15.2 Examples of Feeds Commonly Used Within U.S. Livestock Production Systems. Source: Capper et al., 2013**

| Feed Source | Examples | Human Edible? |
|---|---|---|
| Forage crops | Pasture grasses, alfalfa, clovers, hays, silages (grass or crop based) | No |
| Cereals | Corn, wheat, barley, millet, sorghum, triticale, oats | Yes |
| Plant proteins | Soybean (meal and hulls), cottonseed (whole & meal), safflower meal, canola meal, peanut meal | Partially |
| Vegetable oils | Soybean, canola, corn, sunflower oils | Yes |
| Grain by-products | Distillers grains (wet and dry), corn gluten, wheat bran, straw, crop residues | No |
| Vegetable by-products | Apple pomace, citrus pulp, almond hulls, pea silages | No |
| Waste fruit/vegetables | | Partially |
| Food industry by-products | Bakery waste, cannery waste, restaurant waste, oils, expired candy, potato chips, bakery products, greases | Partially |
| Sugar industry by-products | Molasses (cane, beet, citrus), beet pulp | Partially |
| Animal by-products | Meat and bone meal, tallow, feather meal, bloodmeal, poultry litter | Partially |
| Dairy by-products | Milk, whey products, casein | Partially |
| Marine by-products | Fish and seafood meal and oils, algae | Partially |

The utilization of food used in animal feed based on estimates by the Food Agriculture Organization is in Table 15.3 (FAOSTAT). Overall, 13.6% of world food production was used in animal feeds. Cereal grains comprise the largest component of animal feeds with 34.9% of production going into animal feed, representing almost 65% of animal feeds. Starchy roots (22.2% of production) and pulses (19.4% of production) were the next largest contributors to animal feed based on production totals. Slightly over 15% (Table 15.2, 15.7%) of fish and seafood production was used for animal feed, but it comprised only

1.86% of total animal feeds. Milk and milk products were a significant contributor at 6.26% of animal feeds and 10.68% of total production. Oilcrops, such as soybeans, cottonseeds, and rape seed, contributed 6.46% of total production to animal feed, but were only 2.82% of animal feed.

**Table 15.3 Major Food Groups Used in Animal Feeds. Source: FAO, 2011**

| Food group | Production | Animal feed | % of production | % used as animal feed |
|---|---|---|---|---|
| | Metric tonnes | | | |
| Cereals | 2,345,593 | 818,837 | 34.9 | 64.9 |
| Sugar crops | 2,092,347 | 51,699 | 2.5 | 4.1 |
| Vegetables | 1,087,523 | 52,565 | 4.8 | 4.2 |
| Starchy roots | 798,181 | 177,246 | 22.2 | 14.0 |
| Milk | 739,114 | 78,962 | 10.7 | 6.3 |
| Fruits | 629,014 | 5,547 | 0.88 | 0.44 |
| Oilcrops | 550,924 | 35,565 | 6.5 | 2.8 |
| Meat | 296,615 | 74 | 0.025 | 0.006 |
| Sugar | 205,114 | 434 | 0.21 | 0.034 |
| Vegetable oils | 159,184 | 773 | 0.49 | 0.061 |
| Fish, seafood | 149,496 | 23,443 | 15.7 | 1.9 |
| Eggs | 70,684 | 73 | 0.10 | 0.006 |
| Pulses | 68,336 | 13,243 | 19.4 | 1.1 |
| Animal fats | 36,480 | 2,065 | 5.7 | 0.16 |
| Aquatic products | 23,127 | 159 | 0.69 | 0.013 |
| Stimulants | 18,398 | 12 | 0.065 | 0.001 |
| Offals | 18,151 | 1,041 | 5.7 | 0.083 |
| Totals | 9,288,281 | 1,261,738 | 13.6 | 100.0 |

Table 15.4 looks more closely at the use of cereals and other animal products in animal feeds. Of cereals used for animal feed, maize was 59.25% of

**Table 15.4 Utilization of Agricultural Products for Animal Feed. Source: FAO, 2011**

| Item | World production | Feed use | % of total production | Feed tonnes | % of feed |
|---|---|---|---|---|---|
| | Metric tonnes | | | Metric tonnes | |
| All cereals | 2,345,593 | 818,837 | 34.9 | 818,837 | 100.0 |
| Rye | 13,029 | 4,845 | 37.2 | | 0.59 |
| Oats | 22,318 | 15,463 | 69.3 | | 1.9 |
| Cereals, other | 25,494 | 18,236 | 71.5 | | 2.2 |
| Millet | 27,137 | 4,134 | 15.2 | | 0.51 |
| Sorghum | 58,093 | 26,697 | 46.0 | | 3.3 |
| Barley | 132,894 | 89,430 | 67.3 | | 10.9 |
| Rice (milled) | 481,177 | 32,838 | 6.8 | | 4.0 |
| Wheat | 699,350 | 142,039 | 20.3 | | 17.3 |
| Maize | 886,101 | 485,155 | 54.8 | | 59.2 |
| Starchy roots | 798,181 | 177,246 | 22.2 | 177,246 | 100.0 |
| Cassava | 246,120 | 77,662 | 31.6 | | 43.8 |
| Potatoes | 374,425 | 49,702 | 13.3 | | 28.0 |
| Sweet potatoes | 103,174 | 40,899 | 39.6 | | 23.1 |

| Item | World production | Feed use | % of total production | Feed tonnes | % of feed |
|---|---|---|---|---|---|
| | Metric tonnes | | | Metric tonnes | |
| Yams | 56,677 | 8,235 | 14.5 | | 4.6 |
| Roots | 17,785 | 748 | 4.2 | | 0.42 |
| Sugar crops | 2,092,347 | 51,699 | 2.5 | 51,699 | 100.0 |
| Sugar cane | 1,814,267 | 34,772 | 1.9 | | 67.3 |
| Sugar beet | 278,080 | 16,927 | 6.1 | | 32.7 |
| Sugar | 8,671 | 320 | 3.7 | | 73.7 |
| Sugar raw | 170,669 | 78 | 0.046 | | 18.0 |
| Sweeteners | 24,196 | 36 | 0.15 | | 8.3 |
| Honey | 1,578 | 0 | 0.0 | | 0.0 |
| Pulses | 68,336 | 13,243 | 19.4 | 13,243 | 100.0 |
| Beans | 22,926 | 2,753 | 12.0 | | 20.8 |
| Peas | 9,739 | 3,330 | 34.2 | | 25.1 |
| Pulses other | 35,671 | 7,160 | 20.1 | | 54.1 |
| Oilcrops | 550,924 | 35,565 | 6.5 | 35,565 | 100.0 |
| Soybeans | 261,892 | 13,404 | 5.1 | | 37.7 |

| Item | World production | Feed use | % of total production | Feed tonnes | % of feed |
|---|---|---|---|---|---|
| | Metric tonnes | | | Metric tonnes | |
| Peanuts | 28,168 | 2 | 0.007 | | 0.006 |
| Sunflower seed | 40,785 | 2,659 | 6.5 | | 7.5 |
| Rape & mustard seed | 63,269 | 5,165 | 8.2 | | 14.5 |
| Cottonseed | 48,814 | 12,199 | 25.0 | | 34.3 |
| Coconuts/copra | 57,194 | 46 | 0.080 | | 0.13 |
| Sesame seed | 4,667 | 3 | 0.064 | | 0.008 |
| Palm kernels | 13,433 | 0 | 0.0 | | 0.0 |
| Olives | 20,417 | 0 | 0.0 | | 0.0 |
| Oilcrops, other | 12,285 | 2,087 | 17.0 | | 5.7 |
| Vegetable oils | 159,184 | 773 | 0.49 | 773 | 100.0 |
| Soy oil | 41,915 | 22 | 0.052 | | 2.8 |
| Peanut oil | 5,703 | 0 | 0.0 | | 0.0 |
| Sunflower oil | 13,354 | 0 | 0.0 | | 0.0 |
| Rape & mustard oil | 22,908 | 750 | 3.3 | | 97.0 |
| Cottonseed oil | 5,194 | 0 | 0.0 | | 0.0 |

| Item | World production | Feed use | % of total production | Feed tonnes | % of feed |
|---|---|---|---|---|---|
| | Metric tonnes | | | Metric tonnes | |
| Palmkernel oil | 6,020 | 0 | 0.0 | | 0.0 |
| Palm oil | 48,543 | 0 | 0.0 | | 0.0 |
| Coconut oil | 3,109 | 0 | 0.0 | | 0.0 |
| Sesameseed oil | 1,073 | 0 | 0.0 | | 0.0 |
| Olive oil | 3,620 | 0 | 0.0 | | 0.0 |
| Ricebran oil | 1,084 | 0 | 0.0 | | 0.0 |
| Maize germ oil | 2,325 | 0 | 0.0 | | 0.0 |
| Oilcrops, other | 4,336 | 1 | 0.023 | | 0.13 |
| Vegetables | 1,087,523 | 52,565 | 4.8 | 52,565 | 100.0 |
| Tomatoes & products | 158,055 | 1,751 | 1.1 | | 3.3 |
| Onions | 84,907 | 82 | 0.097 | | 0.16 |
| Vegetables, other | 844,561 | 50,732 | 6.0 | | 96.5 |
| Fruits | 629,014 | 5,547 | 0.88 | 5,547 | 100.0 |
| Oranges & mandarins | 95,569 | 10 | 0.010 | | 0.18 |
| Lemons, limes & products | 15,555 | 0 | 0.0 | | 0.0 |

| Item | World production | Feed use | % of total production | Feed tonnes | % of feed |
|---|---|---|---|---|---|
| | Metric tonnes | | | Metric tonnes | |
| Grapefruit & products | 7,773 | 0 | 0.0 | | 0.0 |
| Citrus, other | 12,610 | 0 | 0.0 | | 0.0 |
| Bananas | 102,172 | 1,543 | 1.5 | | 27.8 |
| Plantains | 35,898 | 2,917 | 8.1 | | 52.6 |
| Apples & products | 76,108 | 633 | 0.83 | | 11.4 |
| Pineapples & products | 21,429 | 0 | 0.0 | | 0.0 |
| Dates | 6,922 | 333 | 4.8 | | 6.0 |
| Grapes & products | 69,979 | 0 | 0.0 | | 0.0 |
| Fruits, other | 185,399 | 111 | 0.060 | | 2.0 |
| Stimulants | 18,398 | 12 | 0.065 | 12 | 100.0 |
| Coffee | 8,299 | 0 | 0.0 | | 0.0 |
| Cocoa beans | 4,691 | 12 | 0.26 | | 100.0 |
| Tea, incl. maté | 5,408 | 0 | 0.0 | | 0.0 |
| Meat | 296,615 | 74 | 0.025 | 74 | 100.0 |
| Beef | 66,359 | 8 | 0.012 | | 10.8 |

| Item | World production | Feed use | % of total production | Feed tonnes | % of feed |
|---|---|---|---|---|---|
| | Metric tonnes | | | Metric tonnes | |
| Lamb & goat | 13,518 | 17 | 0.13 | | 23.0 |
| Pork | 107,893 | 0 | 0.0 | | 0.0 |
| Poultry | 102,466 | 0 | 0.0 | | 0.0 |
| Meat, other | 6,379 | 49 | 0.77 | | 66.2 |
| Offal | 18,151 | 1,041 | 5.7 | 1,041 | 100.0 |
| Animal fats | 36,480 | 2,065 | 5.7 | 2,065 | 100.0 |
| Butter & ghee | 9,404 | 4 | 0.043 | | 0.19 |
| Cream | 3,327 | 0 | 0.0 | | 0.0 |
| Fats, raw | 22,710 | 1,592 | 7.0 | | 77.1 |
| Fish, body oil | 1,022 | 462 | 45.2 | | 22.4 |
| Fish, liver oil | 17 | 7 | 41.2 | | 0.34 |
| Eggs | 70,684 | 73 | 0.10 | 73 | 100.0 |
| Dairy | 739,114 | 78,962 | 10.7 | 78,962 | 100.0 |
| Fish & seafood | 149,496 | 22,443 | 15.7 | 23,443 | 100.0 |
| Freshwater fish | 50,546 | 585 | 1.2 | | 2.5 |

| Item | World production | Feed use | % of total production | Feed tonnes | % of feed |
|---|---|---|---|---|---|
| | Metric tonnes | | | Metric tonnes | |
| Demersal fish | 20,761 | 1,168 | 5.6 | | 5.0 |
| Pelagic fish | 34,088 | 19,908 | 58.4 | | 84.9 |
| Marine fish | 10,855 | 1,439 | 13.3 | | 6.1 |
| Crustaceans | 11,869 | 174 | 1.5 | | 0.74 |
| Cephalopods | 3,850 | 151 | 3.9 | | 0.64 |
| Molluscs, other | 17,527 | 18 | 0.10 | | 0.077 |
| Aquatic products | 23,127 | 159 | 0.69 | 159 | 100.0 |
| Meat, aquatic mammals | 0 | 0 | 0.0 | | 0.0 |
| Aquatic animals, other | 1,351 | 0 | 0.0 | | 0.0 |
| Aquatic plants | 21,776 | 159 | 0.73 | | 100.0 |

all cereals. Wheat (17.35% of cereals used in animal feeds) and barley (10.92% of cereals in animal feeds) were the next major cereals used in animal feeds. A little over half of maize production (54.75%) was used for animal feed, whereas 69.3% of oat production and 71.5% of other cereals were used for animal feed, but these cereals were less than 4% of all cereals used as animal feeds. The cereal grains barley (67.294%) and oats (69.285%) were primarily used for animal feed but their total contribution to animal feed was only 10.92% and 1.89%, respectively. Cassava, followed by potatoes and sweet potatoes were the most common starchy roots used as animal feed. Of oil seeds, soybeans and cottonseeds were the major co-products used in animal feeds.

For vegetables, fruits, and meat small percentages of total production were used in animal feed (Table 15.4). These amounts would represent cull products that were excluded from processing or sale from retail markets. Very little of these products were used as animal feed of that which was produced. Most of the animal feed that results from these products is a result of co-products derived from food processing.

Table 15.5 presents the utilization of by-products from the food processing industries and cereal grains used for animal feeds in the U.S. in 2012 (Anon., 2012). As in the FAO data, cereal grains are the major item used as animal feed with food processing residues comprising 25.8% of livestock feeds. The majority of co-products come from the vegetable oil industries producing oil seed meals, which typically are high protein feeds (Table 15.5). However these estimates do not include vegetable, citrus, bakery, candy, chip and other food waste which comes from retailers and juice and vegetable processors.

## CO-PRODUCTS OF PROCESSING INDUSTRIES

Data from Crawshaw (2001), Macgregor (1989), and the NRC (2001) is summarized in Table 15.6 for co-product feeds typically used in ruminant and swine diets. As Crawshaw has pointed out, the production of feed grade co-products inevitably increases as higher standards are imposed on foods for human consumption. Thus, we may expect feeds from these co-products to increase in production as more rigid standards are applied based on appearance, uniformity in size, absence of defects, and lower contamination thresholds. Presented in Table 15.6 are values for dry matter (DM), crude protein (CP), neutral detergent fiber (NDF, plant cell wall), starch, fat, ash, and metabolizable energy for ruminants (ME, Mj/kg, (Mjoules/kg). For comparison, fine-ground corn has an energy value of 14.4 Mj/kg on a DM basis whereas alfalfa hay has a value of 7.8

**Table 15.5 Statistics for Products Used in Animal Feeds in 2012. Source: Anon., 2012**

| Feed group<br>By-products | Utilization<br>Metric tonnes | Cereals<br>grains | Utilization<br>Metric tonnes (US) |
|---|---|---|---|
| **Oilseed meals** | | **Cereals** | |
| Soybean meal | 27,487,698 | Corn | 116,845,395 |
| Cottonseed meal | 2,290,641 | Sorghum | 1,360,777 |
| Linseed meal | 178,715 | Oats & barley | 2,630,836 |
| Peanut meal | 86,183 | Wheat | 5,352,390 |
| Sunflower meal | 326,587 | Rye | 90,718 |
| **Animal proteins** | | | |
| Tankage & meat meal | 2,131,884 | | |
| Fish meal | 181,437 | | |
| Dried milk | 226,796 | | |
| **Mill products** | | | |
| Wheat mill feeds | 5,805,982 | | |
| Gluten feeds & meal | 4,603,963 | | |
| Rice mill feeds | 521,631 | | |
| Alfalfa meal | NA | | |
| **Totals** | | | |
| By-products | 43,841,517 | Cereals | 126,280,116 |
| By-products used as a percent of total feeds | | **25.8** | |

Mj/kg. Most co-products have an energy value slightly lower than ground corn, unless they have a significant fat content, but have a higher energy value than most forages. However, many co-products have low DM content and, unless excess water is removed, present a problem in transport and storage due to low aerobic stability (see Table 15.6). Removal of water is necessary to economically

**Table 15.6 Major Co-products of Food Processing Used as Animal Feed. Sources: Macgregor, 1989; Crawshaw, 2001**

| Industry | Product | DM | CP | NDF | Starch | Fat | Ash | ME |
|---|---|---|---|---|---|---|---|---|
| | | % | %DM | | | | | |
| Almond nuts | Almond hulls | 91 | 4.2 | 28-42 | 6 | 3.5 | 7.6 | 10.4 |
| Apple processing | Cider apple pomace | 20-28 | 6.7 | 43-56 | 2-7 | 2.3-3.2 | 2.3 | 9-12 |
| | Culinary apple pomace | 19 | 4.5 | 34 | NR | NR | 3.0 | 11 |
| | Apple pomace | 21 | 7.6 | 42-52 | 4 | 4.8 | 2.2 | 11 |
| Bakery products | Bread | 65/90 | 14.0 | . | 70-73 | 3.0 | 2.8 | 14 |
| | Cakes | 68-87 | 5-14.7 | . | 18-32 | 6-35 | NR | 14-19 |
| | Cookie dough | 90-94 | 5-10 | . | 21-76 | 10-25 | 4.5 | 15-17 |
| | Breakfast cereal | 90-93 | 9-13 | . | 49-62 | 1-2.5 | 2.5 | 13.5 |
| | Bakery waste | 91 | 12.1 | 7.0 | 51 | 11.0 | 4.4 | 14 |
| Candy | Candy products | 97 | 8.5 | 10.9 | 15 | 24.4 | NR | 15 |
| Corn (maize) | Screenings | 87 | 9.2 | 16 | 72 | 3 | 0.6 | 13.7 |
| | Corn steep liquor | 45-50 | 40-45 | 0 | 2-5 | 0.1-2 | 9-22 | 11.5-13 |
| | Corn gluten feed | 86-91 | 20-25 | 33-45 | 9-28 | 3-7 | 4-10 | 11.3-14 |
| | Corn gluten meal | 89 | 66 | 7.5 | 15.5 | 6 | 2 | 14.5 |
| | Corn germ meal | 88 | 25-26 | 37 | 23 | 3.4 | 3 | 13.6 |

| Industry | Product | DM | CP | NDF | Starch | Fat | Ash | ME |
|---|---|---|---|---|---|---|---|---|
| | | % | %DM | | | | | |
| | Corn cannery waste | 22 | 8.6 | 53.2 | NR | 5.2 | NR | 10.5 |
| | Corn steep liquor dry | 94 | 33 | 19 | 1 | 1.3 | NR | 12.5 |
| Cottonseed | Whole cotton seeds | 93 | 23 | 44 | 5 | 20 | 4.8 | 15.1 |
| | Cotton seed hulls | 91 | 4.1 | 85 | 15 | 2.4 | 2.8 | 6.5 |
| Malting, brewing & vinegar | Malt powder | 88-95 | 6-17 | 12-71 | NR | 1.5-4.5 | 2-11 | 8-13 |
| | Malt screenings | 88 | 10-13 | 16-21 | NR | 3.8 | 2.6 | 13 |
| | Malt culms | 92-96 | 20-32 | 48.4 | 3-19 | 2.9 | 5.7 | 11.1 |
| | Malt residue pellets | 90 | 23 | 48 | 15.8 | 2.5 | 6.2 | 11.5 |
| | Brewers' grains | 18-25 | 19-31 | 50-64 | 2-8 | 8.5-11 | 3.5-9 | 11-12.5 |
| | Mash filter grains | 24-30 | 20-25 | 49-57 | 4-11 | 9-12 | 4.8 | 12-12.7 |
| | Vinegar grains | 21.5 | 21.7 | 50-59 | 7-13 | 8.6-13 | 4.2 | 11.7-13 |
| | Malt extract grains | 21-27 | 21-24 | 43-57 | 6-18 | 7.5-10 | 4.6 | 11.9-13 |
| | Black grains | 27.4 | 21-29 | 45-55 | 6-11 | 6-8 | 3-17 | 9-10.7 |
| | Grains pressings | 9 | 38 | 17 | 23 | 7.9 | 3 | NR |
| | Brewers' yeast | 12-16 | 36-50 | 3-10 | 2-20 | 2.5-4.5 | 6-10 | 13.5 |

| Industry | Product | DM | CP | NDF | Starch | Fat | Ash | ME |
|---|---|---|---|---|---|---|---|---|
| | | % | %DM | | | | | |
| | Beer | 5.5-9.5 | 4-7 | | | | 9-16 | NR |
| | Vinegar still bottoms | 67.4 | 33.1 | 0.1 | 2 | 0.15 | 10.8 | 14 |
| | Liquid malt extract | 76-78 | 6.4 | 0 | 0 | 0 | 1.6 | NR |
| | Brewer's grains wet | 20 | 26 | 42 | 10.3 | 5 | 4.8 | 10.5 |
| Milk products | Whey | 5-7 | 13-15 | 0 | 0 | 1 | 10 | 13.5 |
| | Whey concentrate | 30-50 | 12.4 | 0 | 0 | 1.4 | 7.4 | 13.5 |
| | Whey permeate | 18,25,45 | 3.8 | 0 | 0 | 0.2 | 11 | 11.6 |
| | Delactosed whey | 38-45 | 24 | 0 | 0 | 1.2 | 16 | 11.2 |
| Distillery co-products | Draff | 20-26 | 20-23 | 60-66 | 0-5 | 9-13 | 3.5 | 11 |
| | Pot ale syrup | 30-50 | 34-38 | 0.6 | 1.3 | 2-3 | 9.5-10.5 | 15.6 |
| | Supergrains | 25 | 29 | 53-64 | 5 | 9-12 | 2.5 | 14.1 |
| | Evaporated spent wash | 27 | 31 | 38-51 | 7 | 6-9 | 2.8 | 13.6 |
| | Distillers' malt | 90 | 27 | 42 | 2.5 | 7.5-9 | 6 | 12.6 |
| | Distillers' wheat | 90 | 32 | 23-46 | 4.5 | 6-7.5 | 5.3 | 13.3 |
| | Distillers' corn | 90 | 29 | 23-51 | 2.5 | 10-11.5 | 4.5 | 14.9 |

| Industry | Product | DM | CP | NDF | Starch | Fat | Ash | ME |
|---|---|---|---|---|---|---|---|---|
| | | % | %DM | | | | | |
| | Distillers' with solubles | 92 | 29.3 | 12 | 12 | 8-14.5 | 4.8 | 13.8 |
| Animal products | Blood meal | 90 | 93 | 0 | 0 | 2 | 2.4 | 14.6 |
| | Feather meal | 93 | 88 | 0 | 0 | 10 | 1.9 | 11.9 |
| | Fish meal | 90 | 68 | 0 | 0 | 9 | 19 | 12.5 |
| | Meat meal | 94 | 58 | 0 | 0 | 12 | 23 | 10.2 |
| | Meat and bone meal | 95 | 50 | 0 | 0 | 12 | 30 | 9.0 |
| Wheat co-products | Wheat bran | 88.8 | 17 | 44 | 22 | 4.5 | 5.8 | 10.8 |
| | Wheat millrun | 89 | 18 | 40.8 | 33 | 4.6 | 2 | 11.4 |
| | Wheat midds | 89 | 18.4 | 38 | 19 | 5 | 6.5 | 12.5 |
| | Wheat shorts | 89 | 15 | 25 | 36 | 3.5 | 2 | 12.9 |
| | Wheat red dog | 88 | 20.1 | 27 | 37 | 4.1 | 3.7 | 13.3 |
| | Wheat germ | 91 | 25 | 15 | 13 | 7 | 2 | 15.1 |
| Potato co-products | Potato feed | 10-14 | 17 | 22.1 | 36.6 | 1.6 | 9.3 | 11.7 |
| | Potato feed permeate | 9-13 | 18.5 | 11 | 43 | 1 | 10.0 | 12.6 |
| | Potato feed solids | 11-15 | 17 | 33 | 30 | 2 | 10 | 10.4 |

| Industry | Product | DM | CP | NDF | Starch | Fat | Ash | ME |
|---|---|---|---|---|---|---|---|---|
| | | % | %DM | | | | | |
| | Abraded peel | 7 | 11.4 | 32-52 | 24-38 | 2.6 | 8.4 | 8.5-11 |
| | Potato skin | 10.6 | 18.4 | 70.1 | 4.2 | 3.9 | 6.6 | 5.5 |
| | Off-cuts potatoes | 21.1 | 7.6 | 10 | 75.3 | 0.9 | 3.2 | 13.3 |
| | Potato slice | 17.4 | 9.3 | 25.5 | 60.1 | 1.9 | 6.1 | 11.1 |
| | Peel and trim | 32.4 | 6 | 18.5 | 66.9 | 3.8 | 10.5 | 12.6 |
| | Potato mash | 21.8 | 8.1 | 7.4 | 69 | 1.4 | 3 | 13.5 |
| | Potato flake | 90.9 | 8.5 | 5.4 | 78.9 | 1 | 4.5 | 13.4 |
| | Potato chips | 34 | 6.9 | 5.8 | 65.3 | 10-20 | 3.2 | 14.8-16 |
| | Fries | 39.9 | 6 | 14.1 | 42.9 | 20.3 | 3.7 | 16.4 |
| | Hash browns | 38.6 | 7.1 | 10.8 | 52.9 | 18.8 | 6.2 | 16.1 |
| | Potato starch | 60 | 1 | 0.7 | 97.7 | 0.2 | 0.3 | 13.9 |
| | Potatoes | 23 | 9.5 | 6 | 71.7 | 0.4 | 4.8 | 13.2 |
| | Potato by-product meal | 93.6 | 10.7 | 12.1 | 62 | 5.9 | 4 | 15.1 |
| Sugar beet co-products | Tails | 9-15 | 7.5 | NR | NR | 0.3 | 7-25 | 10-11.5 |
| | Dried molasses feed | 88 | 11 | 32.1 | 6.5 | 0.4 | 8.8 | 12.5 |

| Industry | Product | DM | CP | NDF | Starch | Fat | Ash | ME |
|---|---|---|---|---|---|---|---|---|
| | | % | %DM | | | | | |
| | Beet pulp | 21-30 | 10 | 52.4 | 0.4 | 0.7 | 8.2 | 12.5-13 |
| | Beet molasses | 74-78 | 10-14 | trace | 0 | trace | 11 | 10.3-12 |
| | Beet pulp dry | 91 | 14.7 | 41.6 | 9 | 1 | 12.2 | 11.7 |
| Citrus/tropical fruit | Citrus pulp | 17-24 | 6.8-9.7 | 19-26 | .1-8.8 | 1.1-3.7 | 4.3 | 13.6 |
| | Orange pulp | 17-24 | 7.9 | 20.8 | 2.5-21 | 1.1-3.5 | 3.9 | 13.6 |
| | Lemon pulp | 16-19 | 6.8 | 28.7 | 5.0 | 1 | 5.0 | 13.0 |
| | Citrus molasses | 71 | 5.8 | 0.0 | NR | 0.3 | 6.6 | 11.3 |
| | Fruit salad | 8.7-9.5 | 11-13 | 33 | 11.5 | 5.2 | 5.3-11 | 11.0 |
| | Citrus pulp pellets | 90 | 6.9 | 23 | 2 | 14.4 | 6.8 | 12.7 |
| Hard nuts | Pistachio culls | 90 | 11.4 | 5.3 | NR | 25.4 | NR | 10.1 |
| | Peanut skins | 90 | 14 | 34 | 22 | 20 | 6 | 9.9 |
| | Peanut hulls | 90 | 15 | 45 | 28 | 2 | 6 | 8.7 |
| Oil meals | Soybean meal | 90 | 55 | 10 | 2 | 3 | 6.7 | 13.9 |
| | Canola meal | 90 | 36 | 30 | 14.3 | 5.7 | 7.3 | 10.6 |
| | Cottonseed meal | 92 | 42 | 30 | 1.7 | 6.1 | 6.9 | 11.1 |

| Industry | Product | DM | CP | NDF | Starch | Fat | Ash | ME |
|---|---|---|---|---|---|---|---|---|
| | | % | %DM | | | | | |
| | Linseed meal | 90 | 32 | 31.4 | 13.1 | 3.5 | 6.5 | 10.1 |
| | Peanut meal | 92 | 52 | 14 | 11 | 1.6 | 7.8 | 13.2 |
| | Safflower meal | 91 | 27.3 | 51 | 4.4 | 1.4 | 5.3 | 6.4 |
| | Sunflower meal | 93 | 49 | 35 | 4.8 | 3.3 | 8 | 9.0 |
| Malting, brewing & vinegar | Malt powder | 88-95 | 6-17 | 12-71 | NR | 1.5-4.5 | 2-11 | 8-13 |
| | Malt screenings | 88 | 10-13 | 16-21 | NR | 3.8 | 2.6 | 13 |
| | Malt culms | 92-96 | 20-32 | 48.4 | 3-19 | 2.9 | 5.7 | 11.1 |
| | Malt residue pellets | 90 | 23 | 48 | 15.8 | 2.5 | 6.2 | 11.5 |
| | Brewers' grains | 18-25 | 19-31 | 50-64 | 2-8 | 8.5-11 | 3.5-9 | 11-12.5 |
| | Mash filter grains | 24-30 | 20-25 | 49-57 | 4-11 | 9-12 | 4.8 | 12-12.7 |
| | Vinegar grains | 21.5 | 21.7 | 50-59 | 7-13 | 8.6-13 | 4.2 | 11.7-13 |
| | Malt extract grains | 21-27 | 21-24 | 43-57 | 6-18 | 7.5-10 | 4.6 | 11.9-13 |
| | Black grains | 27.4 | 21-29 | 45-55 | 6-11 | 6-8 | 3-17 | 9-10.7 |
| | Grains pressings | 9 | 38 | 17 | 23 | 7.9 | 3 | NR |
| | Brewers' yeast | 12-16 | 36-50 | 3-10 | 2-20 | 2.5-4.5 | 6-10 | 13.5 |

| Industry | Product | DM | CP | NDF | Starch | Fat | Ash | ME |
|---|---|---|---|---|---|---|---|---|
| | | % | %DM | | | | | |
| | Beer | 5.5-9.5 | 4-7 | | | | 9-16 | NR |
| | Vinegar still bottoms | 67.4 | 33.1 | 0.1 | 2 | 0.15 | 10.8 | 14 |
| | Liquid malt extract | 76-78 | 6.4 | 0 | 0 | 0 | 1.6 | NR |
| | Brewer's grains wet | 20 | 26 | 42 | 10.3 | 5 | 4.8 | 10.5 |
| Milk products | Whey | 5-7 | 13-15 | 0 | 0 | 1 | 10 | 13.5 |
| | Whey concentrate | 30-50 | 12.4 | 0 | 0 | 1.4 | 7.4 | 13.5 |
| | Whey permeate | 18,25,45 | 3.8 | 0 | 0 | 0.2 | 11 | 11.6 |
| | Delactosed whey | 38-45 | 24 | 0 | 0 | 1.2 | 16 | 11.2 |

transport co-products to farms or feed mills which may be located at some distance from the processing plant. In addition a dry product is necessary to improve stability and handling.

Co-products from processing industries are an important source of revenue and comprise the greatest food waste source for animal feed. An advantage is they are collected in large volumes at the site of processing. A disadvantage is they contain large amounts of water and may be located at some distant from livestock farms. Co-products in Table 15.6 include items such as pomace or pulp from the canning and juice industries, secondary products of the ethanol, spirits, brewing or oil industries, expired bakery, vegetable, and candy products, and residues from the rendering industry. Most of these items would only be partially consumable by humans, and would not be preferred food items. The livestock industries provide a valuable resource for disposal of these items as opposed to municipal disposal or incineration. High quality animal protein for human consumption in meat and milk can be produced from these disposal items by use in livestock diets.

## FOOD WASTE FROM RETAIL, RESTAURANT, AND HOUSEHOLD SECTORS

The Waste & Resources Action Programme in the United Kingdom estimated that only 4.15% of available food is diverted to animal feed in the UK. This would be comparable to estimates of world food production in Tables 15.3 and 15.4 from the FAO. Based on FAO estimates this number is about 13.6% of production. The WRAP estimate would reflect data just for the UK, whereas the FAO data is from across the globe. Using FAO estimates for total domestic supply of food for the U.S. in 2011, 616,645,000 tonnes (679,734,758 U.S. tons) of food was available from all products. An estimate of 133,390,000 tonnes (147,037,306 U.S. tons) of food went for animal feed (21.6%) and 243,717,000 tonnes (268,652,006 U.S. tons) was available for human food (39.5% of total domestic supply) and 4,549,000 tonnes (5,014,414 U.S. tons) was wasted (0.7%). Diversion of food to animal use includes poor quality, outdated shelf life, changes in consumer preferences leading to over-inventory, mislabeled packaging, and overstocking. These totals do not include residues from processing industries.

The Food Waste Reduction Alliance (FWRA) was formed in 2011 under the auspices of the Grocery Manufacturers Association (GMA), the Food Marketing Institute (FMI), and the National Restaurant Association (NRA) (http://www.foodwastealliance.org). The alliance includes more than 30 manufacturing, retailing and foodservice companies and expert partners from the anti-hunger community and waste management sectors (http://

www.foodwastealliance.org). The FWRA commissioned Business for Social Responsibility (BSR), an international non-profit promoting sustainability issues in the business community, to conduct and analyze a survey they designed to assess food waste among food manufacturers, retailers, and wholesalers (BSR, 2013, 2014). Two surveys were completed, Tier I and Tier II.

In the FWRA surveys food waste from the manufacturing sector was 20.09 million tonnes (44.3 billion U.S. lbs) and only 1.09 million tonnes (2.4 billion U.S. lbs) was disposed in MSW. The majority of food waste was diverted to animal feed, 13.88 million tonnes (30.6 billion U.S. lbs). Next greatest diversion from landfills was land application at 3.81 million tonnes (8.4 billion U.S. lbs). Only 0.32 million tonnes (0.7 billion lbs) of manufacturing food waste was donated for food consumption. A survey of the retail sector estimated food waste was 1.72 million tonnes (3.8 billion U.S. lbs) and 0.77 million tonnes (1.7 billion U.S. lbs) was diverted to MSW. Only 0.24 million tonnes (0.53 billion U.S. lbs) was recovered as animal feed from the retail/wholesale sector, but 0.72 million tonnes (1.59 billion U.S. lbs) was used for food donation and composting from this sector (Tier 1 and Tier II BSR survey, 2013). Restaurant respondents donated or recycled 15.7% of food waste, largely recycled cooking oil (BSR, 2014). However, 84.3% of restaurant waste went to landfills with an annual food waste of about 14,969 tonnes (33 million U.S. lbs) (BSR, 2014).

Constraints for utilization of food waste as human food or as animal feed from manufacturing, retail and restaurant industries include safety and nutrient quality, particularly with animal products. Plant by-products may be a source of mold toxins, bacterial toxins, and other contaminants. Mold toxins may be produced prior to harvest or post-harvest if grains are not properly stored at appropriate moisture content. Typically, moisture content below 20% will protect most grains from post-harvest mold growth. Pre-harvest fungal invasion of the grain may occur from insect damage, drought conditions and other stresses to the plant in the field. Mycotoxin(s) may be present in the grain at harvest from conditions in the field which encourage spore invasion and growth within the grain head. Threshold levels of mycotoxin concentration in grains make it illegal to use these grains in human foods and in animal feeds. Bacterial contamination is possible when moisture content is above 30% in grains and food waste. Bacterial growth under aerobic conditions causes spoilage and loss of nutrient content and is a significant problem in utilizing high-moisture food residues. Moisture content may be 70% to 80% in residues and in food waste, which makes them extremely unstable at ambient temperatures. Food waste may be ensiled under anaerobic conditions to stabilize and preserve the material, but this requires large volumes delivered to a farm at one time to make ensiling a feasible storage option. Restaurant plate and kitchen waste would include a significant proportion of meat in addition to high moisture content, which creates further problems with handling and safety. High proportions of animal meat residues in food waste are a risk for transmission of animal diseases, and often high meat content is associated with a high fat content, which may be prone to rancidity.

Animal products pose a significant feed safety risk, as was dramatically evident in the Bovine Spongiform Encephalopathy (BSE) outbreak in Great Britain in 1986-1997. This led to a ban on all animal meat and bone meal as a feed ingredient in animal feeds. In addition to prion disease, animal products may harbor food-borne bacteria, such as *Salmonella, E. coli, Listeria* and *Corynebacteria* spp. Rules are in place to either ban certain animal products from diets fed to specific animals, such as no ruminant meat and bone meal may be fed to ruminants in the U.S., or specific processing procedures must be employed to ensure safety of the material. The USDA has regulations for using food waste as an animal feed to reduce the risk of disease for inclusion in swine rations. The Swine Health Protection Act mandates that food waste from restaurants and cafeterias must be heated to 100°C for 30 minutes throughout the material to be safely offered to swine. Only licensed facilities may handle food waste for swine feeds, and there are only 2,722 licensed facilities in the U.S. These facilities produce 266,104 tonnes (293,330 U.S. tons) of swine feed, which is only 0.6% of all swine feed sold in the U.S.

The EU has a total ban on feeding animal co-products to animals so other methods of disposing of animal processing residues must be pursued, such as incineration. Regulations are in place concerning feeding animal components to ruminants to minimize the risk of BSE. Basically, ruminant meat and meat and bone meal may not be fed to a ruminant. Ruminant blood meal may be fed to ruminant animals and meat, bone, and blood from non-ruminant animals may be fed to ruminants.

An additional problem with food waste is that it can be extremely variable in nutrient content (Table 15.5). Westendorf and Myer (2004) and Westendorf et al. (1999) reported on the nutrient variability in food plate waste for swine. Coefficients of variation for nutrient content for dry matter (19.3%), crude protein (27.5%), fat (30.4%), acid detergent fiber (41.2%) and ash (35.3%) were extremely large. This could lead to significant variation in animal performance when food waste is incorporated in swine diets. As the proportion of food waste increased, animal performance decreased. This was not only due to the nutrient content, but the high moisture content leads to a reduction in average daily gain as food waste replaced corn/soy bean meal in a typical swine grower diet due to reductions in DM intake (Westendorf and Myer, 2004). The high moisture content of food waste limited feed intake, reducing performance in addition to potential effects due to variation in nutrient content.

## CONCLUSION

Food waste can be used as a significant source of animal feed and can be a reasonable energy or fiber source and replace cereal grains to a variable extent in animal diets. Food processing waste is easier to incorporate into animal feed due to concentrated points of collection and processing to remove water. Restaurant and retail food wastes are more difficult to include in animal feed due to dispersed points of collection, and higher animal food content, which requires special processing or exclusions for it to be used in animal feeds. High moisture content increases transportation and handling costs. Aerobic spoilage is a greater risk with high moisture content. High moisture content can limit feed intake. Variable content of sources of waste products incorporated into food waste can cause significant variation in nutrient content which limits extent of inclusion in animal diets.

## REFERENCES

Alltech. 2014. 2013 Alltech® Global Feed Summary. http://www.alltech.com/sites/default/files/2013-feed-tonnage-report.pdf.

Anonymous. Agricultural Statistics. 2012. National Agricultural Statistics Service. United States Department of Agriculture. US Government Printing Office. Washington D.C. Internet:bookstore.gov.

Food and Agriculture Organization of the United Nations (FAO). 2011. Food Balance: Food Balance Sheets. FAO, Rome, Italy.

Business for Social Responsibility (BSR). 2013. Analysis of U.S. food waste among food manufacturers, retailers, and restaurants. http://www.foodwastealliance.org/wp-content/uploads/2013/06/FWRA_BSR_Tier2_FINAL.pdf.

Business for Social Responsibility (BSR). 2014. Analysis of U.S. food waste among food manufacturers, retailers, and restaurants. http://www.foodwastealliance.org/wp-content/uploads/2014/11/FWRA_BSR_Tier3_FINAL.pdf.

Capper, J.I., L. Berger, M.M. Brashears, and H.H. Jensen. 2013. Animal Feed vs. Human Food: Challenges and Opportunities in Sustaining Animal Agriculture Towards 2015. Council for Agricultural Science and Technology (CAST) Issue Paper 53.

Crawshaw, R. 2001. Co-product Feeds: Animal Feeds from the Food and Drink Industries. Nottingham University Press, Nottingham, UK. pp 232-249.

Food Waste Reduction Alliance. http://www.foodwastealliance.org/.

Macgregor, C.A. 1989. Directory of Feeds and Feed Ingredients. W.D. Hoard and Sons Company, Fort Atkinson, WI.

National Research Council. 2001. Nutrient Requirements of Dairy Cattle. National Academy Press, Washington D.C.

Westendorf, M.L., and R.O. Myer. 2004. Feeding Food Wastes to Swine. University of Florida IFAS Extension Publ. No. AS143. http://edis.ifas.ufl.edu.

Westendorf, M.L., T. Schuler, and E.W. Zirkle. 1999. Nutritional quality of recycled food plate waste in diets fed to swine. Prof. Anim. Sci. 15(2):106-111.

# Chapter 16

## *The Last Resort Before Landfill*

### *Food Waste Composting in the U.S.: An Overview*

**Nora Goldstein**

## ABSTRACT

After reducing the quantity of food wasted, and maximizing recovery and distribution of edible food, the next step in the food supply chain is composting the nonedible food to replenish soils – and in turn growing more food to put back into the supply chain. The optimum scenario is to close that loop as close to the sources of consumption. Increasingly, this is happening in communities via urban agriculture, community gardens and local farms and composting operations. To optimize this closed loop of food production, recovery and recycling, the food waste generated needs to be free of contaminants, especially plastic and glass. This requires an investment in training and education at the source of waste generation, and having the proper tools to maximize separation efficiency. This is also a point in the food supply chain where wasting of food can be identified, with information fed back into food purchasing, food preparation and donation programs. The final step is to optimize composting systems to produce high quality compost that can be incorporated back into soils for fertility and organic matter. Currently in the U.S., according to data collected by *BioCycle* magazine, there are over 4,900 composting operations; about 70 percent compost only yard trimmings (leaves, grass, brush, tree trimmings), and about 7-8 percent compost food waste, usually mixed with yard trimmings, and in some cases, livestock manure and crop residuals. An increasing number of anaerobic digestion facilities located on farms, and at wastewater treatment plants or stand-alone commercial operations, receive source separated food waste. Addition of food waste to anaerobic digesters increases the production of biogas, which can be used to generate electricity and produce a vehicle fuel. This paper focuses on food waste composting in the U.S.

## INTRODUCTION

In the United States, Americans generate over 254 million tons of trash annually (USEPA, 2015). Food waste comprises 37.06 million tons of that total, of which the U.S. EPA reports only 1.84 million tons, or 5.0 percent, are recovered. The U.S. EPA's definition of municipal solid waste only includes waste from the residential, commercial and institutional sectors. Therefore, food waste tonnages generated by processors and manufacturers, as well as in the agricultural sector, are not included in that reported tonnage. *This paper focuses on composting food waste from the commercial and institutional sectors.* For readers interested in the residential sector, *BioCycle* does biannual surveys on residential food waste composting. Its latest report (January 2015) identified 198 communities in the U.S. that provide households with curbside collection of source separated organics, including food waste. This represents 2.74 million households (2% of total) in the U.S. (Yepsen, 2015).

The majority of waste disposed in the United States ends up in landfills. And it is now widely recognized that food waste decays in a landfill and generates methane before that greenhouse gas is typically captured. (According to the U.S. EPA's Landfill Methane Outreach Program, there are 645 operational landfill gas energy projects in the United States as of March 2015 (USEPA LMOP, 2015).) According to *BioCycle*'s 2010 "The State of Garbage In America" Report, there were 1,098 MSW landfills in the U.S. (van Haaren et. al, 2010). The recently revised U.S. EPA WARM Model reflects this reality. The U.S. EPA created the WARM (WAste Reduction Model) to help solid waste planners and organizations track and voluntarily report greenhouse gas (GHG) emissions reductions from several different waste management practices. The original model underestimated the greenhouse gas emissions from food waste. The revised model now calculates that food waste in landfills generates about 7.13 MT $CO_2e$ (million metric tons of carbon dioxide equivalent) per dry ton of food waste (or 1.75/wet ton; Brown, 2014). Faster decay rates for food waste (how quickly a material decays and generates methane) have also been recognized. The faster the decay rate, the greater the chances are that the methane will escape into the atmosphere before any gas collection systems are operational.

Recognition of the greenhouse gas emissions from landfilling of food waste has been one key factor in development of programs around the country to divert food waste from landfills to composting and anaerobic digestion facilities. It has been an underlying factor in several states' recent restrictions on disposal of food waste in landfills. Another motivating factor is waste diversion goals adopted by states and municipalities, which are typically based on tonnages diverted. Food waste is by nature wet and heavy, thus diverting it from disposal helps boost recycling rates. Finally, generators of food waste also recognize that keeping that material out of their trash and diverting it to composting – along with their wet and waxed corrugated and soiled paper – reduces their overall cost of waste disposal.

## STATE OF COMPOSTING IN THE U.S.

In July 2014, the Institute for Local Self-Reliance (ILSR), in collaboration with *BioCycle*, released a report, "State of Composting In The U.S." (ILSR, 2014). *BioCycle* conducted a state-by-state survey of composting activity in all 50 states and the District of Columbia. Data (2012) on composting infrastructure in each state, and tonnages diverted to composting, were requested. Composting infrastructure includes municipal and commercial facilities, as well as composting on farms and at institutions (e.g., universities, correctional facilities). A total of 4,914 composting sites were reported (44 states reporting). Of that number, about 3,500 compost only yard trimmings (70%). About 350 (7%) compost food waste. (An update of the composting facilities in *BioCycle*'s online directory, www.findacomposter.com, identified *over 500 composting operations in the U.S. that process food waste* (Krossovitch et al., 2014).)

Composting facilities receiving source separated food waste (i.e., separated by the generator or consumer at the source, e.g., store, restaurant, household) typically require a permit (or an official exemption from a permit) to accept and compost that material. These permits are issued by state solid waste agencies and/or local health or air quality agencies. Food waste has a high moisture content and is high in nitrogen, therefore a carbon source is required for it to be composted effectively. As a result, the majority of food waste composting sites in the U.S. also process yard trimmings and in many cases, wood waste. To encourage more composting of food waste, a number of states have or are streamlining their composting rules to enable yard trimmings composting sites to receive food waste streams. Revised rules also are facilitating composting of small quantities of food waste at urban farms, community gardens and neighborhood composting sites.

Organic waste streams have to travel from the point of generation to the point of processing (composting, anaerobic digestion, livestock feeding, etc.). Ideally, the distance between those two points is as minimal as possible to lessen the carbon footprint of organics diversion. Composting where the organics are generated is ideal, whether that is in a residential backyard, a neighborhood community garden, school grounds, or institutional and corporate campuses. *BioCycle* and ILSR have co-organized two national forums on community composting, as the number of small-scale, and often urban, locations has been growing rapidly. New York City, for example, has more than 200 community composting sites (primarily at community gardens) and 8 to 10 medium-scale operations in the five boroughs. Composting methods range from a single tumbler or 3-bin systems at community gardens to windrows and aerated static piles at some of the medium-scale sites. The latter are receiving upwards of 5 tons/week of household food scraps (Goldstein, 2013). A detailed analysis of community composting in the U.S. is provided in a recent report, *Growing Local Fertility: A Guide to Community Composting* (ILSR/Highfields, 2014). The report describes successful initiatives, their benefits, tips for replication, key start-up steps, and profiles 31 model programs in 14 states and the District of Columbia.

For a host of reasons (e.g., space and labor constraints, urban density) composting on-site or in the neighborhood where the organics are generated isn't always feasible. At this point, the only way to compost the organic waste streams is off-site. Overall, *the pace of developing composting infrastructure to manage food waste has not kept up with the demand to divert this material to composting.* Efforts to create this capacity have been stymied by facility siting and permitting challenges, inadequate financing, and for existing yard trimmings composting facilities, the need to retrofit the sites to receive and process food waste. The situation is exacerbated by the perception, and frequently the reality, that source separated organics are contaminated with plastic and glass. Composting facilities processing only yard trimmings and clean wood can be reluctant to introduce a feedstock with contamination into their facilities. That said, there are many yard trimmings sites in the U.S. that have successfully introduced commercial, institutional and residential source separated food waste into their operations.

## SOURCE SEPARATION STEPS

The most effective source separated organics programs start with source reduction, and then donation of edible food. In almost all cases, source separated organics diversion programs will be replacing traditional trash disposal practices where the generators merely throw everything into the same container inside, which is then taken outside to a dumpster or trash compactor. Many businesses and institutions have been doing some sorting of recyclables, so are already engaged in a limited amount of source separation behavior. For the most part, however, *initiating a source separation program will be a new behavior for everyone involved,* from top management to the food and custodial services – and to the waste haulers servicing these establishments.

Anything that ends up in the source separated organics stream that is not an organic material in origin (or in the case of compostable products, manufactured to biodegrade as an organic material) is a contaminant. Common contaminants include film plastics, packaging, twisty ties, latex gloves used in food service, and glass. There are costs associated with contaminants in the organics stream, primarily related to their removal and the impact on compost product quality. Some programs allow generators to include wet and soiled paper, waxed corrugated and compostable products. Some only allow food waste. All programs spend a lot of time and effort on training kitchen and custodial staff and collection services about source separation.

An effective tool for training is to photograph source separated organics that have contaminants and/or materials not accepted (photos are taken of the contents in the cart or outside container or upon unloading at the composting site) and immediately email them to the generator. This enables the manager at that establishment to identify the source and do follow up training with employees. One composting company makes sure that any new generator being added to the program is serviced last on the collection route for the first several weeks so that those loads are easy to identify (first off the truck) and examine for contaminants. Other collection and composting companies may reject loads, or else charge the generator a premium for directing a contaminated load to a disposal facility.

Continual training is necessary to ensure that employees are properly separating out contaminants/materials not accepted and that they are recovering as much of the food waste that is generated. Other factors that need to be addressed with employee training are creating signage in multiple languages, and maximizing the use of pictures of allowed/not allowed materials on the signage. Also important is positive reinforcement and recognition by management to reward the source separation/participation behavior. Bringing employees to the composting site, where they can see how the food waste they separate is being transformed into compost, pays huge dividends in enthusiasm for the program and proper separation. This also provides generators a first-hand look at the negative impact of contaminants.

## COMPOST UTILIZATION

Amending soils with compost adds organic matter, as well as nutrients. Increasing the organic matter concentration of soils increases the amount of pore space, which in turn allows water to flow into soils more quickly and provides surfaces that can hold onto the water once it is in the soil. These water infiltration and water retention benefits have led to compost use in green infrastructure to manage storm water, e.g., bioswales, green roofs, rain gardens, etc. They also are key benefits in food production, along with the nutrients, including nitrogen and phosphorus.

Use of compost in agriculture can be limited as the nutrients available in the compost may not be sufficient to meet the needs of the crops being grown. In addition, farmers may not have the equipment needed to spread compost. However, local and regional farms participating in food waste composting programs are able to use the compost they produce in food production – and often are willing to receive and compost food scraps if they can keep a percentage of the compost to use on their soils. This "win-win" situation is leading to an increase in on-farm composting of off-farm food waste – providing needed composting infrastructure and closing the loop as close to the point of food waste generation as possible.

## CASE STUDIES

The following case studies illustrate the variations in scale and scope of food waste composting infrastructure.

### Red Hook Community Farm, Brooklyn, New York City

Red Hook Community Farm accepts and processes food scraps as part of the NYC Compost Project (Goldstein, 2013). The composting program processes over 225 tons/year of organic material. Started in 2003, the Red Hook Farm has sought to close the loop with its own organics left over from food production (weeds, spoiled produce, spent crop material), as well as organics generated by its farmers market customers, CSA members, and community partners, including the citywide Greenmarkets compost collection program (ILSR/Highfields, 2014). Red Hook Farm's composting program began with hot compost bins and worm bins, and evolved to a solar-powered aerated static pile (ASP) system.

### Gainesville Compost, Gainesville, Florida

This community composting company collects food scraps from about 20 commercial customers and over 30 households in Gainesville using bicycles and bike trailers, and brings the food scraps to a network of community gardens for composting. Gainesville Compost staff manage the composting operations at the gardens. Finished compost is used in the gardens as well as brought back to the restaurants and businesses diverting their food waste for sale in those establishments (Clark, 2015).

### Two Particular Acres and Weis Markets

Weis Markets has over 160 stores in five states (Pennsylvania, Maryland, New Jersey, West Virginia and New York). In late 2009, Weis initiated a composting pilot using 65-gallon wheeled carts in nine of its stores; however, it found that the cost for the service was prohibitive. About a year later, it decided to test organics separation and collection again. A store audit conducted with the Pennsylvania Recycling Markets Center after the initial pilot found that 50 to 70 percent of its waste stream was packaged and unpackaged food waste. For the second pilot, Weis worked with Two Particular Acres (TPA) in Royersford, Pennsylvania, which has an on-farm composting permit and collects and processes commercial organics. Four stores close to TPA were put on the program, using 65-gallon totes that are collected weekly. Acceptable items include all produce, bakery waste, deli meats and salads, all floral plants, cut flowers and soil, and coffee grounds and filters. Each department within the stores has its own collection containers that can be stored in a cooler.

On average, stores are diverting 1.5 tons/week of food waste (this quantity is post food donation). In addition, the Weis Market in Carlisle, Pennsylvania is working with the student organic farm at Dickinson College to divert the food scraps. Several stores with 40 cubic yard (cy) compactors were added to the program. By diverting food waste, these stores were able to reduce the size of the compactor from 40 cy to 20 cy. The cost of composting with the reduced trash service cannot be more expensive than what the store currently is paying for disposal (Goldstein, 2013). In 2013, a store-brand bagged compost, named "Weis Choice" was introduced in Weis Market stores. The sales copy stated: "Did you know that compost is the best natural soil additive for your garden? Compost increases soil health, improves water retention, and enhances plant growth! Our Weis Choice compost is recycled from our food scraps and organic waste that we generate in stores and is turned into compost. This eliminates the amount of trash we send to the landfill and helps reduce our carbon footprint."

**Onondaga County, New York**

The Onondaga County Resource Recovery Agency (OCRRA), a nonprofit waste management organization, manages the waste for 33 out of 35 municipalities in the county and operates two composting facilities. Its Amboy Compost Facility receives source separated organics from commercial facilities and institutions. OCRRA has three recycling specialists that educate the food waste generators and the community about proper sorting practices. Generators hire haulers to bring the food and yard waste to the Amboy Facility, where trucks are weighed in across an automated scale. Although there is no mandate in Onondaga County to divert organics from disposal or combustion, OCRRA's competitive pricing provides a financial incentive that attracts customers to its program. It charges \$35/ton for organic waste compared to the \$79/ton rate for standard municipal solid waste – a difference of \$44/ton for composting food scraps.

Currently about 44 businesses and institutions in Onondaga County, including Syracuse University, Wegmans Supermarkets and the local shopping mall, participate in the food waste composting program. The Amboy Compost Facility receives approximately 6,000 commercial food waste and yard trimmings deliveries each year. The site is permitted to handle approximately 9,600 tons/year of food waste and 48,000 cy/year of yard trimmings (Siegrist, 2014).

## CONCLUSION

The Last Food Mile Conference focused on how to reduce food wastage "post harvest" and along the food supply chain. The Conference also addressed the challenges of feeding the expected 9 billion people on the planet by 2050, and the reality that about 50 million Americans live in food-insecure households. *BioCycle* magazine connects to the themes of the Last Food Mile Conference when food, no longer edible, needs to be managed sustainably via composting and anaerobic digestion. This includes making sure that the end products derived from these organics recycling processes, especially the compost, become an amendment to build healthy soils to grow more food. In a nutshell, many of the speakers at the Last Food Mile focused on food for people, whereas *BioCycle*'s primary focus is on food for soil.

The reasons to divert unavoidable food waste to composting and anaerobic digestion and back to soil are compelling, including: 1) The U.S. currently disposes over 95 percent of the food waste generated, primarily in landfills; food waste decays and generates methane, which is emitted before the methane is captured in gas collection systems; 2) Agricultural and urban/suburban soils in the U.S. need organic matter; compost, which can be made from unavoidable food waste, is a common vehicle to provide organic matter. Connecting the dots between wasted food, hunger and nutrition, drought and soil health (via organics recycling) and local food production is critically important to address food insecurity and the reality of feeding the world's growing population.

## REFERENCES

Brown, S. 2014. A better warm. BioCycle 55(8):85.

Clark, N. 2015. The business of community composting. Biocycle 56(1):32.

Goldstein, N, 2013.Community composting in New York City. BioCycle 54(11):22.

Goldstein, N. 2013. Trimming costs with composting. BioCycle 54(1):22.

Institute for Local Self-Reliance and Highfields Center For Composting (ILSR). 2014. Growing Local Fertility: A Guide to Community Composting. http://ilsr.org/wp-content/uploads/2014/07/growing-local-fertility.pdf.

Krossovitch, A., S. Katsaros, and N. Goldstein. 2014. Found: Composters taking food scraps!, BioCycle 55(1):24.

Platt, B., N. Goldstein, C. Coker, and S. Brown. 2014. State of composting in the U.S. Institute for Local Self-Reliance. http://ilsr.org/state-of-composting/.

Siegrist, C. 2014. Smooth transition to food waste composting. BioCycle 55(10):20.

USEPA Landfill Methane Outreach Program (USEPA LMOP). 2015. http://www3.epa.gov/lmop/.

USEPA. 2015. Advancing Sustainable Materials Management: Facts and Figures 2013. http://www.epa.gov/epawaste/nonhaz/municipal/pubs/2013_advncng_smm_rpt.pdf.

Van Haaren, R., N. Themelis, and N. Goldstein. 2010. The state of garbage in America. BioCycle 51(10):16.

Yepsen, R. 2012. Residential food waste collection in the U.S. BioCycle 56(1):53.

# *Chapter 17*

## *Food Waste Composting on a Bahamas Resort*

### *Barriers Encountered and Lessons Learned*

**Kathleen Sullivan-Sealey and Jarrell Smith**

**ABSTRACT**

The ability for small islands to meet sustainability goals is exacerbated by the costs of transporting goods onto, and then, solid waste off the islands. Tourism-based industries rely largely on imported foods, and tourism developments often destroy coastal resources (e.g. mangroves) which can diminish the ability of local populations to feed themselves. For the island of Great Exuma (Exuma), The Commonwealth of The Bahamas, solid waste has accumulated faster in recent years than the island solid waste management can absorb, thus threatening coastal fisheries with land-based sources of pollution. The removal of food and organic waste by composting would both reduce waste and provide valuable compost products for small-scale agriculture. This chapter outlines the costs, available resources and barriers to food waste recycling on the small island of Exuma, then explores possible solutions through food composting for the tourism industry.

## CHALLENGES TO RECYCLING AND COMPOSTING FOOD WASTE ON ISLANDS

The management of solid waste represents one of the most difficult challenges to the environment of small island developing states (SIDS) such as The Commonwealth of The Bahamas (Bahamas Environment, 2002; SENES Consultants, 2005). For the wider Caribbean, tourism is critical for both employment and balance of trade. However, both overnight and cruise ship tourists are sources of environmental impacts and resource consumption with consequent public health problems (Mateu-Sbert et al., 2013). One of the most important impacts of tourism is the increased generation of municipal solid waste (MSW; Holden, 2008). Many studies have reported MSW increases with both seasonal tourist populations and opening of destination resorts (Shamshiry et al., 2011). The long-term profitability of island tourism depends on responsible collection, transportation, processing as well as the final deposition of the MSW in an environmentally-sound and cost-efficient way (Chen et al., 2005).

The critical challenge to recycling and composting food waste on islands centers on three areas: 1) high and hidden *costs* for waste management, 2) lack of *benchmarking* in solid waste accumulation with landfill performance, and 3) lack of *standards* for environmental protection.

## 1) Cost of Waste Management on Islands

Solid waste management should include recycling or reuse of material for cost savings, environmental protection, and revenue enhancements. Unfortunately for Caribbean islands, labor is costly and limited in supply. Governments of small islands lack a dedicated revenue source and expertise to manage dumps and landfills. The long-term environmental and public health costs of unsorted wastes are not calculated, including pest control costs, groundwater protection or mitigating air pollution. These waste management costs (Figure 17.1) are often not accounted for in considering the value of recycling or composting. Removal of food and compostable waste could be the most effective way to reduce solid waste associated with resorts, and improve sanitation.

In the Bahamas, businesses and residents have no concerns with solid

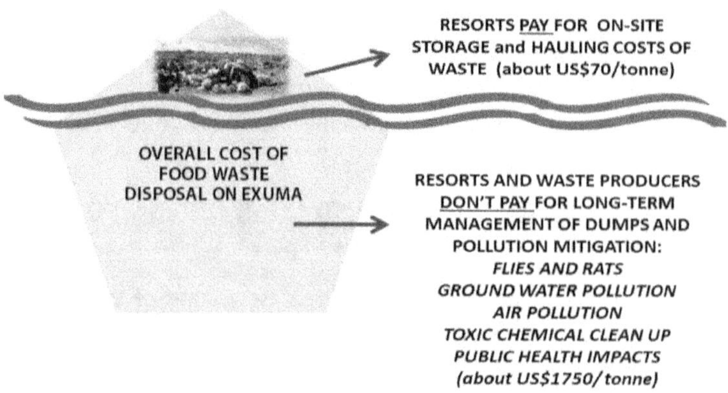


**17.1** Solid waste management costs are often hidden from the resort operator and tourists in The Bahamas. Resorts pay for the storage and hauling of trash, but seldom pay tipping fees to manage landfills. The "underwater" or hidden costs of solid waste on islands include control of flies and rats, groundwater pollution, air pollution and public health impacts (see Sullivan-Sealey and Smith, 2014)

waste beyond having the waste hauled away. Trash haulers get paid to move trash, and then the central government (Department of Environmental Health and Safety) is responsible for the management and the long-term consequences of waste management (Mateu-Sbert et al., 2013). Costs for waste removal should include hauling *and* disposal/ recycling expenses. Recycling is going to be a key component of island tourism into the future, though the implementation will be slow, with setbacks and many "lessons learned" in the process.

As of 2015, there were no Governmental requirements to manage food wastes outside of the immediate food receiving and processing areas of a resort. Resorts pays no waste disposal charges, thus waste disposal costs are tied to the cost of equipment and hauling fees. The costs of solid waste disposal and management on Exuma can be broken down into four areas: a) Resort labor costs to collect and store waste (resort stewarding staff), b) Resort bin and compactor rental, then waste hauling charges from the commercial Exuma Waste Management Company, c) Public environmental and health costs associated with solid waste dumping and disposal, and d) Public long-term costs of management and mitigation of island dumps. Recycling initiatives have to address the labor, hauling, environmental and long-term management costs of solid waste production; these long-term environmental management costs are now assigned to the Government. From this initial study it was estimated that Exuma generated over 21 kg solid waste per capita per week, a significant increase over the 16.6 kg per week documented in 2002, with over 80% of the refuse being placed in open dumps (Firdaus and Ahmad, 2010). Recycling and composting are attractive options for islands to reduce the costs of managing landfills, the pollution impacts of accumulated solid waste and the public health threats.

## 2) Benchmarks and Boundaries for Waste on Islands

All small island tourism is dependent upon the quality of the country's natural resources, namely its native flora and fauna, beaches and coastal waters; thus efforts to promote environmentally-responsible tourism have become widely embraced (Sullivan-Sealey and Cushion, 2009). Ideally, the most profitable and easiest items to remove and reuse/resell are targeted for solid waste reduction. However, food waste in tropical environments presents a particularly pressing problem. Improper disposal of food waste is linked to ground water contamination, coastal hypoxia as well as the multiplying of pests and sporadic solid waste fires (Heileman and Walling, 2005). Modest reductions of up to 20% of the solid waste entering landfills can save millions of dollars in future landfill management and pollution mitigation costs. Integrated solid waste management, particularly large-scale composting, can also be linked to water reclamation and sewage/wastewater processing (Hernández and Martín-Cejas, 2005; Stan et al., 2009).

Food waste makes up a significant component of the solid waste generated in island tourism. In a 2014 study of Bahamian tourism, food waste accounted for as much as 85% of unsorted hotel waste for a single all-inclusive resort (Sullivan-Sealey and Smith, 2014). However the cost of sorting and hauling organic wastes for composting was ten times the cost of hauling unsorted trash. Composting can also offer linked economic opportunities in agriculture and landscaping (David, 2012). However, composting has been shown to be both capital- and labor-intensive in tropical settings, needing management and a market for the compost (Sullivan-Sealey and Smith, 2014). The costs associated with solid waste disposal on islands are especially challenging with the limitations of scale, labor costs, increasing energy costs, as well as the repair and maintenance of machinery (Cooney, 2012). For SIDS, determining the composition and weight of the food waste and organic components of the solid waste stream can be a critical first step to formulating an integrated waste management plan that includes recycling (Gidarakos et al., 2006).

The direct benefits of recycling are reduction in solid waste entering the local landfill, and elimination of flies and rats associated with unsorted solid waste. Indirect benefits include consumption of fewer plastic bags in waste management, reduction in pest management costs, and improved morale in employees seeing the foreign-owned corporation investing in the Bahamian island environment. The Bahamas represent a hot and dry climate, with a sedimentary carbonate geology. All materials deposited on land eventually end up in the ocean through the porous limestone substrate. Food waste benchmarking becomes particularly important if food waste and organic material is disposed of in open dumps. Ideally, food would not be wasted in resorts; surplus or waste food would be used to feed people, then animals, and as a last choice, composted to recycle nutrients for agriculture or landscaping. Food waste is a particular problem on hot, tropical islands as spoilage can limit food waste acceptable for human consumption. With the long distance food travels to get to the island, and challenges in storage and transport to the kitchen, spoilage can account for significant losses to resorts.

### 3) Standards for Environmental Protection

Most large Caribbean hotels do rely on the EarthCheck[1] certification process (formally "Green Globe") to provide technical expertise on the benchmarking and management of solid waste. EarthCheck certification provides assurance to visiting tourists that the hotel management is operating in an environmentally-responsible manner, including the management of food waste. This Earth Check certification requires at least two years of benchmarking documenting the amount of solid waste generated by a resort; however, the resort

[1] See http://earthcheck.org/ for a full description of Earth Check certification for hotels.

must invest in the necessary equipment and personnel. Sandals Emerald Bay Resort (SEB) determined after three months of initial observations in 2012 that action was needed to reduce the flies in the loading dock area. Two recycling methods were developed with the SEB Director of Environmental Management, the kitchen staff and the steward department staff. First, food wastes and compostable materials were to be sorted in the kitchens into plastic seven-gallon (26.5 liters) buckets with sealing lids and second, waste grease and oils were drained into metal buckets for recycling.

The very core of the Bahamian quality of island life is access to clean beaches, healthy reefs and coasts, and successful fishing. But what is "sustainability" for island tourism development? While broadly defined by the Brundtland Commission,[2] much of what is written about sustainability in tourism refers to the operations of a given facility, namely the quality of its infrastructure (e.g. solid waste management, water and waste water management, and energy generation management). With limited resources to manage solid waste accumulating on islands, the Government of The Bahamas has considered, "Could the benefits of recycling solid waste and long-term pollution reduction outweigh the convenience and current low cost of unsorted solid waste disposal?" To address this question, an easy and reliable system of benchmarking solid waste accumulation should be developed, with a particular focus on food waste generated by tourism.

## CASE STUDY OF FOOD WASTE AT SANDALS EMERALD BAY RESORT

Sandals Emerald Bay is a large destination, all-inclusive resort with multiple restaurants and dining facilities. The hotel area includes about 11.25 hectares of outdoor areas, and 245 guest rooms. The original solid waste management plan for the SEB resort had all kitchen, housekeeping, and bar waste enter one central 20-cubic yard (15.3 cubic meters) trash compactor. Unsorted waste enters the single compactor which was emptied on an "as needed" basis, up to three times per week. The undesirable features of the compactor include frequent over-filling of the compactor with dense food wastes, which often resulted in costly break-downs as well as flies and rats attracted to unsorted waste. The pickup of the compactor occured when the compactor was full, which means several days can pass between pick-ups, and the entire service area was malodorous and fly-infested. More specifically, the off-loading of food supplies

[2] Formally the World Commission on Environment and Development (WCED).

on the loading dock adjacent to the compactor room increases the probability of public health hazards (see Figure 17.2).

Exuma is the largest island in the Exuma island chain, with 7,314

**17.2** Visual illustrating the location of the trash compactor in relation to the food and supply delivery station for Sandals Emerald Bay. The unsorted trash adjacent to the food loading dock creates challenges with food safety and hygiene

residents living on the island in six major settlements (Department of Statistics, 2011). There are approximately 546 hotel rooms on Exuma, divided between 16 properties. Sandals Emerald Bay (SEB) resort is the largest, while other luxury condominium properties (Grand Isle and February Point) account for 114 units. Eight resorts have ten rooms or less (e.g. boutique resorts), and five hotels have 10 to 40 rooms. Tourism produces an estimated 50 to 60 per cent of the Gross Domestic Product of The Bahamas. The hospitality industry directly or indirectly employs 60 per cent of the total workforce (Bahamas Ministry of Finance, 2011), and tourism is the primary economic focus for Exuma.

A seven-year study of the solid waste management on Exuma was conducted by collecting records of waste accumulated in the Exuma Regional Sanitary Landfill (ERSL) and photographic documentation of the amount and composition of solid waste generated for the overall island of Great Exuma. The ERSL is the prime facility that accommodates all solid waste generated for the entire island. The opening of SEB in 2010 on Exuma, and the Sandals corporate commitment to EarthCheck certification provided an ideal opportunity to explore cost-effective means to reduce solid waste, particularly the problems associated with food waste. The largest single source of food waste on the island of Exuma is SEB. Removal of food and organic wastes could be the single most important step in solid waste management.

The case study focused particularly on food waste from SEB in 2012 and 2013. The question was, "How much food is thrown away, and can this food waste be economically composted?" Records were kept on solid waste arrivals (tonnes per week, month and year) from 2006 until May 2015. Total solid waste arrivals could be compared to the solid waste hauling from SEB, based on private waste management hauling records of weight and billing (BD$/per tonne). Lastly, a detailed photographic survey of food waste both at the point of origin (waste in garbage bins in SEB, other resorts and private businesses), and at the landfill was used to determine the amount of food waste dumped on the island.

Figure 17.3 illustrates the amount of solid waste entering the ERSL. The increase in waste production from 2010 has posed a serious challenge to both the hauling capacity on the island, and landfill capacity. The ERSL and a transfer

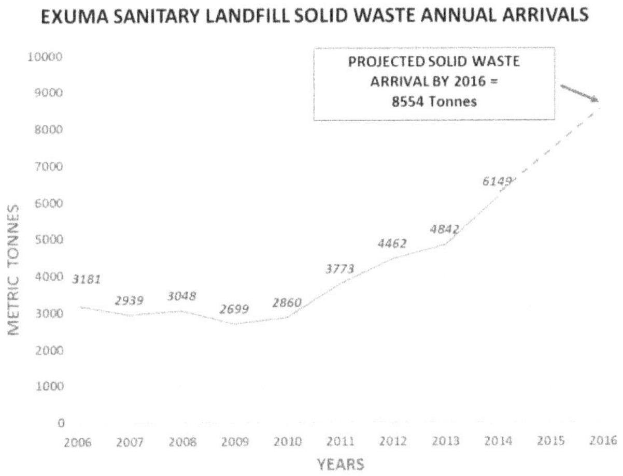

EXUMA SANITARY LANDFILL SOLID WASTE ANNUAL ARRIVALS

PROJECTED SOLID WASTE ARRIVAL BY 2016 = 8554 Tonnes

METRIC TONNES

10000
9000
8000
7000
6000
5000
4000
3000
2000
1000
0

3181
2939
3048
2699
2860
3773
4462
4842
6149

2006 2007 2008 2009 2010 2011 2012 2013 2014 2015 2016

YEARS

**17.3** Exuma Regional Sanitary Landfill solid waste annual arrivals from 2006 projected through 2016. Sandals Emerald Bay purchased the property in 2010, and after remodeling, began operating in 2011. High food waste composition of the solid waste results in about 320,000 kg of nitrogen added to ground water and adjacent wetlands

station have been under contract to private companies on a fixed budget since 2006. Currently, solid waste accumulation in the ERSL exceeds 6000 tonnes per year. Solid waste accumulation has increased as anticipated with the re-opening of the Emerald Bay Resort by Sandals. Sandals Emerald Bay has had a significant impact on solid waste streams since 2010. Most notably, waste streams are higher and consistent throughout the year, and waste streams have a much higher content of food waste. Food waste represented about 37% of the solid waste generated by SEB. The year-round operations of Sandals with a larger number of rooms have produced a more consistent trash stream, unlike the previous situation (prior to 2010) when waste volume would drop sharply over the summer (off-season). More uniform waste streams mean greater opportunities for recycling. Recycling is dependent on consistent waste streams, so there has never been a better time to invest in recycling infra-structure.

A pilot project to recycle food waste from SEB was carried out after several months of benchmarking. Buckets were provided in the resort kitchens to sort compostable materials from the solid waste flow to the compactor. Although the food waste sorting was quickly adopted by the kitchen staff, a serious barrier to food waste recycling was the labor and cost of composting off-site. Although a number of online training videos were developed for SEB, there were few resources to re-train and re-design service collection areas for sorted waste streams. The kitchen staff could make the change to sort food waste into the buckets relatively quickly, while the stewarding staff had to keep up with removal of full food waste buckets and re-supplying empty buckets. The food waste buckets required an increased work load.

The stewarding staff was less enthusiastic about additional work with limited resources, citing the problems of storage of empty buckets, and the additional time needed to collect and distribute empty buckets to all the kitchens. SEB management was not in a position to re-negotiate labor relations with stewarding staff; thus food waste management initiatives were difficult to retro-fit. The message from SEB management was, "Sandals wants to be EarthCheck certified, but not put any new resources into this effort." There are other alternatives to sorting the waste stream that require capital investment; mechanical grinders installed in kitchens could masticate all food waste in a slurry system that can reduce the handling of food waste, and allow storage of food waste slurry in an outdoor tank for bulk removal.

Other labor issues associated with island tourism include the dependence of SEB on employees brought in from other countries or other islands in The Bahamas. Outsiders would have little interest or investment in the local environmental quality of Exuma. Most workers associated with solid waste management are paid minimum wage in The Bahamas, meaning that a full time worker (40 hours per week) earns about $11,429 per year.[3] However the annual cost of living on the Out Islands of The Bahamas is well over $37,500.[4] The declining purchasing power of Bahamians (especially on out islands, referring to islands beyond the capital, Nassau on New Providence, or the second largest city, Freeport on Grand Bahama island) is made worse by the decline of near shore fisheries resources as a supplementary food and income source (Sullivan-Sealey, 2004; Sullivan-Sealey et al., 2014). Although The Bahamas is widely considered the wealthiest independent country in the wider Caribbean, the escalating cost of imported food, utilities and transportation (fuel) costs on the islands has outstripped the earning power of the population. This problem is exacerbated by the increase in land-based sources of pollution and declining coastal fisheries and natural resources. Moreover, the recent implementation of a Value Added Tax (VAT) as of January 1, 2015, further strains the income of Bahamians. The proximity of SEB, the ERSL and coastal wetlands is illustrated in Figure 17.4.

The greatest barrier to island composting of food wastes is cost – the short-term costs and re-training required changing old habits and embracing new technologies. Recycling would be ideally implemented in steps, and issues of environmental health would mandate a timetable for integrated solid waste sorting by location. Currently, a small amount of food waste from the SEB kitchens is used to feed domestic pigs confined to small offshore islands (Figure 17.5). The phenomenon of the "Exuma swimming pigs" has become a popular, if misguided, tourist attraction (MacNaull, 2012). The amount of untreated food waste fed to pigs accounts for less than 60 kg per week or about 0.006% of the food waste generated. Feral pigs are a known conservation issue on islands, as they destroy slow-growing vegetation, eat native reptile species, excavate sea turtle nests, and potentially transmit disease/ parasites. Feeding the swimming pigs is popular and lucrative (up to US$250 per person for the boat trip and "feeding" experience); this type of "agro-tourism" could help fund more significant food waste composting efforts if managed properly.

In essence, the Government of The Bahamas has been giving away something for free (solid waste management and processing) and now SEB is being asked to pay or contribute to food waste processing costs through green certification programs such as EarthCheck. Employees are not likely to "donate" their time or work on recycling initiatives that they perceive to benefit primarily the (subsidized) resorts. The broader impacts of improper solid waste disposal

---

[3] See Database on Labour Legislation, Bahamas Minimum Wage Act at http://www.ilocarib.org.tt/cariblex/bahamas_act4.shtml

[4] http://www.numbeo.com/cost-of-living/country_result.jsp?country=Bahamas

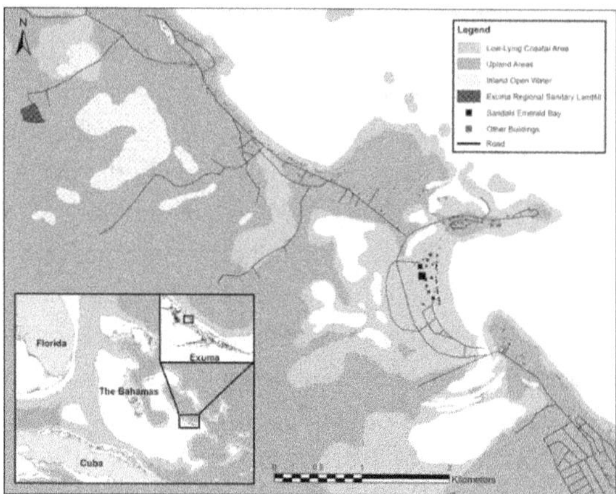


**17.4** Map illustration the footprint of the Sandals Emerald Bay resort, and Exuma Regional Sanitary Landfill. The area is surrounded by coastal wetlands and ponds, all part of a sensitive island environment supporting coastal fisheries, seagrass beds and coral reefs

**17.5** "Swimming pigs and food waste" are becoming a popular attraction in the Exuma Cays. Pigs confined to small off shore islands swim to an approaching boat to be rewarded with food scraps. These same pigs are then used in pig roasts and tourism restaurants

linked to declining fisheries resources puts The Bahamas on an unpleasant trajectory to increased poverty and environmental degradation. The pollution from food waste dumping has in fact contributed to local poverty by decreasing fish habitat and availability of fish near shore (Sullivan-Sealey, 2004). The ability of island populations to supplement their diet with local fish has been key to Bahamian culture and island economies.

Tourists coming to The Bahamas predominantly from North America or Europe expect recycling to occur, and would self-sort high value items like aluminum cans or compostable waste (Gidarakos et al., 2006). The EarthCheck Standards incorporate an on-going system of benchmarking for each resort property. The goal-setting within EarthCheck aims to reflect the best science and technology available for a sustainable tourism and travel industry (Ball and Taleb, 2011). The Government of The Bahamas can actually encourage and facilitate the EarthCheck certification process by installing and maintaining scales at all regional landfills and transfer stations, and then requiring weight records to be maintained and reported.

## CONCLUSION

Tourism on islands is a trashy business, and there are challenging issues in providing luxury food items on an island. Food waste is generated from spoilage, over-preparation for buffets and events, and even serving specialty foods with limited storage life. Successful recycling and environmental protection require three actions: government oversight, industry responsibility and consumer awareness. The Government of The Bahamas has outlined a systematic approach to the monitoring of tourism developments for environmental impacts once the resorts are in operation in a national Environmental Management Action Plan (NEMAP) (Wells-Moultrie, 2006). An assessment of pollution and environmental threats to The Bahamas lists the contamination of freshwater lenses (ground water) and near-shore marine waters by nutrients and organic wastes as the greatest danger, based on the carbonate geology of the islands (Buchan, 2000). The Bahamas has signed numerous international environment agreements, clearly highlighting the adoption of global environmental responsibilities.

Lastly, the issues of outreach and education are crucial to managing expectations and encouraging compliance in recycling. Work-place education needs to be implemented with department heads and unions to ensure a smooth implementation of sorted solid waste streams from resorts and hotels. Resorts such as SEB as well as island residents need to embrace the consumer limitations of island life, and reduce food waste accumulation. Any discussion of agriculture must include solid waste management and the availability of compost as a fertilizer alternative. Large resorts can facilitate or fund industrial-scale composting, including waste water sludge, which could reduce reliance on imported fertilizers, allow water recovery, and reduce the threat of nutrient loading to critical coastal and marine environments.

## ACKNOWLEDGEMENTS

Special thanks to Exuma Waste Management (Mr. Cyril Rolle) and Island Waste Management (Mrs. Velthia Rolle); these companies also provided commercial waste hauling, weight and billing records. Assistance in recording weights came from Louis Grace, Anastasia Gibson and Paulette McPhee for their work with Recycle Exuma. The food waste bucket experiment would not have been possible without the assistance of Ms. Charlene Reid and Ms. Annette Carey, Sandals Emerald Bay. Jacob Patus completed data entry of the landfill records and GIS work. Funding was provided by the Earthwatch Institute Coastal Ecology Project, University of Miami College of Arts and Science and by Recycle Exuma.

## REFERENCES

Anonymous. 2005. Solid Waste Management Loan to The Bahamas from IDB. Caribbean Update 21(1):4.

Bahamas Environment, S. a. T. C. B. 2002. Bahamas Environmental Handbook. Nassau, The Commonwealth of The Bahamas.

Bahamas Ministry of Finance. 2011. Labour Market Information Newsletter. Nassau, The Bahamas.

Ball, S., and M.A. Taleb. 2011. Benchmarking waste disposal in the Egyptian hotel industry. Tourism Hospitality Res. 11(1):1-18. http://dx.doi.org/10.1057/thr.2010.16.

Buchan, K.C. 2000. The Bahamas. Marine Pollut. Bull. 41(1):94-111. http://dx.doi.org/0.1016/s0025-326x(00)00104-1.

Chen, M.C., A. Ruijs, and J. Wesseler. 2005. Solid waste management on small islands: the case of Green Island, Taiwan. Resour. Conserv. Recy. 45(1):31–47. http://dx.doi.org/10.1016/j.resconrec.2004.12.005.

Cooney, S. 2012. The economics of recycling in Hawaii. Hawaii Business:58. http://www.hawaiibusiness.com/the-economics-of-recycling-in-hawaii/.

David, S. 2012. Table scraps to farm food; Johns Island man sets up business to haul food waste for compost. *The Post and Courier,* May 7 p. D.3. http://www.postandcourier.com/article/20120507/PC05/120509374.

Firdaus, G., and A. Ahmad. 2010. Management of urban solid waste pollution in developing countries. Int. J. Environ. Res. (4):795-806. http://dx.doi.org/10.1605/01.301-0012935403.2011.

Gidarakos, E., G. Havas, and P. Ntzamilis. 2006. Municipal solid waste composition determination supporting the integrated solid waste management system in the island of Crete. Waste Manage. 26(6):668-679. http://dx.doi.org/10.1016/j.wasman.2005.07.018.

Heileman, S., and L.J. Walling. 2005. Caribbean Environmental Outlook. United Nations Environment Programme (UNEP), Earthprint, Nairobi, Kenya.

Hernández, M. G., and R.R. Martín-Cejas. 2005. Incentives towards sustainable management of the municipal solid waste on islands. Sustain. Develop. 13(1):13-24. http://dx.doi.org/10.1002/sd.241.

Holden, A. 2008. Environment and Tourism (second ed.). Routledge, New York, NY.

MacNaull, S. 2012. Plethora of porkers: Swimming with swine: Meeting four giant pigs a highlight of Bahamas adventure tour. *The Spectator,* p. T.4.

Mateu-Sbert, J., I. Ricci-Cabello, E. Villalonga-Olives, and E. Cabeza-Irigoyen. 2013. The impact of tourism on municipal solid waste generation: The case of Menorca Island (Spain). Waste Manage. 33(12):2589-2593. http://dx.doi.org/10.1016/j.wasman.2013.08.007.

SENES Consultants, Ltd. 2005. National Environmental Management Action Plan (NEMAP) for The Bahamas. Ministry of the Environment, Nassau, The Bahamas.

Shamshiry, E., B. Nadi, M.B. Mokhtar, I. Komoo, H.S. Hashim, and N. Yahaya. 2011. Integrated models for solid waste management in tourism regions: Langkawi Island, Malaysia. J. Environ. Public Health article no. 709549. http://dx.doi.org/10.1155/2011/709549.

Stan, V., A. Virsta, E.M. Dusa, and A.M. Glavan. 2009. Waste recycling and compost benefits. Not. Bot. Horti. Agrobot. Cluj 37(2):9-13.

Sullivan-Sealey, K. 2004. Large-scale ecological impacts of development on tropical islands systems: Comparison of developed and undeveloped islands in the central Bahamas. Bull. Marine Sci. 75(2):295-320.

Sullivan-Sealey, K., and N. Cushion. 2009. Efforts, resources and costs required for long term environmental management of a resort development: the case of Baker's Bay Golf and Ocean Club, The Bahamas. J. Sustain. Tour. 17(3):375-395. http://dx.doi.org/10.1080/09669580802275994.

Sullivan-Sealey, K., and J. Smith. 2014. Recycling for small island tourism developments: Food wsate composting at Sandals Emerald Bay, Exuma, Bahamas. Resour. Conserv. Recycl. 92:25-37. http://dx.doi.org/10.1016/j.resconrec.2014.08.008.

Sullivan-Sealey, K., V.N. McDonough, and K.S. Lunz. 2014. Coastal impact ranking of small islands for conservation, restoration and tourism development: A case study of The Bahamas. Ocean Coast. Manage. 91:88-101. http://dx.doi.org/10.1016/j.ocecoaman.2014.01.010.

Wells-Moultrie, S. 2006. The National Environmental Policy for the Commonwealth of The Bahamas. Ministry of Health and The Environment Nassau, Bahamas.

# *Part Four*

## *Save food reduce waste – policies, the law*

# Chapter 18

## *Targeting Federal Programs and Policies to Combat Food Loss and Waste*

**Elise Golan**[1]

## ABSTRACT

The amount of food loss and waste (FLW) in the United States – estimated by USDA at approximately 133 billion pounds or $162 billion at the retail and consumer levels in 2010 – has spurred federal analysts and policymakers to consider the root causes of food loss and waste and how to design effective policy responses. Not all food loss and waste is indicative of inefficiencies in the food supply system. However, the societal and environmental ramifications of food loss and waste can create a wedge between societal and private costs and benefits. As a result, government interventions have the potential to stimulate reductions in FLW that reap net benefits to society. USDA and other federal interventions targeting food loss and waste have focused on five primary objectives: 1) support efficient markets for agricultural commodities and food; 2) increase incentives to donate wholesome, otherwise wasted food; 3) cultivate innovative technologies and systems for reducing, recovering or recycling; 4) strengthen USDA programs and policies to minimize food loss and waste; and 5) raise awareness about food loss and waste to motivate private-sector action.

## INTRODUCTION

Reducing food loss and waste (FLW) [2] is integral to improving the sustainability of food systems around the world. High levels of FLW – estimated

---

[1] The views presented here are those of the author and not necessarily those of the U.S. Department of Agriculture.

[2] The FAO defines food loss as "the decrease in the quantity or quality of food," and food waste as the part of food loss which is fit for consumption but is removed from the food supply chain "by choice, or which has been left to spoil or expire as a result of negligence by the actor – predominantly, but not exclusively the final consumer at household level." While the term food loss encompasses food waste, FAO uses the term food loss and waste (FLW) "to emphasize the importance and uniqueness of the waste part of food loss" (FAO, 2014a). Similarly, USDA defines food loss as the edible amount of food, postharvest, that is available for human consumption but is not consumed for any reason. It includes cooking loss and natural shrinkage (for example, moisture loss); loss from

by the Food and Agricultural Organization of United Nations at about 30 percent (FAO, 2014b) of the global food supply – raise questions about the efficiency of global food systems. They raise doubts about our ability to sustainably nourish our growing population while safeguarding the earth's natural resources.

The amount of food loss and waste in the United States – estimated by the U.S. Department of Agriculture (USDA) at approximately 133 billion pounds or $162 billion at the retail and consumer levels in 2010 (Buzby et al. 2014) – likewise raises questions about the efficiency of the U.S. food supply system. It has spurred federal analysts and policymakers to consider the root causes of FLW in the United States and investigate effective policy responses. This chapter examines some of the basic economics of food waste, the potential for market failure and inefficiencies in FLW generation and management, and the policy response of the federal government, with a focus on USDA policy and programs.

## FLW Could be Cost-effective for Individual Businesses and Consumers but not Society

Food is a logistical nightmare. Most is highly perishable, necessitating real time delivery for fresh, untreated products, or processing and/or controlled atmosphere storage for products distributed over time and space. Variability in the quality and quantity of production at the farm level further complicates the challenge of delivering specified amounts of consistent quality products to manufacturers and consumers.

The difficult logistics of food delivery are further evidenced by the fact that food delivery is one of the last frontiers for e-commerce. Dot-com vanguard WebVan shut down its Internet grocery store in 2001 after a disastrous six years. Amazon and Google are only slowly entering the business. As quoted in a 2013 Reuter's article (Reuters, 2013), Roger Davidson, a former grocery executive at Wal-Mart, Whole Foods and Supervalu, said Amazon will struggle to make money from AmazonFresh because fresh produce can easily go bad in storage warehouses and get damaged during delivery. "Will it work? I would bet against it," Davidson said. "The reasons these businesses have failed in the past have not gone away."

The difficult logistics of food production, storage and distribution contribute to the expense of building flexible, fail-safe processing and distribution systems that minimize food waste. As a result, there are situations all along the chain where FLW may be the less expensive option. At the farm level, for

---

mold, pests, and inadequate climate control; food discarded by retailers due to color or appearance and plate waste by consumers.

example, the net benefits of leaving an over-abundance of "seconds" in the field may outweigh those of hiring extra labor and selling product into secondary markets. Processors may throw away edible scraps because food recovery or recycling is more expensive than landfilling. Restaurants may find it less expensive to throw some food away rather than lose customers for lack of reliable menu options. Consumers may over buy and waste produce because it is less expensive to stock up and waste than to make multiple runs to the grocery store.

While FLW might sometimes be the least-cost solution for individual businesses and consumers, it may not be the least cost solution for society. This is the case when the net private benefits of FLW are greater than societal net benefits, either because societal costs are greater or societal benefits are less than those of the individual business or consumer. In the case of FLW, three observations suggest that individual and societal costs and benefits may not be well aligned:

- Wholesome food that could have helped feed people in need is sent to landfills. FAO estimates that 805 million people were chronically undernourished in 2012-2014 (FAO, IFAD, WFP, 2014). USDA estimates that in 2013, 3.8 million American households were unable to provide adequate, nutritious food for their children at some time during the year (Coleman-Jensen et al., 2014).

- The land, water, labor, energy and other inputs used in producing, processing, transporting, preparing, storing, and disposing of discarded food are pulled away from uses that may have been more beneficial to society – and generate impacts on the environment that may endanger the long-run health of the planet.

- FLW also has ramifications for climate change. In the United States, FLW (including the inedible parts of food) is the single largest component going into municipal landfills, where it quickly generates methane; helping to make landfills the third largest source of methane in the United States (EPA, 2012).

The societal and environmental ramifications of FLW drive a wedge between social and private costs and benefits. They open the door to market failure and the possibility of efficiency-enhancing government interventions. Well-targeted interventions have the potential to stimulate reductions in FLW that reap net benefits to society. It is important to note, however, that this is not the case in every instance. In some cases, food waste may represent the least-cost option for society and the environment. For instance, a low-efficiency refrigeration system for a secondary apple harvest or a non-recyclable polystyrene container for scant leftovers from a restaurant may be more costly for the environment than the wasted food.

## U.S. Policy Targets FLW Across the Food Supply Chain to Final Disposition

Economic theory advises that the best-targeted policy to correct market failure aims to better align societal and private costs and benefits. Well-targeted FLW policy should therefore aim to align societal and private FLW costs and benefits. If successful, such policy would motivate individual businesses and consumers to make FLW decisions that are optimal for society and the environment. The U.S. Federal Government has initiated a number of programs and policies to reduce or better manage FLW, many of which will result in a better alignment of societal and private costs and benefits.

U.S. policy can be grouped into five main objectives:

### 1. Support Efficient Markets for Agricultural Commodities and Food

Efficiently functioning markets are key for reducing FLW. Such markets include robust storage and transportation infrastructures to distribute commodities and food products to consumers with minimal deterioration in safety and quality. They include efficient market mechanisms and credit programs to direct the flow of commodities and food products to high-valued uses (and out of landfills). They include outlets for different qualities of produce, such as markets for freezing, canning, drying, or juicing. They include clear safety standards, product definitions and labeling standards to minimize FLW due to safety or quality issues.

The U.S. Department of Agriculture supports a wide variety of programs to strengthen agricultural and food markets. USDA's Farm Service Agency (FSA) administers the Department's credit and loan programs and manages its conservation, commodity, disaster and farm marketing programs through a national network of offices. USDA's Food Safety and Inspection Service (FSIS) works to ensure that the Nation's commercial supply of meat, poultry, and egg products is safe, wholesome, and properly labeled and packaged. FSIS guidelines and outreach on proper food storage and food safety, such as the "Be Food Safe" program, help to directly reduce FLW. USDA's Agricultural Marketing Service (AMS) administers programs that facilitate the efficient, fair marketing of U.S. agricultural products, including food, fiber, and specialty crops. It is an active participant in setting national and international standards. USDA's Economic Research Service (ERS) provides market analyses and research reports to improve the efficiency of agricultural and food markets.

**2. Increase Incentives to Donate Wholesome Otherwise-Wasted Food**

Some FLW is composed of unmarketable but still wholesome food. This wholesome food can be recovered and donated to people in need. The U.S. Environmental Protection Agency (EPA) and USDA prioritize source reduction and feeding people as the top activities to address FLW, as shown in the U.S. hierarchy of action (Figure 18.1).

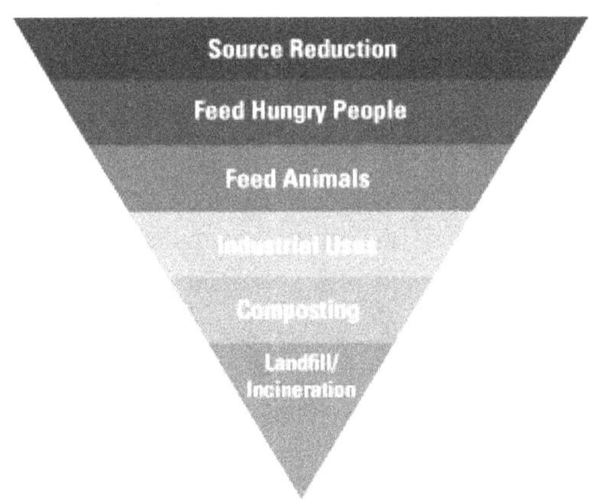


**18.1** U.S. Hierarchy of Action Against Food Loss and Waste

This prioritization hierarchy reflects efficiency as well as food-security objectives. It reflects the observation that it is generally inefficient to use food produced to feed people as animal feed or as inputs to industrial uses or composting; the same or fewer resources could have been used to produce nutritionally balanced animal feed, well-targeted feed stocks for industrial use, and high-value soil amendments.

The U.S. Federal Government has enacted two key legislative actions to incentivize greater donation of wholesome otherwise wasted food:

- The Bill Emerson Good Samaritan Act, which was enacted in 1996, was created to encourage the donation of food and grocery products to qualified nonprofit organizations. Under this Act, as long as the donor has not acted with negligence or intentional misconduct, the donor is not liable for damage incurred as the result of illness.
- Internal Revenue Code 170(e)(3) provides federal tax deductions to businesses to encourage donations of fit and wholesome food to qualified nonprofit organizations serving the poor and needy. Qualified business taxpayers can deduct the cost to produce the food and half the difference between the cost and full fair market value of the donated food.

These government incentives are helping to support and expand the culture of giving and food donation in the United States. Feeding America, a leading U.S. hunger-relief charity, estimates that in 2014, food donations to its food banks diverted 2.5 billion pounds of food from landfills, supplying approximately 2.1 billion meals to feed communities across the United States (Hanner, 2014). Feeding America argues that tax incentives help support this level of giving by noting that when federal tax incentives for food donations were temporarily expanded to cover more businesses in 2006, food donations across the country rose by 137 percent (Feeding America, 2012).

**3. Cultivate Innovative Technologies and Systems for Reducing, Recovering or Recycling FLW**

By increasing the feasibility or reducing the cost of better FLW management, innovation helps to make reducing, recovering and recycling food waste economically viable for businesses, organizations and households. Through its research agencies, USDA invests in research on new technologies for reducing spoilage of fresh foods and the development of new products from agricultural by-products and waste materials.

Recently, USDA researchers, often in collaboration with industry and academic partners, have undertaken a number of FLW-reducing research projects:

- Development of a fruit- and vegetable-based powder to inhibit spoilage of fresh-cut produce.

- Investigation of genetic/breeding options for inhibiting sprouting of potatoes during storage.

- Development of active packaging to extend fruit and fresh-cut produce shelf life.

- Development of a technology to utilize olive-mill waste-water in body-care or beverage products.

- Development of a process to produce new oils and dietary-fiber products from fruit and vegetable seed byproducts.

- Development of a grape-seed flour, a byproduct of wine making, as a healthy food ingredient that helps to lower the risks of heart disease and obesity.

- Development of a 2-stage anaerobic digestion process for potato-processing waste to produce a substitute for peat moss, an imported non-renewable matrix for potting and garden soils.

- Development of a digester process for coffee grounds to substitute for peat moss and imported non-renewable materials.

- Development of a new design and operational procedures for retail grocery store open-air, lighted and refrigerated produce display cases to reduce spoilage, and extend shelf-life and nutritional quality.

- Development of new food and feed ingredients from fish processing waste.

- Development of a small scale peanut dryer and peanut sheller that is suitable for use in remote areas of Haiti, where as much as 50% of the peanut crop is lost due to poor moisture control.

- Development and commercialization of novel nutritious gluten-free fruit and vegetable wraps.

- Development of a licensed technology for forming 100% fruit health bars.

- Commercialization of an ARS-developed process to create a product from sweet potato culls.

- Development of an optical property analyzer to help growers assess crop maturity and quality of food products, and thus help determine optimal harvest time and appropriate postharvest handling/processing procedures to minimize food loss and enhance marketability.

- Development of an automated in-orchard apple sorting technology to enable apple growers to remove inferior fruit in the orchard and better manage the harvested fruit in postharvest handling, thus avoiding potential devastating product loss during postharvest storage/handling.

- Development of value-added food products from rice hulls, including products (such as ground beef and catfish patties) utilizing antioxidants from rice hulls to reduce lipid oxidation.

**4. Strengthen USDA Programs and Policies to Minimize FLW**

In June 2013, USDA and EPA joined together to launch the U.S. Food Waste Challenge, calling on businesses and organizations across the food chain to step up their efforts to reduce, recover and recycle FLW. As part of its commitment to the Challenge, USDA initiated a number of activities within its mission areas to contribute to reducing FLW. By the end of 2014, USDA had delivered on a number of these commitments, including those to:

*Minimize Food Waste in the School Meals Programs.* USDA has taken a number of steps to measure plate waste in the school meal programs and to develop innovative approaches to reducing it.

- Completed the design and school recruitment for a study on the amount of plate waste in schools with respect to the type of foods wasted and student and school characteristics. This research is on track for completion in 2017.
- Worked with the Cornell Behavioral Economics Center to develop and then disseminate Smarter Lunchroom training to over 2,400 school professionals. Funded 12 subgrants to university researchers examining the impact of behavioral economics approaches in school cafeterias.
- Conducted approximately 60 trainings from January to October 2014 to school food service personnel to reduce in-kitchen food loss, reaching over 3,500 participants.

*Generate and Disseminate Information about FLW.* USDA has conducted a variety of activities to educate consumers about safe food storage, package dating, and the benefits of and steps to successfully reducing, recovering, and recycling food waste.

- Updated the safe-storage and date-labeling information on the website of USDA's Food Safety and Inspection Service's website.
- Updated and expanded the 10-year-old online FoodKeeper Resource, which provides food storage information on a wide range of products (in partnership with the Food Marketing Institute).
- Developed (in partnership with the Food Marketing Institute and Cornell University) a FoodKeeper App to provide consumers with an option to access clear, scientific information on food storage, proper storage temperatures, food product dating, and expiration dates. The App is on track for delivery in June 2015 in conjunction with a nationwide consumer education campaign.
- USDA's National Institute of Food and Agriculture (NIFA) funded research examining childhood obesity, eating patterns and food waste. Of particular relevancy to this volume, NIFA co-sponsored a conference at the University of Pennsylvania (December 8-9 2014) entitled "The Last Food Mile".

- Purchased a demo composter for composting food waste for the garden at headquarters and provided composter demonstrations as part of the educational tours of the USDA headquarters garden.

*Recover or Recycle Food that has been Removed from Commerce.* USDA completed a variety of activities to increase the recovery or recycling of wholesome food that is removed from commerce.

- Streamlined procedures for donating wholesome misbranded meat and poultry products by making changes to allow establishments to donate such product without temporary label approval provided that the bills of lading for the product include certain information for Agency verification activities. (*Notice 68-13 Verifying Donation of Misbranded and Economically Adulterated Meat and Poultry Products.*)

- Conducted a one-year meat composting pilot program for meat samples submitted for chemistry analysis, diverting 8,800 pounds of meat from solid waste disposal to recycling (at the Food Safety and Inspection Service's Western Laboratory). This successful program is now in use at two of FSIS's three labs (Western Laboratory and Eastern Laboratory). The labs will continue to track monthly quantities of meat composted and explore options for expanding and enhancing the Meat Sample Composting Program to include the Midwestern Laboratory.

- Connected fresh produce importers with charitable institutions to help increase donations of wholesome fresh imported produce that is subject to destruction or rejection because it does not meet the same or comparable federal marketing order standards as the domestic product.

- Began working with the California Desert Grape Administrative Committee to specify alternative exempted outlets for fresh table grapes that are not inspected or that do not meet certain federal marketing order requirements.

*Update Estimates of Food Loss in the United States.* USDA's Economic Research Service (ERS) calculates and maintains the U.S. Food Availability data system, including the Loss Adjusted Food Availability (LAFA) data series. This data series was primarily designed to estimate daily per capita calorie availability and food-pattern equivalents of the five major food groups plus the amounts of added sugars and sweeteners and added fats and oils. These data include the widely cited estimates of food loss at the retail and consumer levels in the United States. USDA is on track to release in June 2016 the results of a study updating the loss estimates for fresh fruit, vegetables, meat, poultry, and seafood at the retail level in the United States.

*Reduce and Recycle Food Waste at USDA Headquarters.* USDA increased the amount of food waste it composts from the USDA headquarters in Washington DC from 2,400 to about 2,650 pounds of food waste per week. This represents about a 10% increase, which is double the 2014 goal of 5%.

**5. Raise Awareness About FLW to Motivate Private-sector Action**

USDA and other federal agencies have been raising awareness about the importance of reducing FLW for a long time. In fact, some of these efforts stretch back to the early 1900s, as demonstrated by the WWI-era poster in Figure 18.2.

**18.2** USDA WWI-Era Poster on reducing FLW

Recent awareness-raising efforts trace back to Secretary of Agriculture Glickman's gleaning initiatives. In 1996, USDA published "A Citizen's Guide to Food Recovery" (revised in 1997) and worked with the National Restaurant Association to publish a food recovery guide for restaurants. In September 1997, USDA joined with a number of non-profit organizations to sponsor the first ever National Summit on Food Recovery and Gleaning. During this period, the Secretary of Agriculture also served as the Chair of the interagency working group on Food Recovery to Help the Hungry, which had the objective of fulfilling the President's directive that all federal agencies donate excess food to the extent practicable. This activity helped lay the groundwork for passage of the U.S. Federal Food Donation Act of 2008. This act specifies procurement contract language encouraging federal agencies and contractors of federal agencies to donate excess wholesome food to eligible nonprofit organizations to feed food-insecure people in the United States.

More recently, in 2010, the EPA launched the *Food Recovery Challenge* to encourage U.S. businesses and industry to make measureable commitments to reduce their food waste. With this Challenge, EPA offers participants access to data management software and technical assistance to help them quantify and improve their sustainable food management practices. Participants enter goals and report food waste diversion data annually into EPA's data management system. They then receive an annual climate profile report that translates their food diversion data results into greenhouse gas reductions as well as other measures such as "cars off the road" to help participants communicate the benefits of activities implemented. EPA provides on-going technical assistance to EPA's Food Recovery Challenge participants to encourage continuous improvement.

In 2013, USDA and EPA joined together to launch the U. S. Food Waste Challenge. This Challenge builds on the EPA challenge but offers businesses and others across the food supply chain a way to participate by reporting their food waste activities, not their food waste amounts and goals. The goal of the greater U.S. Food Waste Challenge is to disseminate information about the best practices to reduce, recover, and recycle food waste and stimulate the development of more of these practices across the entire U.S. food chain. The inventory of activities will also provide a snapshot of the country's commitment to – and successes in – reducing, recovering, and recycling food waste. By the end of 2014, the joint U.S. Food Waste Challenge had over 1,800 participants, well surpassing its goal of 400 participants by 2015.

## CONCLUSION

Entities across the food chain are stepping up to take action to reduce, recover and recycle food loss and waste. Government policy is helping to support and incentivize further action through a wide variety of programs ranging from improving market efficiency to educating consumers about proper food storage. Well-targeted government interventions are helping to stimulate reductions in FLW that reap net benefits to society, including putting food on the table for families in need and helping to conserve our precious natural resources. The U.S. Food Waste Challenge website provides information about federal and other activities to improve the reduction, recovery and recycling of food loss and waste. Stay tuned to learn about future policies and activities.

## REFERENCES

Buzby, J.C., H.F. Wells, and J. Hyman. 2014. The estimated amount, value and calories of postharvest food losses at the retail and consumer levels in the United States. Economic Information Bulletin Number 121. Economic Research Service/USDA.

Coleman-Jensen, A., M. Nord, M. Andrews, and S. Carlson. 2012. Household food security in the United States in 2011. ERR-141. U.S. Dept. of Agriculture Economic Research Service, Washington, DC.

FAO. 2014a. Definitional Framework of Food Loss. http://www.fao.org/fileadmin/user_upload/savefood/PDF/FLW_Definition_and_Scope_2014.pdf.

FAO. 2014b. Global Initiative on Food Loss and Waste Reduction. http://www.fao.org/3/a-i4068e.pdf.

FAO, IFAD and WFP. 2014.. The State of Food Insecurity in the World 2014. Strengthening the enabling environment for food security and nutrition. FAO, Rome. http://www.fao.org/3/a-i4037e.pdf.

EPA. 2012. Municipal solid waste generation, recycling, and disposal in the United States: Facts and figures for 2012. http://www.epa.gov/epawaste/nonhaz/municipal/pubs/2012_msw_fs.pdf.

Feeding America. 2012. Feeding America Urges Swift Vote on Extended Tax Provisions http://www.feedingamerica.org/hunger-in-america/news-and-updates/press-room/press-releases/feeding-america-urges-swift-vote-on-expired-tax-provisions.html.

Hanner, K. 2014. Feeding America: Rescuing Wasted Food to Feed People. *The Last Food Mile Conference*. http://repository.upenn.edu/thelastfoodmile/sessions/session/15/.

Reuters. 2013. Amazon Plans Major Move Into Grocery Business. *CNBC* June 4. http://www.cnbc.com/id/100789495.

# Chapter 19

## *Food Recovery, Donation, and the Law*

**Nicole M. Civita**

At the end of a long, dry summer in the Ozarks, Susan Schneider got to worrying about some peaked possums on her property. Droughts don't discriminate: they make it awfully hard for all critters – great and small – to fill their bellies. A true animal lover, Susan laid out a little buffet of food scraps for the wildlife that make their home around her own. Her offering disappeared without delay. Sensing a need greater than the odd carrot peelings and wilted lettuce she had on hand, she set about sourcing some surplus from her local grocers. Surely, she thought, they must have a few bruised and brown bananas, some soggy spinach – the sort of deteriorated food that is hardly suitable for donation to a food bank – that they would be more than happy to part with.

Susan was correct on one count: poorly produce was in staggering supply. But, to her surprise, no one was willing to share the bounty with her woodland friends.

*We can't give that to you. It's against the law*, said a produce manager with a stern and disapproving shake of his head.

Susan resisted the urge to press her business card into the man's palm. She's terribly polite, you see, and she did not want to embarrass him by revealing that she happens to be an expert in food law. Instead, she tried a different grocery chain. Same question, same sort of response. Again and again, retailers raised a red flag:

*It's just too risky to put something we can't sell out there. You say you want it for the critters, but what if you ate it or fed it to another person and someone got sick. We don't want to be responsible for that. We'd surely get sued.*

Her concern grew: If food retailers are afraid to help some wild animals, they're probably not allowing charities to collect the edible excess either. With a few follow up questions, Susan confirmed, to her dismay, that all the grocers in her area – from the biggest of the box stores to the natural foods coop – were dumping their edible-but-unsalable produce. Immobilized by unwarranted fears of legal risk, these food businesses were paying hefty tipping fees to dispose of food that, with just a little bit of extra effort and attention, could be turned into healthy meals for people and animals in need.

The idea that food businesses – even those with robust sustainability plans and active community engagement programs – would conduct themselves in such a wasteful manner in an increasingly "green" era boggles the mind. Oftentimes, insensible behavior can be traced back to a law or policy that has unintended consequences. When that is the case, good lawyers investigate and expose the problem and press for reform. This time, though, the problem did not seem to be the law itself – there is no federal or state law that categorically prohibits the donation of food or mandates disposal of food that did not sell. Rather, lack of knowledge about the law – specifically, about the express legal protections for those who donate food – was driving wasteful behavior.

Fortunately, legal education is what Susan does best. As the William H. Enfield Professor of Law at the University of Arkansas and the Director of the School of Law's LL.M. Program in Agricultural and Food Law,[1] Susan has trained hundreds of lawyers to support a more sensible and sustainable food system.[2] The next task, it seemed, was to teach the public that they were not legally required to waste food. With a critical infusion of start-up funding from the Women's Giving Circle[3] and the aim of creating and disseminating legal-informational resources to support businesses in developing and implementing food recovery programs, Susan founded the Food Recovery Project (FRP).[4] Under her initial direction and with the help of James Haley, a diligent and accomplished

---

[1] For over 30 years, the LL.M. Program in Agricultural & Food Law at the University of Arkansas School of Law has been leading the nation in agricultural and food law education, research, and outreach. It was the first to offer an advanced legal degree program in agricultural and food law, first to publish a specialized journal devoted to food law and policy issues, and first to offer a fully integrated opportunity for face-to-face and distance education for law students, attorneys, and graduate students in related disciplines.

[2] http://law.uark.edu/directory/?user=sschneid

[3] The Women's Giving Circle (WGC) members are alumnae and friends of the University of Arkansas who encourage and create opportunities for women to become philanthropic leaders. The WGC leverages collective resources to support their University and community by funding timely and innovative projects. http://wgc.uark.edu/12216.php

[4] For further information about the Food Recovery Project, see http://law.uark.edu/food-recovery-project/

graduate research fellow,[5] the Food Recovery Project began shedding some much needed light on the law pertaining to unsold food.[6]

## RISK ASSESSMENT: BETTING THE HOUSE OR BESTED BY A BOOGEYMAN

As demonstrated by Susan's initial effort to access added calories for armadillos, many food businesses hesitate to donate their excess food for fear that doing so will expose their enterprise to unnecessary risks associated with foodborne illnesses, allergen exposure, and any other negative consequences for the ultimate consumers of such food. It is hard to fault food-sector businesses for wanting to dissociate themselves from scary-sounding and potentially devastating outbreaks of *Salmonella*, *Shigella*, *E. coli*, *Listeria*, *Campylobacter* and norovirus. Absolutely no producer, retailer, or restaurant wants to sicken its customers or any other consumers of its food. Plus, developing a reputation as a purveyor of tainted food is not exactly a good business strategy.[7]

Although several well-respected leaders in the area of food recovery and charitable feeding report that they have never heard of a lawsuit alleging harm from donated food,[8] the fear of litigation and costly damage awards have a chilling effect on donations. Indeed, a national survey conducted by America's Second Harvest in 2002 (now Feeding America), demonstrated that more than 80 percent of the responding companies identified the threat of liability for food related injuries as the greatest deterrent for donating excess food.[9] A 2014 report prepared for the Food Waste Reduction Alliance (FWRA) indicates that this fear is eroding very slowly: 67% of food manufacturers, 54% of food wholesalers and retailers, 67% of small restaurants, and 56% of large restaurants that responded to the

---

[5] James Haley is the author of a comprehensive legislative history and statutory analysis of federal liability protections related to donated food, which he produced for the Food Recovery Project, THE LEGAL GUIDE TO THE BILL EMERSON GOOD SAMARITAN FOOD DONATION ACT, available at http://media.law.uark.edu/arklawnotes/2013/08/08/the-legal-guide-to-the-bill-emerson-good-samaritan-food-donation-act/

[6] A few months after the FRP was founded, this author took over as the FRP's Director and created the FRP's first publication: *Food Recovery: A Legal Guide*, available at http://law.uark.edu/documents/2013/06/Legal-Guide-To-Food-Recovery.pdf

[7] Jonathan Bloom, AMERICAN WASTELAND, 222 (2010).

[8] Jonathan Bloom, AMERICAN WASTELAND, 223 (2010)("[L]iability for [a] donation suit has yet to be seen. Nobody I spoke with can recall any such lawsuit. Not Robert Egger, who has run D.C. Central Kitchen for the past twenty years, nor retired executive director of Second Harvest Heartland Richard Goebel, a man Egger described as 'the godfather of food recovery.' Neither Dan Glickman, the secretary of agriculture from 1995 to 2001, nor Joel Berg, the USDA's food-recovery coordinator at the time, could think of one.")

[9] David. L. Morenoff, Lost Food and Liability: The Good Samaritan Food Donation Law Story, 57 FOOD & DRUG L.J. 107, 2002, at 116-17.

FWRA's survey identified liability concerns as a barrier to food donation.[10] To facilitate a more honest assessment of risk, the Food Recovery Project researched the history of lawsuits in this space. Our exhaustive national search of case law turned up zero claims of illness or injury from consuming recovered or donated food. We also found absolutely no examples of a food donor or charitable feeding organization invoking federal liability protections for food donation (detailed below) as a defense to such a claim. No complaint filings, no briefs, no judicial opinions. With no case law to analyze, our lawyers found themselves in the uncommon position of having very little to say. Nevertheless, we were delighted to get the word out: Food recovery is a very low risk, high benefit activity. There are no publically available records of anyone in the United States being sued or having to pay damages because of harms related to donated food.

In many respects, this finding is not a surprise. Socially-responsible businesses and well-intentioned persons who take the time and make the effort to provide food for the hungry are uncommonly sympathetic defendants – the kind of defendants with whom juries feel compelled to side. As long as no one has acted in a totally reckless or deliberately destructive manner, lawyers are not interested in sticking it to people who make sure the needy do not starve. Likewise, the media is not eager to vilify those who give or distribute food to the hungry. What is more, the very people who depend on donated food – the potential plaintiffs and interviewees – hesitate to bite the hands that feed them.

Of course, we cannot reduce the risk of food borne illness to zero. A low-level of litigiousness does not mean that donated food never sickens recipients. Nor does it mean that an outbreak of foodborne illness among recipients of charitable food will not generate a bit of negative attention. Indeed, in July 2012, mishandling of donated food sickened at least 60 diners at a Denver Rescue Mission dinner; this incident was, quite appropriately, covered by the local news. [11] Encouragingly, the tone of the reporting was generally measured, factual, and low on outrage. Moreover, the Denver Rescue Mission, after conducting its own investigation into the incident, admitted that there had been a lapse in time and temperature protocol. This lapse created an opportunity for *Staphylococcus aureus* bacteria to taint pre-prepared smoked turkey that the Mission re-served to its clients. The Mission owned up to the error and shielded its generous donors

[10] BSR, Analysis of U.S. Food Waste Among Food Manufacturers, Retailers, and Restaurants (2014) at http://www.foodwastealliance.org/wp-content/uploads/2014/11/FWRA_BSR_Tier3_FINAL.pdf

[11] http://www.denverpost.com/news/ci_21141863/denver-health-investigators-probing-tainted-meal-that-sickened

from undue reputational harm by declining to disclose the sources of the donated food to reporters.[12]

The Denver Rescue Mission's own conclusions about causation and their role in the outbreak were later validated by Denver Environmental Health food safety investigators and Denver Public Health epidemiological staff, who determined that some of the turkey had been insufficiently cooled and re-heated and that proper hygiene practices had not been assiduously followed.[13] Because these missteps all violated the Mission's established and ordinarily used handling practices, the city agencies concluded that there was "no reason to believe [the outbreak] was anything other than an isolated incident where organizational practices in handling of the food were not followed" and noted that the lapse was "documented and immediately corrected by the Denver Rescue Mission's leadership and staff."[14]

Although this incident was undoubtedly unfortunate and frightening for the sickened consumers and the well-meaning agents of the Mission alike, it demonstrates that an outbreak does not spell certain doom for either charitable food providers or their donors. The Denver Rescue Mission continues to serve the community and in 2013 and 2014 it offered well over a million meals to people in need.[15] Rather than discouraging food donation this story provides a strong reminder for all food donors and charitable food providers to develop and consistently implement safe food handling protocol, carefully train and supervise volunteers, and nurture mutually-supportive relationships. Putting good procedures in place, maintaining sanitary food storage, preparation and service spaces, and cooperating with investigating authorities in the event of an outbreak, undoubtedly allowed the Denver Rescue Mission to emerge relatively unscathed and continue its good work.

## TAKING LIABILITY OFF THE TABLE: GOOD SAMARITAN LAWS

The absence of litigation and the minimal, measured news coverage of the occasional mishap with donated food demonstrates that legal and reputational

---

[12] Denver Rescue Mission president and CEO Brad Meuli simply stated, "We get pre-prepared food from a lot of locations. Rather than me pointing the blame, we're just taking responsibility." http://www.denverpost.com/ci_21158871/denver-charity-did-not-handle-food-donation-properly?source=infinite

[13] http://foodpoisoningbulletin.com/2012/denver-rescue-mission-outbreak-caused-by-improper-food-handling/

[14] http://foodpoisoningbulletin.com/2012/denver-rescue-mission-outbreak-caused-by-improper-food-handling/

[15] http://www.denverrescuemission.org/2013-annual-report and http://www.denverrescuemission.org/2014-annual-report

risks are overstated and largely illusory barriers to food recovery. Of course, noting that something has not yet happened is different than proving that it will not be a problem in the future. Prospective food donors want to know both that the socio-legal dynamics disfavor derision and that they are legally protected from liability.

Beyond reputational harm and the burdens of being sued, there are harsh legal consequences for being the source of a foodborne illness. All 50 states impose "strict liability" – liability even in the absence of negligence – on those who make and sell defective food products that cause injury.[16] A person injured by a food product need only prove that: (1) the product was defective, (2) it was used as intended, and (3) the defect caused the injury.[17] In other words regardless of how careful, reasonable, or responsible, if a food producer sells a product that is contaminated and causes harm, it will be on the hook for significant damages. In many states, this practically-automatic form of liability attaches to all entities in the chain of distribution, even if they received the product in tainted form from a supplier and had no reason to suspect it was unfit to eat.[18] Thus, the specter of strict liability produces a strong disincentive to putting less-than-pristine food into the stream of consumption. Without the countervailing promise of profit, the risk of trading in food is hardly worth it.

Fortunately, the liability standard is far less draconian when it comes to food *donation*. In fact, there are express statutory protections from liability designed to protect donors and charitable distributors of excess food. Since 1977, state laws have shielded food donors and charitable food providers from much of the potential liability associated with their donations.[19] These laws are helpful but

---

[16] http://www.marlerclark.com/pdfs/product-liability-strict.pdf

[17] See *Greenman v. Yuba Power Products, Inc.*, 59 Cal.2d 57 (1963) (establishing strict liability in tort as the standard for product liability claims, displacing contract-based warranty theories of liability); see also, Restatement (Second) Torts § 402A ("One who sells any product in a defective condition unreasonably dangerous to the user or consumer or to his property is subject to liability for physical harm thereby caused to the ultimate user or consumer, or to his property, if (a) the seller is engaged in the business of selling such a product, and (b) it is expected to and does reach the user or consumer without substantial change in the condition in which it is sold.). http://www.marlerclark.com/pdfs/intro-product-liability-law.pdf

[18] Twenty-five states and Washington DC apply the rule of strict liability to all in the chain of distribution. Nine states have so-called "pass-through" statutes that protect sellers from liability where they had both no knowledge of the defect and no reason to suspect it was present. Thirteen states require proof of negligence to hold a non-manufacturing seller liable, unless the manufacturer is insolvent or beyond the reach of a lawsuit. A mere three states the reject the rule of strict liability and require proof of negligence. http://www.marlerclark.com/pdfs/chain-of-distribution-liability.pdf

[19] California was the first state to pass a Good Samaritan law in 1977.

widely variable in their requirements and scope of coverage. [20] As a result, before 1996, businesses that operated in more than one state, made donations across state lines, or contributed food to a bank that supplied emergency feeding programs in a multi-state region, had to hire lawyers to conduct a comprehensive survey of the law in all states where the food might travel and adopt jurisdiction-specific recovery practices. This patchwork of well-intentioned laws discouraged rather than facilitated food recovery.

Recognizing that consistency was needed to simplify access to liability protection,[21] Congress passed the Bill Emerson Good Samaritan Food Donation Act of 1996, 42 U.S.C. § 1791 (the "Bill Emerson Act" or the "BEA"), which creates a national minimum standard of liability protection for food donation-related activities.[22] When then-President Bill Clinton signed the Bill Emerson Act into law, he praised the United States Department of Agriculture (USDA) for undertaking "a national initiative to help Americans "rescue" food of a highly perishable, but nutritious nature," most of which was "prepared in restaurants, hotels, cafeterias, and other institutional settings and would otherwise have been thrown away." He noted that through this important USDA-led effort, "thousands of hungry people [were] fed at no cost to the Federal taxpayer." At the same time, President Clinton acknowledged that liability concerns impede food recovery: even though "many States have enacted their own 'Good Samaritan' laws..., many businesses have advised that these varying State statutes hinder food donations." Clinton, echoing the statements of many legislative proponents of the Bill Emerson Act, declared that the legislation would "end the confusion regarding liability for food recovery and donation operations through uniform definitions in one national law" and "encourage the charitable and well-intentioned donation of food to the needy, while preserving governmental authority to protect health and food safety."[23]

Shortly after the Bill Emerson Act went into law, the USDA requested and received an opinion from the U.S. Attorney General's Office regarding the extent to which the BEA preempted state laws. The opinion confirmed that the Bill Emerson Act impliedly and partially preempted state-level Good Samaritan laws that provide less liability protection. In other words, the BEA federal liability

---

[20] Attached as an Appendix is a 50 State Compilation of Good Samaritan Laws to Facilitate Food Donation.

[21] 143 Cong. Rec. H7477-79 (daily ed. July 12, 1996).

[22] Preemptive Effect of the Bill Emerson Good Samaritan Food Donation Act, 21 Op. O.L.C. 55, 1997 WL 1188104 (discussing legislative history).

[23] 1996 U.S.C.C.A.N. 3400 (Signing Statement to P.L. 104-210, Oct. 1, 1996).

protection functions as a floor not a ceiling. [24] States remain free to offer greater liability protection or to extend protection to a wider range of covered activities and personnel. [25] Because no state can offer less protection than the BEA, most who engage in the recovery and charitable distribution of excess food should tailor their activities to fit within the parameters of the Bill Emerson Act (explained in the next section). However, smaller or more contained food businesses and feeding charities with a purely local, wholly intra-state geographical scope may fare better by reviewing their state's own Good Samaritan law and determine if it accommodates greater programmatic flexibility or offers greater protection.

**Keeping Liability Off the Table**

To take advantage of the Bill Emerson Good Samaritan Food Donation Act's liability protection, donors and non-profits must first know how this law works. Because it is important to confirm that the statute applies to both the actors and items involved in a contemplated donation, this section will take a very close look at the BEA, its coverage, requirements, and limits. Because the BEA has never been relied on or challenged in court, there is no case law interpreting the statute. As such, we must reasonably interpret the text of law in light of its purpose and legislative history.

The Bill Emerson Act absolves those involved in the donation of food and grocery products from civil and criminal liability for injuries and harms – up to and including death – that arise from the nature, age, packaging, or condition of donated items.[26] The Act extends this same liability protection to the nonprofit organizations that receive and distribute such donations.[27] Its broad sweep encompasses a range of food conservation activities including making, receiving, and distributing donations, field gleaning, perishable produce and prepared food rescue or salvage, and nonperishable processed food collection.[28]

The Bill Emerson Act provides liability protection when qualifying donated items – either *apparently wholesome food* or *apparently fit grocery products* – are donated in *good faith* by a *non-profit organization* for ultimate distribution to *persons in need.* As long as these requirements, which are discussed in detail below, are met the BEA provides strong protection.

---

[24] See Preemptive Effect of the Bill Emerson Good Samaritan Food Donation Act, 21 Op. O.L.C. 55, 1997 WL 1188104 (discussing legislative history).
[25] See Preemptive Effect of the Bill Emerson Good Samaritan Food Donation Act, 21 Op. O.L.C. 55, 1997 WL 1188104 (discussing legislative history).
[26] 42 U.S.C. § 1791(c)(1)
[27] 42 U.S.C. § 1791(c)(2)
[28] Cheryl Maclas, Citizen's Guide to Food Recovery.

**Who is Covered**

"Persons," "gleaners," and "nonprofit organizations" all receive protection from the Bill Emerson Act. If you've got excess food and you want to donate it, chances are, you will fall within the protective embrace of the BEA. Those who wish to accept donations of food and distribute them to those in need are probably covered as well. The very broad category of covered *persons* encompasses individuals, corporations, partnerships, organizations, associations or governmental entities. It expressly includes retail grocers, wholesalers, hotels, motels, hospitals, manufacturers, restaurants, caterers, farmers, and nonprofit food distributors. Protection from liability extends to officers, directors, partners, deacons, trustees, council members, or other elected or appointed individuals responsible for governance of covered entities.[29] The BEA further defines a *gleaner* as a "person who harvests for free distribution to the needy, or for donation to a nonprofit organization for ultimate distribution to the needy, an agricultural crop that has been donated by the owner."[30]

In 2008, the Federal Food Donation Act expanded the coverage of the BEA to include both federal executive agencies and contractors hired by those agencies that have excess food.[31] Four years later, thanks to an under-the-radar legislative maneuver in an appropriations bill, the protection of the Bill Emerson Act was also expressly extended to schools and local educational agencies that participate in federally-funded school meal programs.[32]

**To Whom Can Donations Be Made**

Emergency feeding organizations perform an array of food recovery-related activities, including donation solicitation, food collection, storage, preparation, and distribution. Before the BEA, most nonprofits received no legal protection for their food-handling activities, which come with inherent and inextinguishable risks, because many early state-level Good Samaritan laws covered donors only. Now, thanks to the BEA, nonprofits are protected when they distribute donations and when they perform other covered activities.

---

[29] 42 U.S.C. § 1791(b)(10)

[30] 42 U.S.C. § 1791(b)(5)

[31] 42 U.S.C. § 1792(b)(2); The Government formally encourages "executive agencies and their contractors, to the maximum extent practicable and safe, to donate excess apparently wholesome food to nonprofit organizations that provide assistance to food-insecure people in the United States." 48 C.F.R. 26.402. To be considered "excess" under the Federal Food Donation Act of 2008, it must be food that is "not required to meet the needs of the executive agencies" and that "would otherwise be discarded." 48 C.F.R. 26.401

[32] P.L. 112-55; 42 U.S.C. 1758(l)

The BEA specifically aims to facilitate donations to nonprofit organizations for ultimate distribution to the needy. Although, it does not protect *direct* donation of food to people in need, it also does not require that the intermediary non-profit organizations be registered, tax-exempt 501(c)(3)s. Instead, the BEA uses that term "non-profit organization" to refer to an incorporated or unincorporated entity that (a) operates for religious, charitable, or educational purposes; and (b) does not provide net earnings to, or operate in any other manner that inures to the benefit of, any officer, employee, or shareholder of the entity.[33] Thus, as a general matter, the BEA will protect a loosely organized group of well-meaning citizens who collect food and distribute it to the needy (and their donors), regardless of whether the group of do-gooders is affiliated with an incorporated or tax exempt organization, provided that the members of the group are not enriching themselves in the process. There are, however, a three important provisos: (1) If the food is being recovered from a federal executive agency or a contractor hired by such an agency, it may only be donated to an organization that is described in section 501(c) of the Internal Revenue Code and exempt from tax under section 501(a).[34] (2) If the food is being recovered from a federally-funded school meal program, it may only be donated to "eligible local food banks or charitable organizations"; this category is limited to food banks and charitable organizations that are exempt from taxation under section 501(c)(3) of the Internal Revenue Code.[35] (3) For any donor of the food to take advantage of enhanced tax deduction available for qualified donations of excess food, the donation must be made to a tax-exempt 501(c)(3) that will use the donation for the care of .the ill, the needy or infants [children]."[36]

**What Can Be Donated**

The BEA covers the donation of both food and grocery products, defined as follows:

- *Food* is "any raw, cooked, processed, or prepared edible substance, ice, beverage, or ingredient used or intended for use in whole or in part for human consumption."[37]

- *Grocery Products* are "nonfood" items typically sold in a grocery store, including a disposable paper or plastic products,

---

[33] 42 U.S.C. § 1791(b)(9)
[34] 42 U.S.C. § 1792; 48 CFR 26.401
[35] 42 U.S.C. 1758(l)(4)
[36] 26 U.S.C. §§ 170(e)(3)(A)(i), 170(e)(3)(C)
[37] 42 U.S.C. § 1791(b)(4)

household cleaning products, laundry detergent, cleaning products, or miscellaneous household item.[38]

The expansive definition of food protects the donation of comestibles in whole, processed, raw, or prepared forms.[39] This comes as a surprise to many caterers, restaurateurs and grocers who stock grab-and-go items and have long believed their only options for handling their excess prepared foods were disposal or composting. Because of the "in whole or part" language, the BEA's definition of food can also reasonably be interpreted to allow donation of somewhat deteriorated food, part of which may still be safely salvaged and consumed (i.e., a bruised watermelon with a sizeable unblemished and edible section or a carton of strawberries with a few mushy berries in the mix). The recovery of prepared, re-heated, partially-edible, and near-expiration food, of course, demands more rigorous attention to time, temperature, and hygienic handling than the donation of highly-processed, shelf-stable, packaged, non-perishables. Provided that both donors and donees handle prepared and perishable foods with due care and comply with public health regulations pertaining to the handling, storage, service, and re-service of food, the BEA offers a valuable safety net.

By also covering grocery products, the BEA facilitates the donation of non-food essentials that people in poverty often struggle to purchase and need to maintain a sanitary home and good personal hygiene – items that many food pantries and caring agencies try to offer their clients. It explicitly includes household cleaners, paper goods (toilet paper, facial tissues, napkins paper towels), and disposable paper and plastic items (dishes, cups, and utensils). Because the BEA separately lists "cleaners" and "household cleaners," it is reasonable to interpret this to encompass personal hygiene products like soap, shampoo, conditioner, and toothpaste. Moreover, the "other household items" category can reasonably be interpreted to cover the donation of kitchen utensils (flatware, cutlery, cookware, kitchen tools, and food storage items) and cleaning implements (sponges, brooms, mops, dish towels). This protection can help charities source many of the related items that the people they serve may need to safely prepare and store food at home and to maintain good health.

A critical caveat: the donated items must be "apparently wholesome," (for food) and "apparently fit" (for grocery products). That a food or grocery product "may not be readily marketable due to appearance, freshness, grade, size, surplus, or other conditions," does not automatically render it unfit or unwholesome. Rather, to meet these standards, the food and grocery products

---

[38] 42 U.S.C. § 1791(b)(6)
[39] 42 U.S.C. § 1791(b)(4)

must "meet all quality and labeling standards imposed by Federal, State, and local law and regulations." This is where things can get a little tricky. As a general matter, most excess unsold food should satisfy applicable regulations imposed by all levels of government; otherwise, it could not legally be offered for sale. Thus, one might assume that excess unsold food – especially that which has been recovered from retail food sources – will generally satisfy this requirement.

Remember, though, that not all food that is available for donation is marketable or has been previously offered for sale. Sometimes food producers cannot market perfectly wholesome food because of packaging and labeling errors, cosmetic damage, or production overruns. Some gleaners collect produce directly from the field; some sportsmen hunt wild game with the intent of donating meat. These items may be subject to special regulation by state and local departments of health or departments of agriculture. It is wise to contact these agencies for guidance before donating or accepting and distributing these types of food. With relatively simple steps, most of these items can be safely and legally prepared or rehabilitated for donation. Indeed, the BEA specifies procedures for reconditioning which facilitate the protected donation of food and grocery items that do not meet all quality and labeling standards imposed by Federal, State, and local laws and regulations.[40] The BEA's "partial compliance" provision allows for the recovery and donation of otherwise edible and wholesome items with technical flaws, such as missing or marred product labels, open or broken packaging, and items that require washing, trimming or other cleaning before they can be provided to the ultimate recipients.

To be certain that nonconforming items are covered, donors and nonprofits must follow three steps: First, the donor must inform the nonprofit of the nonconforming nature of the item. Second, the nonprofit must agree to recondition the item so that it is compliant. Finally, the nonprofit must know the standards for reconditioning the item.[41] Technical guidance and legal requirements for food product reconditioning can be found in the Comprehensive Guidelines for Food Recovery[42] and in state food salvage codes.

**What Exactly is a Donation?**

Though most of us have an intrinsic appreciation of what it means to *donate* something, for purposes of the BEA, the term is expressly defined as

[40] 42 U.S.C. § 1791(e)
[41] 42 U.S.C. § 1791(e)
[42] Conference for Food Protection, Comprehensive Guidelines for Food Recovery Programs (2007) available at http://www.foodprotect.org/media/guide/Food-Recovery-Final2007.pdf

"giv[ing] without requiring anything of monetary value from the recipient."[43] Donations must be made in good faith. Although the statute does not specifically address what constitutes good faith, this familiar legal concept embraces conduct that is motivated by a sincere and honest intention to deal fairly with others.[44]

Even though the concept of donation is at odds with the concept of payment, the BEA does allow some money to change hands, provided that the transfer of cash is only a "nominal fee" and that it only passes through the proverbial "hands" of nonprofit organizations. This carve-out is designed to defray costs incurred by food banks, which provide critical collection and aggregation services for the front-line emergency feeding programs that do direct food provisioning. Of course, to remain within the bounds of the BEA, the end consumer or recipient of the food must never be charged.

**Protection for Injuries Not Related to Eating**

Beyond merely quelling fears of liability for foodborne illness, the BEA removes most of the risk associated with injuries and deaths that occur while the injured party is collecting donations on the donor's property. This waiver of premises liability makes it far less frightening to invite third-party volunteers onto one's property – and into the slightly hazardous places where food is most safely held, such as commercial kitchens, walk in coolers and freezers, and designated cold or dry storage pantries.

For a donor to take advantage of the BEA's premises liability protection, four conditions must be met: The property must be (1) owned or occupied by a covered person who (2) permits the gleaners or representatives of a nonprofit (paid or unpaid) to enter his property (3) for the specific purpose of collecting donations; and (4) the collected donations must ultimately be distributed to needy individuals.[45]

Note that the premises liability protection is limited to the property "where gleaning or donation collecting occurs."[46] This means that the premises liability portion of the BEA will not waive liability for injuries that occur during transit or at the facilities to which the donated items are delivered. Thus, food recovery facilitating organizations and receiving non-profits should consider

---

[43] 42 U.S.C. § 1791(b)(3)
[44] Black's Law Dictionary 762 (9th ed. 2009) (defining good faith as 1-honesty in belief or purpose; 2-faithfulness to one's duty or obligation; 3-observance of reasonable commercial standards of fair dealing in a given trade or business; or 4-absence of intent to defraud or to seek unconscionable advantage.)
[45] 42 U.S.C. § 1791(d)
[46] 42 U.S.C. § 1791(d)

taking several protective steps, including (1) consulting a trusted insurance agent to purchase general liability insurance policies and making sure they understand the scope and limits of their insurance coverage; (2) assuring that the personnel involved in collecting, transporting and unloading donations are properly trained, have clean driving records, and maintaining adequate automobile insurance; (3) working with their attorneys to develop a tailored, jurisdiction-specific waiver of liability form for volunteers to sign; and (4) obtaining knowing waivers of liability from volunteers.

**Facilitating Donation, Protecting the Needy**

By both taking charitable food out of the strict liability realm and waiving liability for negligence – the failure to use reasonable care – the BEA removes most of the legal risk associated with food recovery and donation. It does not, however, provide a license for food handlers (on either the donating or receiving ends) to act recklessly or to deliberately cause harm. If an injured party can prove that he was injured by food that became contaminated as the result of "gross negligence" or "intentional misconduct", the BEA will not shield the party at fault.[47] This gross negligence liability floor also applies to the premises liability protection.[48] The BEA defines *gross negligence* as "voluntary and conscious conduct (including a failure to act) by a person who, at the time of the conduct, knew that the conduct was likely to be harmful to the health or well-being of another person."[49] In other words, gross negligence in food handling amounts to a willful disregard for the safety of the ultimate consumers of the donated food. House Report 104-661, which accompanied the enactment of the BEA, counsels that a finding of gross negligence requires nuanced consideration of numerous factors, including the type of food involved, the recommended sell by date, and the ways in which the intended end-user is anticipated to consume the donated food (i.e., raw or cooked, on-site or at a later date).[50] The House Report also makes it clear that food product dates (often erroneously referred to as "expiration dates"), which are neither federally required for most products nor indicative of when a product may be safely consumed, do not function as a bright line test for gross negligence.[51] Accordingly, donors and non-profits are encouraged to make

---

[47] 42 U.S.C. § 1791(e)
[48] 42 U.S.C. § 1791(d)
[49] 42 U.S.C. § 1791(b)(7)
[50] H.R. REP. NO. 104-661, at 5 (1996).
[51] H.R. REP. NO. 104-661, at 5 (1996).

product and use-specific determinations about whether, when, and for how long "out-of-date" food may be donated and consumed.[52]

The BEA defines *intentional misconduct* as "conduct by a person with knowledge (at the time of the conduct) that the conduct is harmful to the health or well-being of another person."[53] That the law does not permit anyone in the charitable food chain to intentionally sicken or injure those who rely on donated food should come as no surprise. Donors might be even more inclined to participate in food recovery – especially of high-quality, nutrient-dense perishable foods that are in short supply at most food pantries – if the BEA's liability protection was absolute. But a complete and unqualified waiver of liability would perversely authorize and could encourage abuse of those in need. Thus, the BEA aims to strike a balance between preventing waste, facilitating an adequate supply of charitable food, and protecting the often vulnerable people (i.e., children, seniors, homeless people, ill and immunocompromised people, and those without ready access to quality medical care) who are food insecure.

Many food business executives are curious about the extent to which they may be legally responsible for the unauthorized acts – both reckless and purposefully harmful – of the employees that participate in food recovery efforts as part of their official duties. Under many circumstances, employers can be "vicariously liable" for the injurious acts of their employees.[54] However, employers are rarely liable for unauthorized criminal acts committed by employees.[55] Liability for both criminal and tortious (non-criminal, but civilly actionable) harmful acts by employees may turn on whether the act was reasonably foreseeable and whether the employer had actual or constructive knowledge (i.e., whether the employer knew or should have known about the employee's dangerous propensity). To decrease the likelihood of being held liable for an employee's unauthorized acts of gross negligence or intentional misconduct (in all matters, not just food recovery), employers should perform a rigorous background check as a standard element of the hiring process; conduct and

---

[52] For an in-depth look at the laws, policies, and practices pertaining to food product dating, past and present, as well as recommendations for waste-reducing and consumer-empowering reforms, see Emily Broad Leib et al., THE DATING GAME: HOW CONFUSING FOOD DATE LABELS LEAD TO FOOD WASTE IN AMERICA (September 2013) at http://www.nrdc.org/food/files/dating-game-report.pdf

[53] 42 U.S.C. § 1791(b)(8)

[54] Restatement Third, Agency § 2.04 ("An employer is subject to liability for torts committed by employees while acting within the scope of their employment.")

[55] Dennis Stearns, INTENTIONAL CONTAMINATION: LIABILITY FOR THE CRIMINAL ACTS OF EMPLOYEES available at http://www.marlerclark.com/pdfs/intentional-contamination-idaccess.pdf

periodically reinforce sufficiently detailed staff training programs to provide and refresh education on important policies and promote safety; document and address any reports or instances of misconduct; and take appropriate action, up to and including termination of employment when the infraction is sufficiently severe. Developing a food recovery protocol that prioritizes food safety and compliance with applicable regulations, providing employees with clear training on and consistent reinforcement of such protocol, and adequately supervising employees are all recommended risk-reducing practices.

Remember that donors, gleaners, and nonprofit organizations must still comply with all state and local health regulations. Though there is no case law on point, it is safe to say that perfect compliance with the pickiest and most technical of regulations is probably not necessary to stay within the protective ambit of the BEA. Generally, failure to comply with a health or safety law or regulation is considered per se evidence of negligence, not of gross negligence.[56] While occasional and small missteps or isolated instances of non-compliance are unlikely to vitiate the BEA's protections, failure to learn the applicable laws and to substantially comply with those that bear upon food safety could support a finding of gross negligence, taking the conduct outside the protection of the Bill Emerson Act. In sum, it is advisable for food donors and receiving non-profits to develop food recovery protocol that promotes safe handling, facilitate good communication, and allows for easy identification and disposal of food that may have become unsafe.

**Planning Recovery, Promoting Safety**

Food sector businesses that wish to take advantage of the Bill Emerson Act's liability protections should begin by developing a formal food recovery plan. Such a plan will guide, organize, and streamline recovery efforts and foster productive relationships with non-profit organizations that collect food. Through developing a food recovery plan, both donating and receiving organizations will be better able to envision the complete recovery process, which includes assessment of excess food streams, identification of possible partner organizations and applicable regulators, development of segregating, handling, training, and record-keeping protocol, and design of efficiency and contingency strategies.

To maximize utility, food recovery plans should identify the typical kinds and approximate anticipated quantities of food to be donated, as well as the periods and frequency of available excess. Building on this foundational

---

[56] Restatement (Third) of Torts § 14 (defining negligence per se); Dennis Stearns, INTENTIONAL CONTAMINATION: LIABILITY FOR THE CRIMINAL ACTS OF EMPLOYEES available at http://www.marlerclark.com/pdfs/intentional-contamination-idaccess.pdf

information, the plan should then identify characteristics of suitable charitable food distribution partners, seeking to optimize between the types of food to be donated and the needs and capacities of the donee organization(s). Next, it should outline proposed terms of the relationship between the donor food business and the food distribution organization(s). If working with more than one donee organization, it may be prudent to think through how donations will be allocated and prioritized.

It is critical to establish food handling protocol and record keeping mechanisms, as well as to set minimum qualifications for the personnel in charge of oversight at both the donating and receiving facilities. Best management practices for food safety and defense should be incorporated into the plan. Doing so will help protect against conduct that could be deemed grossly negligent. The methods used to store and transport donated food and grocery items should be worked out and any necessary equipment or vehicles procured.

The Conference for Food Protection, which established Memoranda of Understanding with the U.S. Food and Drug Administration (FDA), the USDA Food Safety and Inspection Service, the U.S. Centers for Disease Control and Prevention, and the Association of Food and Drug Officials, among others,[57] produced a very detailed and technical set of *Comprehensive Guidelines for Food Recovery Programs*. These guidelines, which were last updated in 2007 and are currently being streamlined,[58] are not an official part of the Food & Drug Administration's Model Food Code; thus, they have not been widely adopted and incorporated into state food codes. As a result, many state and local regulators with authority over food have requirements that differ from these guidelines. Despite the fact that the Conference for Food Protection has no formal regulatory authority, its voluntary guidelines may be a useful tool for organizations seeking to develop a food recovery plan. The guidelines can be presented to local regulators who are unfamiliar with and wary of food recovery or who impose what appear to be unreasonable and unduly burdensome limits on food recovery and donation activities.

The exercise of careful planning will also surface in staff training needs. Because the types of food safety practices are generally the same as those that

---

[57] See, e.g., FDA, MOU 225-93-4006 at http://www.fda.gov/AboutFDA/PartnershipsCollaborations/MemorandaofUnderstandingMOUs/OtherMOUs/ucm118390.htm; CDC MOU at http://www.foodprotect.org/media/site/CDC_MOU.pdf

[58] Conference for Food Protection, Comprehensive Guidelines for Food Recovery Programs (2007) available at http://www.foodprotect.org/media/guide/Food-Recovery-Final2007.pdf

apply to the sale and service of food, additional training and education needs should not be overwhelming. That said, because local departments of health may impose more stringent standards on certain categories of rescued, recovered, or reserved food and because some excess items may be "non-conforming" and in need of reconditioning, a small amount of additional training may be in order.

Finally, because the success of a food recovery can hinge on the quality of the working relationship between donor and donees, it is advisable to develop clear communication protocols (especially when non-conforming items are being donated). It is also wise to establish an unintimidating, neutral means of addressing unsatisfactory interactions, raising concerns about improperly handled food, breaches of protocol, or complaints about the quality and suitability of donated food.

The planning process also provides an opportunity for a donor business to meaningfully integrate food recovery into its business plan. Engaging in food recovery can have a significantly salutary effect on a business's bottom-line by reducing disposal costs (tipping fees), providing access to valuable federal and state tax incentives, and identifying existing areas of inefficiency and waste generation. Food recovery should be integrated into an enterprise's overall sustainability, social responsibility, and community engagement portfolios. Active promotion of this environmentally and socially sensitive practice may yield a significant return on investment as consumer purchasing behavior is increasingly driven by values beyond taste, price, freshness and accessibility.[59] Food businesses looking for a formal way to promote their work in this regard may wish to seek third-party certification through Food Recovery Certified.[60]

**Bottom-line Bonus: Tax Incentives for Donation of Excess Food**

The Bill Emerson Act is not the only tool that the federal government uses to incentivize the donation of surplus food. Hidden within our massive maze of federal tax law, the Internal Revenue Code, is a special rule for contributions of food inventory.[61] It provides an enhanced – and uncommonly generous – deduction for the donation of food inventory.

The amount of the enhanced tax deduction for food donations is the lesser of (a) twice the basis of the donated food or (b) the basis of the donated food plus one-half of the food's expected profit margin, if sold at fair market

---

[59] See USDA Agricultural Marketing Service, FOOD VALUE CHAINS: CREATING SHARED VALUE TO ENHANCE MARKETING SUCCESS (May 2014) at http://www.thegreenhorns.net/wp-content/files_mf/1404014092foodvaluechains.pdf

[60] Food Recovery Certified at http://www.foodrecoverycertified.org/

[61] 26 U.S.C. § 170(e)(3)

value. (i.e., one-half the appreciated value of the food).[62] As used here, basis is simply the cost of producing or acquiring the food.[63] To qualify for this enhanced deduction, the donation must be made to a 501(c)(3) organization, a public charity, or a private operating foundation that operates domestically.[64] Furthermore, the donated property must be used only for the care of the ill, the needy, or infants, and in a manner related to the receiving organization's tax-exempt purpose.[65] Compliance with these requirements must be confirmed by the receiving organization in a written statement provided to and kept on file by the donor.[66]

In December 2015, after several years of lobbying by anti-hunger and food recovery advocates,[67] including the Food Recovery Project,[68] Congress finally made the enhanced tax incentive for food donations more robust, equitably available, and easy to claim. Thanks to changes made in the omnibus federal budget bill for fiscal year 2016,[69] the federal enhanced tax deduction for food donation, as a permanent part of the tax code, is now available to all businesses regardless of their corporate form.[70] (In years past, it was permanently authorized only for C corporations.[71]) This change will help incentivize food donation among

---

[62] 26 U.S.C. § 170(e)(3)(B) available at http://www.law.cornell.edu/uscode/text/26/170.

[63] 26 C.F.R. 1.170A-4A(c)(2) (2015), available at http://www.law.cornell.edu/cfr/text/26/1.170A-4A.

[64] 26 U.S.C. §§ 170(c)(2)

[65] 26 U.S.C. § 170(e)(3)(A)(i); Reg. 1.170A-4A(b)(2)

[66] 26 U.S.C. § 170(e)(3)(A)(iii); Reg.1.170A-4A(b)(4)

[67] Council on Foundations, COALITION URGES HOUSE TO PASS AMERICA GIVES MORE ACT OF 2015 (FEB. 11, 2015) available at http://www.feedingamerica.org/hunger-in-america/news-and-updates/press-room/press-releases/coalition-urges-house-to-pass-Act.html

[68] Food Recovery Leaders Send Sign-On Letter to U.S. Senators, Led by Food Law and Policy Clinic and Professor from University of Arkansas School of Law (Dec 12, 2014) at https://blogs.law.harvard.edu/clinicalprobono/2014/12/12/food-recovery-leaders-send-sign-on-letter-to-u-s-senators-led-by-food-law-and-policy-clinic-and-professor-from-university-of-arkansas-school-of-law/

[69] *See* Consolidated Appropriations Act, 2016, Pub. L. No. 114-113, § 113 available at https://www.congress.gov/bill/114th-congress/house-bill/2029/text#toc-HCC4297736B6849FDA0DB4D36332A480B. (N.B. Tax professionals often refer to § 113 as the PATH Act because the provisions were originally introduced as the Protecting Americans from Tax Hikes Act of 2015.)

[70] Pub. L. No. 114-113, § 113(a)

[71] Before December 2015, the Section 170(e)(3) deduction, as a permanent part of the tax code, was consistently available only to C corporations. The Katrina Emergency Tax Relief Act (KETRA), a package of tax incentives designed to promote charitable giving in the wake of Hurricane Katrina, extended the deduction to all businesses regardless of their corporate form, allowing many smaller food and farm businesses to benefit from the enhanced tax deduction for 2005. After that, the extension had to be re-authorized

a much wider group of potential donors. Typically, large players in the food sector are organized as C-Corporations, but many smaller, independent food entities, including farmers, ranchers, value-added food producers, restaurants, and specialty food purveyors are not. The size and corporate form of a food business have little bearing on whether that business generates excess food or whether that food is suitable for donation. However, for smaller, lower-margin food businesses, the availability of an enhanced tax deduction for food donation – as opposed to the more modest general deduction for charitable contribution that is limited to basis value – can make all the difference in terms of a food recovery program's financial viability. By making the enhanced deduction broadly and permanently available, Congress is encouraging most businesses to establish long-term food donation programs, build relationships with community partners, and promote the highest and best use of excess healthy, wholesome food.[72] The Food Recovery Project is gratified that our efforts at education and advocacy in this arena have, in some small ways, contributed to this important step forward.

In addition to the much-needed permanent eligibility expansion, three other significant changes to the federal food donation tax incentive were made at the end of 2015. First, the cap on the deduction amount was increased to 15% of the donor's net income.[73] Increasing the cap from 10 to 15% will allow food businesses to realize greater financial benefit each tax year and may encourage businesses to donate even more surplus food, not just the "low hanging fruit." Moreover, after 2015, the businesses will also be allowed to carry forward any food-donation related deductions in excess of the 15% of income limitation for up to five years.[74] This provides a way to realize future benefits of food donation even if a business has a year in which income is disproportionately low or its donation value is especially high.

Second, certain taxpayers were given the option of using a new, streamlined formula to calculate the enhanced deduction. To calculate the

---

annually. Congress often waited until the last days or hours of the year to do this. Because non-C-corporations were never certain that the enhanced deduction would be available to them, many were reluctant to undertake or invest in food recovery and donation programs.

[72] *See* Jim Larson, Food Donation Connection, Statement for the Record, House Ways and Means Subcommittee on Oversight Joint Hearing on Food Banks and Front-Line Charities: Unprecedented Demand and Unmet Need, Nov. 19, 2009 available at http://waysandmeans.house.gov/media/pdf/111/ovhearing111909_larson.pdf (demonstrating a 137% increase in food donations after the KETRA expansion of the enhanced deduction).

[73] Pub. L. No. 114-113, § 113(b)(ii)

[74] Pub. L. No. 114-113, § 113 (b)(iii)

deduction enhancement, a taxpayer must know the basis value of its donation. Under the old formula, businesses that used the cash balance method of accounting (tracking cash in and cash out) rather than the accrual method of accounting had a hard time determining basis value for their food donations. The tax code now permits those taxpayers who are exempt from accounting for inventories and not required to capitalize indirect costs to calculate basis as 25% of a product's Fair Market Value.[75] Thus, donors who use different accounting methods can now easily claim the enhanced deduction by choosing to use a fixed basis value for their goods.

Third, the formula for determining fair market value (FMV) of food inventory was updated. For years, donors have also been perplexed about how to determine the Fair Market Value (FMV) for wholesome but non-conforming or unmarketable food products, those that they are unable to sell because the products do not meet internal brand standards, lack a market, are past a quality date, or are missing labeling information. In the past, it was not clear that such food was still worth its original FMV for purposes of claiming the deduction or whether its FMV had decreased because of the market-facing defect. Some businesses were disadvantaged because they were calculating the deduction for unsalable food based on a significantly reduced FMV. The diminution in the recognized value of the donated food and the reduction value of the resulting deduction often made donation uneconomical. To avoid this unintentional disincentive, the new legislation clarifies that the FMV of such food can be calculated with reference to the price of the same or substantially similar food items sold by the business.[76] This updated, more accurate FMV standard will encourage businesses to recover and donate food that is unsalable yet wholesome and safe by helping to defray the costs of separation and donation.

To help donors and donees navigate the process of making qualifying donations, the FRP and the Harvard Food Law and Policy Clinic co-published a legal guide called *Federal Enhanced Tax Deduction for Food Donation* in November 2015.[77] We are revising this guide to reflect the December 2015 changes to the law and will publish an updated version on our websites in early 2016. The Internal Revenue Service also publishes detailed guidance documents about the various special charitable donation deductions and worksheets that aid

---

[75] Pub. L. No. 114-113, § 113 (b)(iv)
[76] Pub. L. No. 114-113, § 113(b)(v)
[77] O. Balkus, N. Civita, et al., *Federal Enhanced Tax Deduction for Food Donation, a Legal Guide* (Nov. 2015) available at https://drive.google.com/file/d/0B3BpfXJ_Lg0QS1ZDZWdMUUk4ZGc/view (N.B., This guide pre-dates passage of the PATH ACT and is in the process of being updated.)

taxpayers in calculating the amount the may deduct.[78] Despite the availability of these explanatory materials, it is advisable to consult a Certified Public Accountant before establishing a significant or ongoing food recovery program. A CPA may be able to help a food donor maximize the value of this deduction.

**Toward Food Conservation**

Since its inception in 2013, the Food Recovery Project has been making good on its mission to educate interested parties about the legal protections for food recovery. As part of this process, we've had the privilege of supporting and dialoging with an array of food businesses, food recovery groups, charitable feeding organizations, and policy makers serving communities throughout the United States. During this relatively brief but intensely active period, we have been inspired by an array of innovative solutions to the food waste crisis being proposed, explored, implemented and refined. We are frequently asked questions about whether the law might protect, incentivize, or even permit a new food enterprise designed to curb or repurpose food waste. Often, we note that current laws pertaining to food recovery are akin to Stone Age tools in the Iron Age or, if you favor a more timely analogy, analog tools in a digital world. In most cases, the applicable laws and regulations were drafted at a time when dialing up an Internet connection was a time-intensive, uncertain task and smart phones were the stuff of sci-fi fantasy. They were debated when the food movement was in its infancy and largely consisted of a few (loveable and prescient) leftover hippies. They were enacted before most people had a handle on the magnitude of the food waste problem, at a time when the common conception of food-to-donate was limited to canned vegetables, and boxes of Hamburger Helper approaching their sell-by dates. Accordingly, our well-intentioned laws are not particularly well-designed to facilitate forward-thinking solutions, to support creative social enterprise, or to encourage risk- and cost-reduced retail. Ill-coordinated regulations and overlapping authority often impede the flow of excess food across jurisdictional boundaries, which seem especially arbitrary when they stand between hungry people and wholesome food. Cramped and shortsighted tax policy sometimes skews economic equations in favor of wasteful solutions.

Within legal circles, the term "conservation laws" refers to legislation that is designed to protect the environment and promote judicious use of

---

[78] See, e.g, IRS Publication 526 (2014) at http://www.irs.gov/pub/irs-pdf/p526.pdf; Ronald Fowler & Amy Henchey, IN-KIND CONTRIBUTIONS, IRS Exempt Organizations Continuing Professional Education publication (1994) available at http://www.irs.gov/pub/irs-tege/eotopice94.pdf. (N.B., These guidance documents pre-date the passage of the PATH Act and are no longer completely accurate. Taxpayers can reasonably expect that the IRS will issue updated guidance, likely in 2016.)

resources. Outside of the legal system, "conservation laws" are fundamental tenets of physics. The law of conservation of energy states that the total amount of energy in a system remains constant. Energy can neither be created nor destroyed, but it can be transformed or transferred. Taking inspiration from this physical reality, the law of food conservation – a much-needed body of human-made policies and laws for a sustainable future – would be based on the premise that the energy in our food system should not be destroyed. Food – which is, after all, encapsulated energy – should be transferred and transformed. Whenever possible, wholesome food should be used to nourish people.

As the agrifood sector and its consumers – which is to say, everyone – continue to creatively confront the food waste problem at all nodes in our food chain, attention must be paid to the legal and policy drivers of food waste and limits on food recovery. We must cultivate an ethos of food conservation and contour a body of law that will promote the careful and most complete utilization of our food resources and of the resources that are embedded in our food.

To this end, in September 2015, Secretary of Agriculture Tom Vilsack made food conservation a national priority by setting a national food waste reduction goal of 50-percent by 2030.[79] Though this was a laudable, tone-setting announcement, it was little more than aspirational. After all, it is nearly impossible to achieve a lofty national goal without a plan of action. The federal government set the bar high, but did not take corresponding steps – such as imposing mandates, allocating resources, setting benchmarks, and establishing methods of measurement – to support progress.

Fortunately, however, Representative Chellie Pingree,(D-Maine) has shown incredible leadership in moving America toward success.[80] Two months after the ambitious national food waste reduction target was set, she introduced legislation designed to tackle America's food waste problem along the entire food chain.[81] The Food Recovery Act of 2015 (FRA) proposes a comprehensive suite of reforms designed to change wasteful practices among consumers, farmers, retailers (grocery stores and restaurants), schools, and the federal government

---

[79] *USDA and EPA Join with Private Sector, Charitable Organizations to Set Nation's First Food Waste Reduction Goals*,Sept 16. 2015 at http://www.usda.gov/wps/portal/usda/usdamediafb?contentid=2015/09/0257.xml&printable=true&contentidonly=true

[80] *See generally*, www.pingree.house.gov/foodwaste

[81] *Pingree Introduces Landmark Legislation Aimed at Reducing Food Waste*, Dec 7, 2015, at https://pingree.house.gov/media-center/press-releases/pingree-introducing-landmark-legislation-aimed-reducing-food-waste

(including Congress and the military).[82] It also seeks to direct wholesome excess food to those in need, fund infrastructure for large-scale composting and food waste-to-energy projects, and fill critical knowledge gaps through expanded research funding.[83] The FRA is so chock full of helpful reforms that it cannot be covered in detail here. However, a quick tour of its most potentially impactful provisions demonstrates how much can be done at the federal policy level to address food waste.

The FRA aims to help farmers and retailers by strengthening federal liability protections for food donation.[84] In parallel, it would assist the charitable food sector by funding public investment in storage and distribution programs, building the infrastructure and capacity that food banks need to safely capture and efficiently distribute available excess food.[85] Moving further down the chain of distribution, this legislation would also empower consumers to conserve food. The FRA calls for funding a national campaign to raise awareness of the impacts of food waste and to share behavioral strategies that everyone can use to decrease wasted food at home.[86] It also aims to reduce consumer confusion over the dates found on food packages by requiring uniform labeling language and clarifying that "use by" and "sell-by" dates are merely manufacturers' suggestions that indicate when a product may begin to lose optimal flavor and texture.[87]

Because Representative Pingree is a farmer, mother, and former school board member, it is not surprising that the FRA pays special attention to schools. Food waste in schools teaches students a very powerful, very negative lesson: that food and the resources embedded within it are expendable. To turn this around and inspire a future generation of food conservationists, the FRA would expand grant programs to educate students about food waste and encourage in-school food recovery.[88] Additionally, to simultaneously reduce waste and improve the affordability and availability of nutritious, whole foods, the FRA proposes incentives for school cafeterias to purchase lower-price "ugly" fruits and vegetables.[89] The FRA also aims to strengthen connections between schools and farms and gives both more resources to combat food waste.[90]

---

[82] *See generally,* H.R. 4184 (2015) available at https://www.congress.gov/114/bills/hr4184/BILLS-114hr4184ih.pdf
[83] *Id.* at §§ 101, 104, & 204
[84] *Id.* at § 202
[85] *Id.* at § 204
[86] *Id.* at § 402
[87] *Id.* at § 401
[88] *Id.* at § 303
[89] *Id.* at § 305
[90] *Id.* at § 304

Finally, the FRA would tweak organization and practices within the federal government to reduce food waste. Most significantly, it proposes the creation an Office of Food Recovery within USDA; this office and its future director would coordinate federal activities related to measuring and reducing food waste and implementing food recovery initiatives.[91] The FRA would also give some teeth to the purely advisory Federal Food Donation Act of 2008 by requiring – not merely recommending – that food service providers to the federal government donate surplus food to organizations like food banks, food pantries, and soup kitchens.[92] This requirement would be imposed in contracts for food service in both houses of Congress and all federal executive agencies, as well as U.S. Army, Navy, and Air Force.[93]

The FRA was drafted in close consultation with a diverse collection of stakeholders who are often left out of the lawmaking process: independent farmers and food producers, environmentalists, food waste and anti-hunger advocates, green waste managers, renewable energy producers, academics, and local leaders. The Food Recovery Project's Director was proud to be one of the experts who advised Rep. Pingree's legislative staff on the content and language of the FRA.[94] The bill developed out of this process embodies a food conservation ethos. If passed, it could be considered the first food conservation law in the United States.

Laws that facilitate connection between consumers and excess edible food and consumers – whether they take the form of donation mandates in select settings, limitations on the rejection of food shipments, protections for dumpster-divers, farmworkers who glean, and others who rescue discarded food – should be explored and vetted, and the best proposals should be swiftly enacted. Food that cannot be safely or appropriately consumed by humans can be directed toward the creation of new food as livestock feed or compost. Bans on directing organic wastes to landfills and incinerators, programs to link livestock producers with nearby generators of suitable food residues, and incentives for composting and anaerobic digestion can all be deployed to make sure that food is rarely destroyed. Using a combination of expanded, updated liability protections and tax incentives, streamlined, modernized health and safety regulations, prohibitions on waste, technical assistance for resource matching, public investment in food-system infrastructure and food conservation education, we can capture the maximum value from our unavoidable excesses and create a cultural shift toward

---

[91] *Id.* at § 301
[92] *Id.* at § 302
[93] *Id.*
[94] The Food Recovery Act and Our Faculty at http://www.agfoodllm.com/2015/12/the-food-recovery-act-and-our-faculty.html

respect and reverence for our food. Though this vision is far grander than the discrete goals that launched the Food Recovery Project, we are excited to be working toward a resilient, conservation-centered food system. We will continue explaining and helping to shape the laws that facilitate the future of food conservation.

## APPENDIX

**50 State Compilation of Good Samaritan Statutes to Facilitate Food Donation**

**Current as of June 2015**

States and Territories Good Samaritan Statutes/ Citations:

- Alabama:
  - Ala. Code 1975 Section 20-1-6
    - http://alisondb.legislature.state.al.us/alison/codeofalabama/1975/coatoc.htm

- Alaska
  - AS Section 17.20.345
  - AS Section 17.20.346
  - Definitions at As 17.20.347
    - All available at
    - http://www.legis.state.ak.us/basis/statutes.asp#17.20.345

- Arizona
  - AZ St. Section 36-916
    - http://www.azleg.gov/FormatDocument.asp?inDoc=/ars/36/00916.htm&Title=36&DocType=ARS

- Arkansas
  - ACA Section 20-57-102
    - http://www.lexisnexis.com/hottopics/arcode/Default.asp
  - ACA Section 20-57-103
    - http://www.lexisnexis.com/hottopics/arcode/Default.asp
  - ACA Section 20-57-201
    - http://www.lexisnexis.com/hottopics/arcode/Default.asp

- California
  - West's Ann. Cal. Civ. Code Section 1714-25

        - http://leginfo.legislature.ca.gov/faces/codes_displayText.xhtml?lawCode=CIV&division=3.&title=&part=3.&chapter=&article=

    - Definitions section in West's Ann. Cal. Food and Agric. Code Section 58501
        - Section 58502
        - Section 58503
        - Section 58503.1
        - Section 58504
        - Section 58505
        - Section 58506
        - Section 58507
        - Section 58508
        - Section 58509
            - http://leginfo.legislature.ca.gov/faces/codes_displayText.xhtml?lawCode=FAC&division=21.&title=&part=1.&chapter=5.&article=

    - Cal. Health and Safety Code Section 114432, Section 114433, and Section 114434
        - Exists at above site on West Law- but not available on Cornell or Cali. State Website. When on Cornell's website it directs you to Cali. State website.
    - West's Ann. Cal. Civ. Code Section 846.2
        - http://leginfo.legislature.ca.gov/faces/codes_displayText.xhtml?lawCode=CIV&division=2.&title=3.&part=2.&chapter=2.&article=

- Colorado
    - CRSA Section 13-21-113
    - CRSA Section 13-21-113.5
        - All available at
        - http://www.lexisnexis.com/hottopics/colorado/

    - CRSA Section 39-22-115 (repealed)
    - CRSA Section 39-22-301 (liability of food donors with corporate tax exemptions)
        - Available at
        - http://www.lexisnexis.com/hottopics/colorado/

- Connecticut
  - CT ST Section 52-557L
    - http://www.cga.ct.gov/current/pub/chap_925.htm#sec_52-557L

- Delaware
  - 10 Del. C. Section 8130
    - http://delcode.delaware.gov/title10/c081/index.shtml

  - 16 Del. C. Section 6820
    - http://delcode.delaware.gov/title16/c068/sc03/index.shtml

- District of Columbia
  - DC ST Section 48-301
    - http://dccode.elaws.us/code?no=48-301

- Florida
  - West's FSA Section 768.135
  - West's FSA Section 768.136
  - West's FSA Section 768.137
  - All found at
  - http://www.leg.state.fl.us/statutes/index.cfm?App_mode=Display_Statute&Search_String=&URL=0700-0799/0768/0768PARTIContentsIndex.html

- Georgia
  - GA ST Section 51-1-31
    - http://www.lexisnexis.com/hottopics/gacode/Default.asp

  - Note Worthy: Ga. Code Ann. Sect. 26-1-1
    - http://www.lexisnexis.com/hottopics/gacode/Default.asp

- Hawaii
  - HI ST Section 663-10.6
    - http://www.capitol.hawaii.gov/hrscurrent/Vol13_Ch0601-0676/HRS0663/HRS_0663-0010_0006.htm

- HRS Section 145D-1
    - http://www.capitol.hawaii.gov/hrscurrent/Vol03_Ch0121-0200D/HRS0145D/HRS_0145D-0001.htm
- HRS Section 145D-2
    - http://www.capitol.hawaii.gov/hrscurrent/Vol03_Ch0121-0200D/HRS0145D/HRS_0145D-0002.htm
- HRS Section 145D-3
    - http://www.capitol.hawaii.gov/hrscurrent/Vol03_Ch0121-0200D/HRS0145D/HRS_0145D-0003.htm
- HRS Section 145D-4
    - http://www.capitol.hawaii.gov/hrscurrent/Vol03_Ch0121-0200D/HRS0145D/HRS_0145D-0004.htm
- HRS Section 145D-5
    - http://www.capitol.hawaii.gov/hrscurrent/Vol03_Ch0121-0200D/HRS0145D/HRS_0145D-0005.htm
- HI ST Section 663-10.7
    - http://www.capitol.hawaii.gov/hrscurrent/Vol13_Ch0601-0676/HRS0663/HRS_0663-0010_0007.htm

- Idaho
    - ID ST Section 5-339
        - http://legislature.idaho.gov/idstat/Title5/T5CH3SECT5-339.htm
    - ID ST Section 6-1302
        - http://legislature.idaho.gov/idstat/Title6/T6CH13SECT6-1302.htm
    - ID ST Section 5-338
        - http://legislature.idaho.gov/idstat/Title5/T5CH3SECT5-338.htm
    - IS ST Section 6-1301

        - http://legislature.idaho.gov/idstat/Title6/T6CH13SECT6-1301.htm

- Illinois
    - 745 ILCS 50/1
    - 745 ILCS 50/2-2.14
    - 745 ILCS 50/3
    - 745 ILCA 50/4
        - All found at
        - http://www.ilga.gov/legislation/ilcs/ilcs3.asp?ActID=2077&ChapterID=58

- Indiana
    - IC 34-30-5-1
        - https://iga.in.gov/legislative/laws/2014/ic/titles/034/

    - IC 34-4-12.5-1 and 12.5-2 repealed

- Iowa
    - ICA Section 672.1
        - https://www.legis.iowa.gov/law/iowaCode/sections?codeChapter=672&year=2015

- Kansas
    - KSA Section 65-687
        - http://www.kslegislature.org/li_2014/b2013_14/statute/065_000_0000_chapter/065_006_0000_article/065_006_0087_section/065_006_0087_k/

- Kentucky
    - KRS Section 413-247
      http://www.lrc.ky.gov/statutes/statute.aspx?id=17886

    - KRS Section 413-248
      http://www.lrc.ky.gov/statutes/statute.aspx?id=17887

- Louisiana
    - LA R.S. 9:2799
        - https://legis.la.gov/legis/Law.aspx?d=107244

- LA R.S. 9:2799.3
  - https://legis.la.gov/legis/Law.aspx?d=107247

- Maine
  - ME ST T. 14 Section 166
    - http://legislature.maine.gov/statutes/14/title14sec166.html

- Maryland
  - MD Code, Health- General Section 21-322
    - http://www.mgaleg.maryland.gov/webmga/frmStatutesText.aspx?article=ghg§ion=21-322&ext=html&session=2015RS&tab=subject5

  - MD Code, Courts and Judicial Proceedings, Section 5-634
    - http://www.mgaleg.maryland.gov/webmga/frmStatutesText.aspx?article=gcj§ion=5-634&ext=html&session=2015RS&tab=subject5

- Massachusetts
  - MA ST 94 Section 328
    - https://malegislature.gov/Laws/GeneralLaws/PartI/TitleXV/Chapter94/Section328

- Michigan
  - MCLA Ch. 691- General Provisions that are relevant
  - MCLA 691.1571
    - http://www.legislature.mi.gov/(S(ajyllpujxfxyuy3buudmbktj))/mileg.aspx?page=getObject&objectName=mcl-691-1571

  - MCLA 691.1572
    - http://www.legislature.mi.gov/(S(ajyllpujxfxyuy3buudmbktj))/mileg.aspx?page=getObject&objectName=mcl-691-1572

  - MCLA 691.1573
    - http://www.legislature.mi.gov/(S(ajyllpujxfxyuy3buudmbktj))/mileg.aspx?page=getObject&objectName=mcl-691-1573

- Minnesota
    - MN ST Section 604A.10
        - https://www.revisor.mn.gov/statutes/?id=604A.10
    - Also Relevant: MSA Section 31.495 (Food Salvage Operation Laws)
        - https://www.revisor.mn.gov/statutes/?id=31.495
- Mississippi
    - Miss. Code Ann. Section 95-7-1
    - Miss. Code Ann. Section 95-7-3
    - Miss. Code Ann. Section 95-7-5
    - Miss. Code Ann. Section 95-7-7
    - Miss. Code Ann. Section 95-7-9
    - Miss. Code Ann. Section 95-7-11
    - Miss. Code Ann. Section 95-7-13
        - All found at
        - https://www.lexisnexis.com/hottopics/mscode/
    - Miss. Code Ann. Section 37-115-47
        - https://www.lexisnexis.com/hottopics/mscode/
- Missouri
    - VAMS 537.115
        - http://www.moga.mo.gov/mostatutes/stathtml/53700001151.html
    - VAMS 192.081
        - http://www.moga.mo.gov/mostatutes/stathtml/19200000811.html
- Montana
    - MCA 27-1-716
        - http://leg.mt.gov/bills/mca/27/1/27-1-716.htm
- Nebraska
    - NE ST Section 25-21, 189
        - http://nebraskalegislature.gov/laws/statutes.php?statute=25-21,189

- Nevada
    - NRS 41-491
        - https://www.leg.state.nv.us/NRS/NRS-041.html#NRS041Sec491

- New Hampshire
    - NH Rev. Stat. Section 508:15
        - http://www.gencourt.state.nh.us/rsa/html/lii/508/508-15.htm

- New Jersey
    - NJSA 24:4A-1
        - http://law.justia.com/codes/new-jersey/2014/title-24/section-24-4a-1/

    - NJSA 24:4A-2
        - http://law.justia.com/codes/new-jersey/2014/title-24/section-24-4a-2/

    - NJSA 24:4A-3
        - http://law.justia.com/codes/new-jersey/2014/title-24/section-24-4a-3/

    - NJSA 24:4A-4
        - http://law.justia.com/codes/new-jersey/2014/title-24/section-24-4a-4/

    - NJSA 24:4A-5
        - http://law.justia.com/codes/new-jersey/2014/title-24/section-24-4a-5/

- New Mexico
    - NM ST Section 41-10-1
    - NM ST Section 41-10-2
    - NM ST Section 41-10-3
    - NM ST Section 41-10-4
    - NM ST Section 41-10-5 (Game Meats Donated)
        - All available at http://law.justia.com/codes/new-mexico/2014/chapter-41/article-10/

- New York
  - N.Y. Agric. And Mkts. Law Section 71-z
    - http://codes.lp.findlaw.com/nycode/AGM/4-D/71-z
  - N.Y. Agric. And Mkts. Law Section 71-y
    - http://codes.lp.findlaw.com/nycode/AGM/4-D/71-y

- North Carolina
  - NCGSA Section 99B-10
    - http://www.ncga.state.nc.us/EnactedLegislation/Statutes/HTML/BySection/Chapter_99B/GS_99B-10.html

- North Dakota
  - ND ST Section 19-05.1-02
  - ND ST Section 19-05.1-03
    - All available at
    - http://www.legis.nd.gov/cencode/t19c05-1.pdf?20150514124952

- Ohio
  - OH ST Section 2305.35
  - OH ST Section 2305.37
    - All available at http://codes.ohio.gov/orc/2305

- Oklahoma
  - OK ST. T. 76 Section 5.6
    - http://www.oscn.net/applications/oscn/DeliverDocument.asp?CiteID=72832

- Oregon
  - OR ST. Section 30.890
    - http://www.oregonlaws.org/ors/30.890
  - OR ST. Section 30.892 (Liability of Donors and Distributors of General Merchandise or Household Items)
    - http://www.oregonlaws.org/ors/30.892

- Pennsylvania
    - 10 Pa. Cons. Stat. 351-358
        - All available at
        - https://govt.westlaw.com/pac/Browse/Home/Pennsylvania/UnofficialPurdonsPennsylvaniaStatutes?guid=N37070715F8634D36881F052F2F94F452&originationContext=documenttoc&transitionType=Default&contextData=(sc.Default)

    - 42 Pa. Cons. Stat. 8338
        - http://law.justia.com/codes/pennsylvania/2010/title-42/chapter-83/8338/

- Rhode Island
    - RI Gen. Laws 1956, Section 21-34-1
        - http://webserver.rilin.state.ri.us/Statutes/TITLE21/21-34/21-34-1.HTM

    - RI Gen. Laws 1956, Section 21-34-2
        - http://webserver.rilin.state.ri.us/Statutes/TITLE21/21-34/21-34-2.HTM

    - RI Gen. Laws 1956, Section 21-34-3
        - http://webserver.rilin.state.ri.us/Statutes/TITLE21/21-34/21-34-3.HTM

- South Carolina
    - SC Code 1976, Section 15-74-10
    - SC Code 1976, Section 15-74-20
    - SC Code 1976, Section 15-74-30
    - SC Code 1976, Section 15-74-40
        - All available at
        - http://www.scstatehouse.gov/code/t15c074.php

- South Dakota
    - SDCL Section 39-4-22
        - http://legis.sd.gov/Statutes/Codified_Laws/DisplayStatute.aspx?Type=Statute&Statute=39-4-22

    - SDCL Section 39-4-23

  - http://legis.sd.gov/Statutes/Codified_Laws/DisplayStatute.aspx?Type=Statute&Statute=39-4-23

- SDCL Section 39-4-24
  - http://legis.sd.gov/Statutes/Codified_Laws/DisplayStatute.aspx?Type=Statute&Statute=39-4-24

- SDCL Section 39-4-25
  - http://legis.sd.gov/Statutes/Codified_Laws/DisplayStatute.aspx?Type=Statute&Statute=39-4-25

- Tennessee
  - TCA Section 53-13-(101-105)
    - All available at
    - http://www.lexisnexis.com/hottopics/tncode/

- Texas
  - VTCA., Civil Practice and Remedies Code Section 76.001-76.004)
    - All available at
    - http://www.statutes.legis.state.tx.us/Docs/CP/htm/CP.76.htm

- Utah
  - UT ST Section 4-34-5
    - http://le.utah.gov/xcode/Title4/Chapter34/4-34-S5.html?v=C4-34-S5_1800010118000101

  - UT ST Section 78B-4-502
    - http://le.utah.gov/xcode/Title78B/Chapter4/78B-4-S502.html?v=C78B-4-S502_1800010118000101

- Vermont
  - VT ST 12 Section 5761
    - http://legislature.vermont.gov/statutes/section/12/197/05761

  - VT ST 12 Section 5762
    - http://legislature.vermont.gov/statutes/section/12/197/05762

- Virginia
    - VA ST Section 35.1-14.2
        - https://leg1.state.va.us/cgi-bin/legp504.exe?000+cod+35.1-14.2
    - VA ST Section 3.2-5129
        - https://leg1.state.va.us/cgi-bin/legp504.exe?000+cod+3.2-5129
    - VA ST Section 3.2-5144
        - https://leg1.state.va.us/cgi-bin/legp504.exe?000+cod+3.2-5144
- Washington
    - West's RCWA 69.80.010
    - West's RCWA 69.80.020
    - West's RCWA 69.80.030
    - West's RCWA 69.80.031
    - West's RCWA 69.80.040
    - West's RCWA 69.80.050
    - West's RCWA 69.80.060
    - West's RCWA 69.80.090
        - All available at
        - http://apps.leg.wa.gov/rcw/default.aspx?cite=69.80
- West Virginia
    - WV ST Section 55-7D-(1-5)
        - All available at
        - http://www.legis.state.wv.us/wvcode/Code.cfm?chap=55&art=7D#07D
    - WV ST Section 19-30-4
        - http://www.legis.state.wv.us/wvcode/ChapterEntire.cfm?chap=19&art=30§ion=4#30#30
- Wisconsin
    - WI ST Section 895.51
        - https://docs.legis.wisconsin.gov/statutes/statutes/895/II/51

- Wyoming
  - WY ST Section 35-7-1301
  - WY ST Section 35-7-1302
    - All available at
    - https://legisweb.state.wy.us/statutes/statutes.aspx?file=titles/Title35/Title35.htm
    -
- Puerto Rico
  - Laws of Puerto Rico Annotated
    - 8 L..P.R.A. Section 802
      - http://www.lexisnexis.com/hottopics/lawsofpuertorico/
- District of Columbia
  - District of Columbia Code Annotated
    - DC ST Section 48-301
      - http://dccode.elaws.us/code?no=48-301

# *Summary and Recommendations*

Twenty thousand years ago humans were hunters and gatherers, and were sparsely distributed over Earth's land areas. Ten thousand years ago humans invented/discovered agriculture, rapidly replacing global floral diversity with intense cultivation of a much smaller number of food plants and global faunal diversity with managed collections of animals for food. Two hundred years ago most humans were still subsistence farmers. Today more than half of us live in cities and know little about how our food is produced.

What has not changed in that time is the fact that food remains the basic necessity of mankind.

In an era of ever-increasing food demand coupled with resource decline, environmental degradation, and climate change, *waste-less-to-feed-more* is an option that must be taken into account in our endeavor to feed the world sustainably.

- More food is wasted than the increases in total food production on an annual basis in the U.S. and worldwide.
- Reducing edible food loss to a reasonably attainable level can feed millions of hungry people.
- Prevention of food wastage offers multiple benefits with few negatives.

## SUMMARY OF FOOD LOSS IN THE U.S. SUPPLY CHAIN

- *The scale of the problem is huge.* More than 300 billion lbs. of food exit the U.S. supply chain annually, of which approximately 150 billion lbs is edible food loss in farm fields, at retail, and at the consumption stage. The amount of annual edible food loss is more than half (54%) of that consumed by Americans; it is also >8-times the average annual increase in U.S. food production in past decades.

- *Wasted food is wasted resources.* Represented in the annual edible food loss are 42 million acres of cropland, 8.6 billion lbs. of fertilizer nutrients, and 4.5 trillion gallons of irrigation water. These massive amounts of resources are spent in vain when the food never reaches a human stomach but is wasted, not to mention other environmental costs associated with the production and provision of that wasted food, such as soil erosion, greenhouse-gas emissions, and water pollution.

- *The amount of food recovered for humans is small.* Food rescued for humans through donation, gleaning, etc. approximates 2 billion lbs annually, which is <2% of edible food loss. Composting diverts about 5% of wasted food away from landfills. Food processing byproducts used in animal feeding amount to roughly 97 billion lbs, which is the most significant beneficial use given the characteristics of the food loss.

- *Efforts to raise awareness and address food waste issues are growing.* In the U.S. and throughout the world, charity organizations and numerous volunteers are at the frontline recovering food to feed food-insecure families. Media coverage of the topic is increasing. Food waste composting is gaining popularity. The food industry has embraced the food waste problem to better understand and address food waste issues. Certain government programs and laws have been initiated, including a newly announced U.S. federal commitment to reduce U.S. food waste at landfills by 50% by 2030. However, the American public is yet to be significantly engaged, as the scope and scale of food waste have not yet registered with the average consumer.

- *Opportunities for food waste reduction, recovery, and recycling depend on the "nature" (composition, form, and distribution) of food loss.* All food loss is not equal. Industry sector food loss is primarily made of food remnants that are unpalatable to humans; it resembles point sources, concentrated at a small number of sites; the economies of scale favor recovery and repurposing, particularly for animal feeding. In contrast, retail and consumption stage food loss consists of consumer food products that are discarded for multiple reasons. As non-point sources, this wasted food is dispersed and the composition and form are variable and random. Once discarded, the food material is usually beyond recovery for human consumption, while other recycling options are subject to technical or logistical challenges. Prevention of food waste (i.e. source reduction) is the best option.

- *Evidence-based data are critically needed.* Current understanding of U.S. food loss is derived from systematic documentation by the USDA Economic Research Service, based on the nation's food-supply inventory, coupled with waste parameters derived via various means. The statistically based data are robust, enabling us to see the big picture. Yet, evidence-based data from ground-up measurements will be essential to help us better understand consumer food behavior and contributing factors. Such understanding is key to finding practical solutions and interventions for change, as consumer-level food wastage is the single largest contributor.

## RECOMMENDATIONS

Humans have always wasted food, but the scale of the problem today is unprecedented in history, as >1/3 of food produced for humans is not eaten but lost while nearly 1 billion people worldwide (and 1 in 6 Americans) are food insecure. The impacts are profound, considering the growing food demand amid accelerating resource and environmental degradation and climate change. Reducing the amount of food loss is a low-hanging fruit for advancing food security and sustainability goals. Seizing the opportunity requires concerted efforts and mobilization of all forces, from public policy support to private enterprise commitment to individual citizen action.

The following recommendations are offered to illustrate the magnitude of what would be required to mobilize the nation to build a sustainable food system. It is by no means exhaustive or exclusive. The order of appearance does not represent any specific priority.

**(Public/Government Services)**

1. Public Education Coordinate national campaigns to engage all stakeholders, consumers in particular, for food conservation, waste reduction, and food behavioral change.
2. Early Intervention Incorporate food, nutrition, and sustainability education into K-12 curriculum. This demographic group represents 16% of the population; the food habits, knowledge, and attitudes they acquire in school would stretch over their own long life-spans and positively affect future generations.
3. Research Support Create funding mechanisms to support evidence-based fact-finding and baseline establishment, consumer food behavior studies, better understanding of socioeconomic factors, and identification of ways and means for positive change.
4. Technological Empowerment Set policies and provide seed money to foster the development of innovative, cost-effective technologies aimed at food waste recovery and repurposing for beneficial uses, particularly for animal feeding, as it contributes to both food security and sustainability goals.
5. Incentive Building Create policies/legislation with reward mechanisms to induce and enable positive change, in addition to regulative (punitive) measures.
6. Holistic Approach Set forth national and local programs and initiatives to foster sustainable food consumption for the long-term goal of building a

sustainable food system, one that integrates sustainable food production with soil and resource conservation and sustainable food consumption.

**(Businesses/Private Sector)**

7. Practical Innovation (for food industries) Reform business practices to facilitate positive changes. For example, smaller and re-sealable packaging can help prolong food freshness, lowering wastage at homes; "buy-one-get-one-free" *in the next purchase instead of now* could help consumers minimize spoilage.

8. Participatory Management (for any institution) Make food waste reduction a component of each organization's sustainability program. Raise awareness of food waste among employees (63% of the population is in the American workforce). Measure your cafeteria's waste streams, set waste reduction goals, and monitor and publicize progress.

9. Entrepreneurship Invest in the development of new technologies for enhancing food uses and reducing wastage, for example, better preservation methods, employment of nanotechnology in food packaging, etc. In particular, technological breakthroughs are crucially needed to repurpose the massive amount of consumer level food waste for animal feeding, which would serve both food security and sustainability purposes. Such technologies would, ideally, sanitize, dehydrate, and homogenize the food waste materials to make safe, healthy and nutritious, and easy-to-use feed supplements. Stationary and mobile operating units are needed to handle diffuse (e.g. neighborhoods) and collective (e.g. restaurants) waste volumes.

**(Households/Individuals)**

10. Self-Appraisal Start a kitchen diary, note the type, amount, and the reasons for food items discarded (e.g. spoiled, expired by date label, no longer wanted, etc.). Tally the waste items weekly. Ask yourself if, and how, you can do better to lower the amount of waste and save money.

11. Better Planning Adopt a (mental or written) planning habit. Outline the week's meals; check the refrigerator and cupboard periodically; have a grocery list before shopping; be mindful of what you purchase and what you can consume in a given timeframe.

12. Common Sense Rules Read date labels AND smell/inspect/touch the food itself if in doubt. Keep perishables at a fixed or ready-to-reach spot. Store food items properly to maximize shelf life.

13. Adventure and Flexibility Give your imagination and creativity a place in the kitchen; for example, try Chinese stir-frying to utilize remaining ingredients from recipe-based cooking (yes, the wok is good for making a delicious dish from a mixture of things). Ask Google for ideas for using leftovers.

14. Ownership Attitude We live on one Earth with shared resources and responsibilities. Make food waste reduction personal. Everyone can be a proud advocate. Engage your family, talk to friends and relatives, reach out to co-workers and neighbors. Ask the waiter to box your leftover food; tell the produce staff that you welcome discounted "ugly" fruits and vegetables – there is no need to be embarrassed about such responsible behavior. Just the opposite!

The challenge of feeding the world cannot be simply met through increases in production – for over half a century, we have experimented with remarkably lopsided strategies and approaches for producing more to feed more. Our intellectual power brought miraculous achievements but simultaneously created havoc. For example, we grow lavish greens in deserts with pumped water while drawing down aquifers far faster than they can be replenished; we greatly enhance crop yields with nitrogen fertilizers fixed industrially from the air but more than half of the nitrogen escapes the agricultural system, polluting waterways and choking off aquatic life. If we are to feed >9 billion people by 2050 without exhausting the basic resources (land, water, energy, etc.) necessary for future generations to survive and thrive, we need a fundamental shift with changes in philosophy, policies, priorities, and everyday actions. Waste-less-to-feed-more offers multifaceted benefits of combating hunger, enhancing food availability, improving resilience of food systems, and improving resource and environmental performance. Clearly, sustainable consumption must be incorporated in the food security and sustainability formula.

Substantially reducing food wastage is attainable, as proven by the UK WRAP experience. Opportunities exist throughout the supply chain. Meaningful progress can be made household-by-household, site-by-site, and step-by-step. These opportunities are not limited to the U.S., but can and must extend across the globe as well.

Zhengxia Dou, James D. Ferguson, David T. Galligan,
Alan M. Kelly, Steven M. Finn, and Robert Giegengack